Human Spaceflight and Exploration

Carol Norberg (Editor)

Human Spaceflight and Exploration

Springer

Published in association with
Praxis Publishing
Chichester, UK

PRAXIS

Editor
Dr. Carol Norberg
Swedish Institute of Space Physics
Kiruna
Sweden

SPRINGER–PRAXIS BOOKS IN ASTRONAUTICAL ENGINEERING

ISBN 978-3-642-23724-9 ISBN 978-3-642-23725-6 (eBook)
DOI 10.1007/978-3-642-23725-6
Springer Heidelberg New York Dordrecht London

Library of Congress Control Number: 2012942470

Cover design: Jim Wilkie
Project copy editor: David M. Harland

Printed on acid-free paper

Springer is part of Springer Science+Business Media (www.springer.com)

Contents

Foreword

Europe plays a key role in all fields of international spaceflight including Earth observation, satellite communication, and navigation, as well as space research, space science, and of course human spaceflight. It therefore stands at the forefront of space exploration. Few regions in the world are able to independently send large satellites and probes into space for exploration purposes; Europe is one of them. We also have our own laboratory on the International Space Station, to which 14 countries contribute. Europe plans and carries out space missions that explore all major celestial bodies—from the Sun to Saturn. Together with our European partners we are building Galileo, a satellite navigation system that will provide a civilian alternative to existing systems and promises great market potential. With GMES, Europe's Global Monitoring for Environment and Security program, we are building extensive infrastructure on Earth, in the air, as well as in orbit, that will allow us to continually observe the changes of the Earth, its atmosphere, seas, polar regions, and its climate.

These numerous achievements also imply a significant gain in knowledge, which needs to be documented in whatever form and made accessible, comprehensible, and understandable. It needs to be emphasized that the ultimate goal of exploration is linked to human presence on other celestial bodies. In addition to the increase of our knowledge, it is curiosity which drives us. It is the young scientists' task to generate new knowledge based on our existing understanding, and to continuously expand the boundary of human action into space. Today's students and scientists- or engineers-to-be will determine the big spaceflight events of tomorrow. Whoever wishes to have an impact on the future must know the past, understand the present, and be motivated to shape the future.

With this in mind, enjoy reading this book!

Thomas Reiter
Director of Human Spaceflight and Operations, European Space Agency
June 2012

Editor's preface

This book is based on the contents of a summer course in "Human Spaceflight and Exploration" given by Umeå University, Sweden. Each of the contributing authors participated in giving the lectures, which were aimed at Swedish and international students in physics and engineering with an interest in space. It proved very popular, and the goal of this book is to encapsulate and expand the material developed by the lectures and transmit it to a wider audience.

Many of those active in the space industry today were inspired by the first Moon landings, which they witnessed in their childhood. Those were inspirational events of significance to all mankind. Space-based applications such as weather forecasting, satellite television, communication, and satellite navigation are now integral to daily life. During a "Relevance of Science Education" study carried out in Sweden, secondary school pupils were asked to rank 108 topics according to how eager they were to learn more about them. The fact that the top 20 contained 5 space-related topics shows the interest that space still holds for young people this millenium. It is important to harness this interest to cultivate a new generation of space scientists, engineers, and astronauts, and space-related educational activities at all levels of the education system can make important contributions to this effort.

Many publications on human spaceflight and exploration focus on the activities of the United States. This book attempts to provide a more balanced presentation of the international contributions to the field, limited only by the lack of publicly available information about the programs of Russia, China, and Japan. As the contributing authors are all Europeans with broad experience of European space activities, the role of Europe in space-related activities is highlighted. In addition, this book is written at a critical time in global human exploration of the solar system. China is forging ahead with its national program, while the way ahead for the traditional spacefaring nations is unclear. We may well be on the brink of a transition from nationally operated space transportation systems to commerically driven systems. It remains to be seen whether commercial operators can deliver low-cost systems with the required degree of safety. If they succeed, space will become accessible to a greater portion of humanity. Vacations in space that have for so long been mere fantasy may be about to become reality. However, it would be a shame if the possibilities of holidaying in low Earth orbit are placed on hold due to the risk posed by all the litter left behind by previous projects. There is no doubt that space exploration is hard, but that is what makes it so satisfying when we make a step forward. I hope this book will increase awareness of global space activities and inspire others to enter the field.

Carol Norberg
Stockholm
May 2012

Contributors

Dr. **Christer Fuglesang**, a particle physicist, joined the European Astronaut Corps in 1992 and became the first Swedish astronaut to fly in space when he was assigned as a Mission Specialist on STS-116 Discovery for a mission to the International Space Station. He flew again on STS-128 in 2009. He has carried out five spacewalks and holds the current European record for spacewalking experience.

Sven Grahn is a space consultant. After participating in the first sounding rocket launches in Sweden in the 1960s as a student, he spent his career in the space industry in Sweden on a broad spectrum of projects. Following his retirement in 2006, he has continued as a part-time advisor in space matters and is a popular lecturer on space topics.

Dr. **Carol Norberg** is a senior lecturer in space physics at Umeå University in Sweden. She has carried out theoretical research in space plasma physics at both the University of Oxford in England and the European Space Agency. She specializes in university physics courses for international students.

Dr. **Åke Ingemar Skoog** is an expert on space suits and life support systems. His career was spent at Dornier System/Astrium GmbH in Germany, where he participated in design and development of life support systems for the European manned space programs Spacelab, Columbus, and Hermes. He was project manager for the development of European spacesuits from 1986 to 1994. He coauthored the book *Russian Spacesuits* published by Springer/Praxis in 2003.

Hansulrich Steimle is a senior advisor to the German Space Agency (DLR) and its former Director of Manned Space Operations. In the 1980s he was responsible for the recruitment of German astronauts and he advised the astronaut recruitment campaign by the European Space Agency in 2008–2009. His career has encompassed astronaut selection and training, human space operations, and sounding rocket programs.

Dr. **Gerhard Thiele** began his career as an environmental physicist. In 1988 he was selected as a German astronaut and subsequently became a member of the European Astronaut Corps. He participated as a Mission Specialist on the STS-99 mission in 2000. In August 2005 he was assigned as the Head of the Astronaut Division of the European Astronaut Centre, and in March 2010 became Resident Fellow at the Space Policy Institute in Vienna.

Acknowledgments

The authors would like to express their gratitude to Umeå University for giving them the great opportunity to participate as lecturers in the international summer courses on Human Spaceflight and Exploration, which led to the production of this text. Thanks are also due to the Swedish Institute of Space Physics in Kiruna for support during those courses and in the production of this manuscript. In particular the assistance of the institute's librarian Carina Kreku is much appreciated.

The authors acknowledge the service of NASA and its excellent website, which has been very useful as a source of historical and current information. Thanks are due to Peter Preu and Peter Turner at the German Space Agency, Hubertus Thomas at the Max Planck Institute for Extraterrestrial Physics, Jennifer Ngo-Anh at the European Space Agency, Natsuhiko Inoue at the Japan Aerospace Exploration Agency, and Asta Pellinen-Wannberg at the Swedish Institute of Space Physics for assistance in researching topics in the text. Thanks to Ebba Kierkegaard and the Chinese Embassy Stockholm in assisting to obtain images for the Chinese space program and to Susan Newsam for help in accessing Virgin Galactic images. Special thanks to the RD&PE Zvezda Joint Stock Company in Tomilino, Russian Federation, and its staff for facts and illustration material on space suits for this book and for making space suit hardware available for hands-on demonstration during lecturing in Kiruna.

We are grateful to Olle Norberg for proofreading the text and assisting in creating diagrams. Finally a very big thank you to Clive Horwood at Praxis for giving us the opportunity to publish the book and for providing the excellent editorial services of David M. Harland.

Figures

Tables

Abbreviations and acronyms

ACE	Advanced Composition Explorer
A/L	AirLock
ALSEP	Apollo Lunar Surface Experiments Package
ALTEA	Anomalous Long-Term Effects in Astronauts' Central Nervous System
AOA	Abort Once Around
APM	Attached Pressurized Module
APU/HYD	Auxiliary Power Unit and HYDraulics system
ARED	Advanced Resistive Exerciser
ASCAN	AStronaut CANdidate
ASI	*Agenzia Spaziale Italiana* (Italian Space Agency)
ATO	Abort To Orbit
ATV	Automated Transfer Vehicle
BFS	Backup Flight Software
BLSS	Biological Life Support System
BPC	Biomass Production Chamber
BRT	Body Restraint Tether
CAPCOM	CAPsule COMmunicator
CAST	China Academy of Space Technology
CCD	Charge-Coupled Device
CDR	Mission CommanDeR
CEBAS	Closed Equilibrated Biological Aquatic System
CELSS	Controlled Ecological Life Support System
CEVIS	Cycle Ergometer
CFR	Carbon Formation Reactor
CMO	Crew Medical Officer
CNES	*Centre National d'Etudes Spatiales* (French Space Agency)

CNSA	China National Space Administration
CODATA	Committee On DATA for Science and Technology
COMM	COMMunication system
CSA	Canadian Space Agency
CSSS	Constellation Space Suit System
CWC	Contingency Water Container
DARA	*Deutsche Agentur für Raumfahrtangelegenheiten* (German Space Agency)
DCS	DeCompression Sickness
DFVLR	*Deutsche Forschungs- und Versuchsanstalt für Luft- und Raumfahrt* (German Research and Development Institute for Air and Space Travel)
DLR	*Deutsches Zentrum für Luft- und Raumfahrt* (German Aerospace Center)
DNA	Deoxyribonucleic acid
DOS	*Dolgovremennaya Orbitalnaya Stanziya* (Long-duration Orbital Station)
EAC	European Astronaut Centre
ECLSS	Environmental Control and Life Support System
EDC	Electrochemical Depolarized Concentrator
EDV	*Emkost Dlya Voda* (water container)
EKON	Russia's Environmental Safety Agency
EMU	Extravehicular Mobility Unit
EPCARD	European Program Package for the Calculation of Aviation Route Doses
EPM	European Physiology Module
ERV	Earth Return Vehicle
ESA	European Space Agency
EVA	ExtraVehicular Activity
FCOD	Flight Crew Operations Directorate
FDF	Flight Data File
FE	Flight Engineer
FGB	*Funktsionalno Gruzovoy Blok* (Functional Cargo Block)
FMARS	Flashline Mars Arctic Research Station
FPEF	Fluid Physics Experiment Facility
GDR	German Democratic Republic
GEO	GEOstationary orbit
GMES	Global Monitoring for Environment and Security
GMK	Chief Medical Commission
GMT	Greenwich Mean Time
GMVK	Joint State Commission
GOST	*GOsudarstvennyy STandart* (Russian Federation State Standards)
GPC	General Purpose Computer

GRAIL	Gravity Recovery And Interior Laboratory
GRC	Glenn Research Center
GSFC	Goddard Space Flight Center
HGS	Hughes Global Services
HRCS	High Rate Communication System
HRF	Human Research Facility
HST	Hubble Space Telescope
HTV	H-II Transfer Vehicle
HUBES	HUman Behavior in Extended Spaceflight
HUT	Hard Upper Torso
HZE	High Z, high Energy
IADC	Inter-Agency Space Debris Coordination Committee
IBMP	Institute for Biomedical Problems
ICE	Isolated and Confined Extreme
ICV	Integrated CardioVascular Alternate Experiment
IKF	*Institut für Kosmische Forschung* (Institute for Space Research)
ILC	International Latex Corporation
IMP	Interplanetary Monitoring Platform
IMS	Inventory Management System
ISRO	Indian Space Research Organisation
ISS	International Space Station
JAMIC	JApan MIcrogravity Center
JAR-FCL	Joint Aviation Requirements–Flight Crew Licensing
JAXA	Japan Aerospace eXploration Agency
JEM	Japanese Experiment Module
JPL	Jet Propulsion Laboratory
JSC	Johnson Space Center
KSC	Kennedy Space Center
KSI	*Kapsula Sbrosa Informatsii* (Information Return Capsule)
LCC	Launch Control Center
LCROSS	Lunar CRater Observation and Sensing Satellite
LCVG	Liquid Cooling and Ventilation Garment
LEA	Launch, Entry, and Abort
LED	Light Emitting Diode
LEM	Lunar Excursion Module
LEO	Low Earth Orbit
LET	Linear Energy Transfer
LLC	Limited Liability Company
LMLSTP	Lunar–Mars Life Support Test Program
LMS	Life and Microgravity Spacelab
LRV	Lunar Roving Vehicle
LSS	Life Support System
MAPHEUS	Material Physics Experiments Under Microgravity

Marangoni	Surface fluid physics experiment
MASER	MAterials Science Experiment Rocket
MAXUS	MAser-teXUS
MCC	Mission Control Center
MDRS	Mars Desert Research Station
MECA	Master Events Controller Assembly
MECO	Main Engine Cut Off
MEDES	*Institut de Médecine et de Physiologie Spatiale* (Institute for Space Medicine and Physiology)
MELiSSA	Micro-Ecological Life Support System Alternative
MERLIN	Microgravity Experiment Research Locker INcubator
MET	Mission Elapsed Tme
MGLAB	MicroGravity LABoratory of Japan
MIM	MIni-Research Module
MIT	Massachusetts Institute of Technology
MLI	Multi-Layer Insulating
MMU	Manned Maneuvering Unit
MORABA	Mobile Rocket Base
MPS	Main Propulsion System
MRM-1	Mini-Research Module 1
MS	Mission Specialist
MSFC	Marshall Space Flight Center
MSG	Microgravity Science Glovebox
MTFF	Man-Tended Free-Flyer
MVL	Man Vehicle Laboratory
NASA	National Aeronautics and Space Administration
NASDA	NAtional Space Development Agency
NBL	Neutral Buoyancy Laboratory
NEAR	Near Earth Asteroid Rendezvous
NEEMO	NASA Extreme Environment Mission Operation
NGA	National Geospatial-Intelligence Agency
NIICHIMMASH	*Nauchno-Issledovatelskii I Konstruktorskii Institut Chimicheskogo Machinostroenija* (Research and Design Institute of Chemical Engineering)
NML	National Microgravity Laboratory
NOLS	National Outdoor Leadership School
NPO	*Nauchno-Proizvodstvennoe Obiedinenie* (Machine Building Scientific Production Association)
NSAU	National Space Agency of Ukraine
OMS	Orbital Maneuvering System
OPS	Orbital Piloted Station
OSIRIS	Optical, Spectroscopic and Infrared Remote Imaging System
OSTP	Onboard Short Term Plan
OSTPV	Onboard Short Term Plan Viewer

PASS	Primary Aviation Systems Software
PKE	*Plasmakristall-Experiment* (Plasma Crystal Experiment)
PLT	PiLoT
pp	partial pressure
PPF	Polar Platform Facility
PSS	Procedures, Specifications and Standards
QUAL	QUALification
RAAN	Right Ascension of the Ascending Node
RCS	Reaction Control System
RELAKSATSIYA	Relaxation (Earth observation experiment)
RMS	Remote Manipulator System
RS	Russian Segment
RSC	Rocket and Space Corporation
RTLS	Return To Launch Site
SAFER	Simplified Aid For EVA Rescue
SAWD	Solid Amine system with Water Desorption
SCOM	Shuttle Crew Operations Manual
SFINCSS	Simulation of Flight of International Crew on Space Station
SLAMMD	Space Linear Acceleration Mass Measurement Device
SLICE	Structure and Lift-off In Combustion Experiment
SMAC	Spacecraft Maximum Allowable Concentration
SNR	Signal-to-Noise Ratio
SPK-U	*Sistemi priema i koncervashii urinje* (System for collection and pretreatment of urine)
SRB	Solid Rocket Booster
SRL	Space Radar Laboratory
SRTM	Shuttle Radar Topography Mission
SRV-K	*Sistema regeneratsii vodi—kondensata* (Water regeneration system—condensate)
SRV-SG	*Sistema regeneratsii vodi—gigieni* (Water regeneration system—hygiene)
SRV-U	*Sistema regeneratsii vodi—mochi* (Water regeneration system—urine; System for collection and pretreatment of urine)
SSC	Swedish Space Corporation
SSRMS	Space Station Remote Manipulator System
STD	STanDard
STS	Space Transportation System
T2/COLBERT	Advanced treadmill
TAL	TransAtlantic Landing
TDRS	Tracking and Data Relay Satellite
TEXUS	Technological Experiments in Weightlessness

TIMES	Thermoelectric Integrated evaporation MEmbrane System
TKS	*Transportniy Korabl Snabzheniya* (Transport Supply Ship)
TMA	*Transportnyi Modifitsirovannyi Antropometricheskii* (Anthropometrically Modified Transport)
TORU	Teleoperator Control System
TSuP	*Tsentr Upravleniye Polyotom* (Mission Control Center)
TVIS	Treadmill
UKSA	United Kingdom Space Agency
URAGAN	Hurricane
USGS	United States Geological Survey
USOS	United States On-orbit Segment
VA	*Vozrashaemiy Apparat* (Return Apparatus)
VCD	Vapor Compression Distillation
VELO	Ergometer bike with load trainer
VPK	*Voenno-Proyshlennyy Komissiy* (Military-Industrial Commission)
WPA	Water Processor Assembly
X-SAR	X-band Synthetic Aperture Radar
ZARM	*Zentrum für angewandte Raumfahrttechnologie und Mikrogravitation* (Center of Applied Space Technology and Microgravity of the University of Bremen)
ZHR	Zenith Hourly Rate

1

Introduction

Carol Norberg

1.1 EXPLORATION

We humans have an inbuilt need to explore. In the Middle Ages sailors made perilous voyages in search of new lands, and in the last half century technology has allowed us to investigate first hand the environment around our planet and even our neighbors in the solar system. Instrumentation is continually being improved to enable us to look farther away to learn more about our universe and its history. The design, construction, and launch of robotic spacecraft for exploration requires manpower, national funding, and often long periods of time. Despite the many setbacks due to launch failures and spacecraft malfunctioning prior to completing their missions, we continue to initiate new projects in the hope of making discoveries with those which succeed. Due to the short time frames of politics, national funding is often difficult to sustain, and projects sometimes have to be redesigned several times in order to accommodate political and budgetary priorities. For a long-term project it is essential to ensure that there will be personnel to support it throughout its lifetime. This means, for example, ensuring that science teams are composed of members at various stages in their careers; it wouldn't do for them all to retire before the spacecraft reaches its destination. A good example is Rosetta, a European mission that was originally intended for Comet 46P/Wirtanen. This was approved in 1993 for launch in 2003, but the failure of the planned type of launch vehicle imposed a postponement which meant that a new target comet had to be identified. The scientists involved selected Comet 67P/Churyumov–Gerasimenko. The mission was finally launched in 2004 and should reach its target in 2014 to initiate a detailed study lasting about 18 months. During the cruise phase, limited measurements are being made to give scientists data to work on. Taking into account the time required to analyze the results from the encounter with the comet, the project will have an overall life-span of around 25 years.

1.2 SPACE

Before discussing the exploration of space it is necessary to first define exactly which region above the surface of the Earth is of interest. Although "space" is a word used in our daily lives, the answer to the question "Where does space begin?" is not obvious. The Fédération Aéronautique Internationale separates aeronautics from astronautics at an altitude of 100 km. Above this height the air is too rarefied for aeronautical flight. Nevertheless, there is still air present, and a satellite must fly tens of kilometers above this altitude to achieve a stable orbit. The 100 km boundary is also referred to as the Kármán line in honor of Theodore von Kármán from Hungary, who proposed it in the 1950s because of the manner in which the properties of the atmosphere change at that height, as is evident from Fig. 1.1.

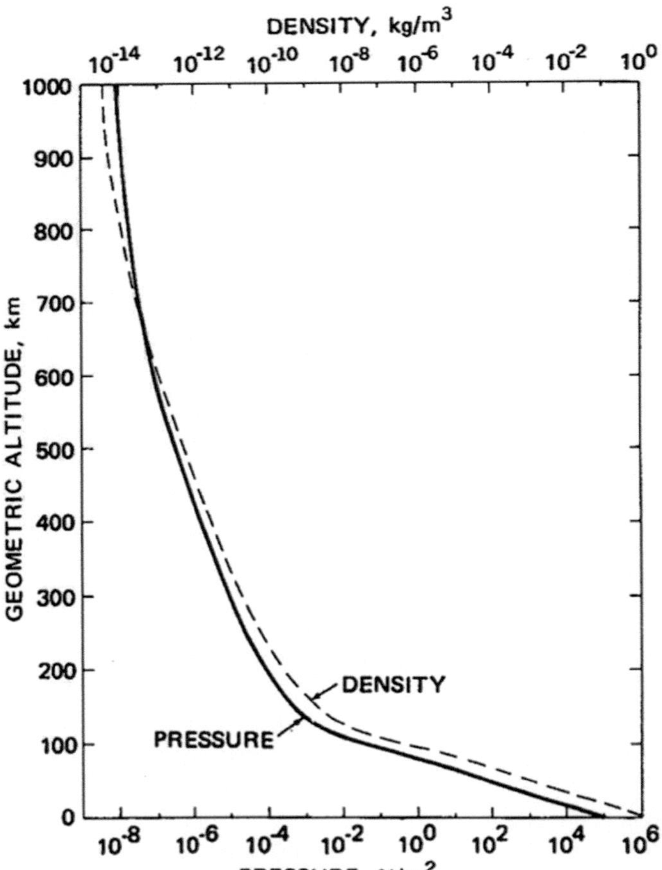

Fig. 1.1. U.S. Standard Atmosphere 1976. It is a model that gives atmospheric parameters based on a limited amount of experimental data (source: NASA).

From the surface of the Earth, the sky appears to be blue to the human eye. This is due to light from the Sun being scattered as it passes down through the atmosphere. At night, when the sunlight is gone, we see a black sky. To astronauts in space their surroundings appear black even in daylight, since there is no atmosphere present to scatter the light. The same is true of the view from an aircraft that flies so high that the local atmosphere is very thin. Aircraft are often described as flying to the "edge of space". Such flights approach but do not reach an altitude of 100 km, and are sufficiently high that it becomes possible to see both the curvature of the Earth and the blackness of space. Passengers on a suborbital flight to space will get a slightly better view, and experience weightlessness for a few minutes as they reach the apex of their parabolic trajectory somewhat above the 100 km boundary. On a night flight, they might also be able to view the aurora at short range, should their trajectory take them close enough to one of the Earth's poles.

Fig. 1.2 illustrates the height distribution of a number of physical phenomena in comparison with the flight paths of vehicles. Passenger planes normally fly above the clouds in the troposphere. The highest clouds in the atmosphere are the noctilucent clouds which appear in the summer at altitudes of about 83 km; the name means "clouds that shine at night". They can be seen when the Sun is below the horizon for an observer at ground level but still shines on the high-altitude clouds, making them appear bright in the twilight. Just how they are produced is still a subject of research. One theory is that there is a thin smoke of nanometer-sized particles at that altitude due to the ablation of meteoroids penetrating the atmosphere and that these particles act as condensation nuclei for

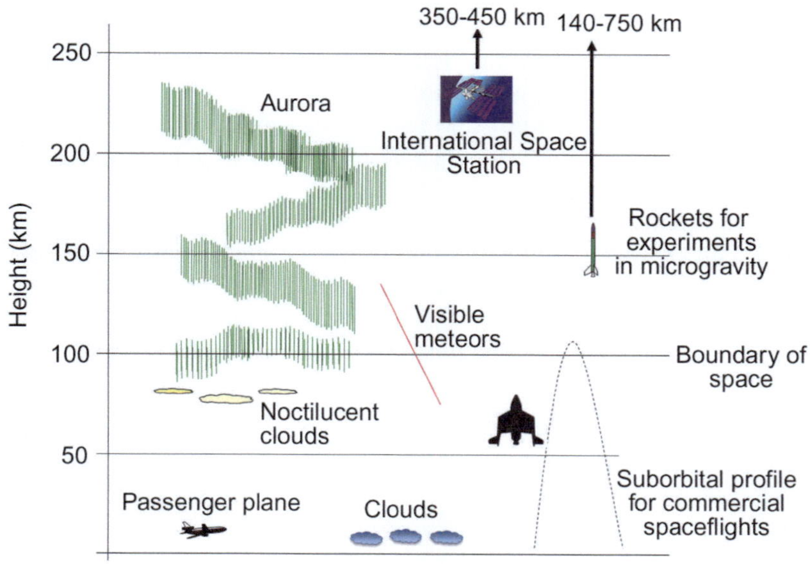

Fig. 1.2. Altitude distribution of physical phenomena and vehicle trajectories.

Fig. 1.3. The solar system (source: The International Astronomical Union/Martin Kornmesser).

cloud droplets. A large number of meteoroids enter the Earth's atmosphere every day, with the larger ones being visible at night as meteors burning up as they descend from around 140 km to 70 km. An aurora has its lower boundary at about 90 km and can extend up into the ionosphere for hundreds of kilometers. The diagram also depicts a suborbital flight profile that peaks just above the official boundary of space. The International Space Station (ISS) operates between 350 and 450 km altitude. Sounding rockets can climb on ballistic trajectories to altitudes approaching 750 km.

Looking out farther from the Earth, our position in the solar system is illustrated in Fig. 1.3, which is not to scale. Our closest planetary neighbor is Venus, but its surface is very hot due to a strong greenhouse effect and it has a toxic atmosphere, so it is not an attractive option for human missions. The next closest planet is Mars, lying about 50% farther from the Sun than Earth. It is chilly, but offers more tolerable conditions and so is a natural target for human exploration. Our exploration of the planet using robotic probes and the challenges posed by sending humans there are considered in a number of sections in this book. The Moon and asteroids constitute potential sources of oxygen, hydrogen, and other resources to enhance space exploration.

1.3 ORGANIZATION OF SPACE ACTIVITIES

We can travel 100 km on Earth in a car, bus, or train in about an hour without much effort, but to climb that distance up from the surface of the Earth is a different matter. Only a very small number of human beings (around 500 in total)

Table 1.1. The world's main space agencies.

Short name	*Full name*	*Nation(s)*
NASA	National Aeronautics and Space Administration	U.S.A.
ESA	European Space Agency	Austria, Belgium, Czech Republic, Denmark, Finland, France, Germany, Greece, Ireland, Italy, Luxembourg, The Netherlands, Norway, Poland, Portugal, Romania, Spain, Sweden, Switzerland, United Kingdom
Roscosmos	Russian Federal Space Agency	Russia
CNES	French Space Agency	France
JAXA	Japan Aerospace Exploration Agency	Japan
DLR	German Aerospace Center	Germany
ISRO	Indian Space Research Organization	India
CNSA	China National Space Administration	China
ASI	Italian Space Agency	Italy

have managed to make the trip, but many more would like to. The difficulties and costs involved in space travel have until now made it an activity carried out exclusively by national and international projects. Many countries have their own governmental space agencies—intergovernmental in the case of the European Space Agency (ESA). The main agencies are listed in Table 1.1, each of which currently has an annual budget of over one billion U.S. dollars.

The European agency has a different character from the other agencies, since it represents cooperation between countries whose governments have realized that the challenges of space exploration can be better met by joining forces. The agency was formed in 1975 to promote and facilitate cooperation between European states in non-military space activities. Its initial membership was nine countries, but over the years it has increased to 20 European countries. Canada also participates in some ESA programs via cooperation agreements. In recent years cooperation with the European Union has increased. There are two classes of ESA activities for each member state—mandatory programs and optional programs. The member states finance mandatory programs in proportion to their gross domestic product. Each member chooses which of the optional programs they wish to support. Human spaceflight is an optional program, and those countries participating (in 2012) are Belgium, Denmark, France, Germany, Italy,

The Netherlands, Norway, Spain, Sweden, and Switzerland. It is evident that expanding human exploration of the solar system is a task too great for any single nation. In 2007, 14 space agencies produced a framework document for a "Global Space Exploration Strategy" in order to promote cooperation and collaboration between nations in space exploration activities and work for a sustained human presence in space. These space agencies are: ASI (Italy), CNES (France), CNSA (China), CSA (Canada), CSIRO (Australia), DLR (Germany), ESA (European Space Agency), ISRO (India), JAXA (Japan), KARI (Republic of Korea), NASA (United States of America), NSAU (Ukraine), Roscosmos (Russia), and UKSA (United Kingdom).

One measure of the development of space activities over time is the rate of global satellite launches. Fig. 1.4 shows the number of satellite launches for each year since Sputnik in 1957. It can be seen from the graph that the launch rate was high at the start of the space age, then stabilized at about 120 per year over a 20-year period, and thereafter decreased to about 60 per year. In Fig. 1.5 satellite launches are divided into those by Soviet Union/Russia and Ukraine; the U.S.A.; and other countries. It can be seen that the Soviets maintained a high launch rate from the end of the 1960s until the end of the Cold War in the late 1980s. Although they invested heavily in optical reconnaissance from space, the performance and lifespan of reconnaissance satellites grew very slowly and the number of launches remained fairly constant in those years. The number of launches from the U.S.A. had a peak in the 1960s when spy satellites were launched to study intercontinental missile sites in the Soviet Union. Following the introduction of the Space Shuttle in 1981, the U.S.A. moved to phase out conventional launchers. The minimum in the launch rate in 1986 came in the aftermath of the loss of Challenger, when the few rockets remaining were assigned to payloads with the

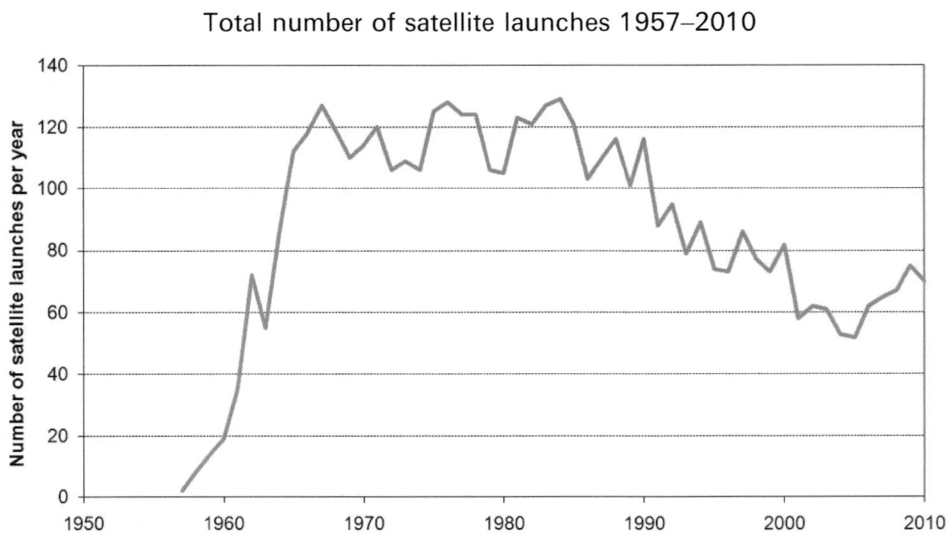

Fig. 1.4. Satellite launches as a function of time (courtesy: Sven Grahn).

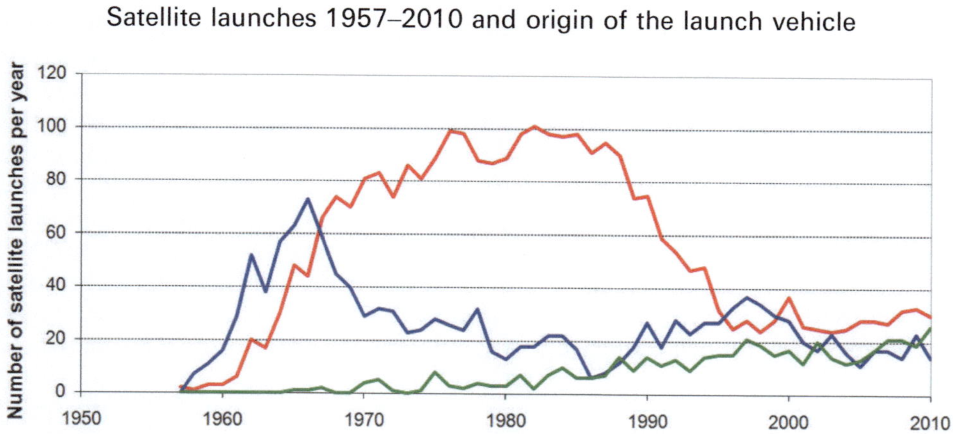

Fig. 1.5. Number of satellites by origin—red represents the Soviet Union/Russia and Ukraine, blue represents the U.S.A., and green other countries (courtesy: Sven Grahn).

highest priorities. From the mid-1990s the number of launches in the three categories are roughly the same, with a slight increase for other countries as, for example, China and India launch their own satellites and build up their space activities.

1.4 CHAPTER SUMMARIES

This section briefly summarizes each chapter. Chapter 2 contains an overview of the history of human spaceflight from the flight of Yuri Gagarin through to the modern era of private space travel. The dominant spacefaring nations to date have been the Soviet Union/Russia and the United States of America. With the exception of China, which has steadily developed its own human spaceflight program, the other countries that have participated in human space missions have done so by collaborating with the two superpowers. The chapter introduces the ISS, the largest international space project yet undertaken, which has hosted an uninterrupted human presence since 2000. Until recently, space travel has been almost exclusively reserved for career astronauts but commercial players are now active. Unburdened by the huge costs of large infrastructures and bureaucracy, the entrepreneurs funding programs to produce new space vehicles are able to quickly transform their innovations into active projects that test new concepts. There are currently several companies working on launch vehicles for suborbital and orbital spaceflights. The chapter concludes with an overview of commercial space travel.

In order for humans to operate spacecraft in popular orbits around the Earth, and to venture beyond, an understanding of the challenges of the space environment is required. Chapter 3 examines the space environment by considering in turn gravity, neutral particles, vacuum, plasma, radiation, micrometeoroids, and space debris. The microgravity environment in space provides a unique laboratory for scientists to carry out experiments for extended periods, but it is detrimental to the human body which has evolved for conditions on the Earth's surface. Neutral particle and space debris populations in orbits close to the Earth are today the most significant factors in limiting the expected life of a satellite, including the ISS. Activities during the space age have left low Earth orbit littered with debris, and this is multiplying over time due to collisions between pieces of debris. The probability of collisions between pieces of debris and operational spacecraft increases relentlessly. The status of the space debris problem is addressed in this chapter. With respect to long-duration missions into deep space, the radiation environment poses a tough challenge to be overcome. The sources of space radiation are described.

Chapter 4 gives an overview of space exploration carried out during the space age, focusing primarily on the Moon and Mars, to which a large number of spacecraft have been sent with varying success. Each mission is briefly described. The current efforts to prepare to undertake a human mission to Mars are presented, including projects to investigate the psychological challenges of such a long-duration mission. In order for humans to survive when removed from their natural habitat, they must take with them life support systems to provide a living environment analogous to that at home. These systems must work flawlessly to ensure survival of the crew. Chapter 5 describes the essential components of such a life support system and different methods of creating one. The popular image of an astronaut is a person in a space suit, especially the kind used for carrying out spacewalks. In effect, such a suit is a tiny spacecraft because it must provide its own life support system. Chapter 6 discusses various types of space suit. This section includes ESA astronaut Christer Fuglesang's personal reflections on carrying out a spacewalk. The principles of space suit design are described, as are the requirements for a suit that could be used on the Moon or Mars.

Chapter 7 reviews national and international astronaut selection campaigns to date. Thorough selection processes are undertaken to find the most suitable candidates to be career astronauts, and these multi-step processes can take over a year. Initially only military pilots were eligible, which led inevitably to there being only male candidates. The Soviets flew Valentina Tereshkova in 1962, but it was almost 20 years before the next woman made it into space. In the 1980s the number of female candidates began to increase, but they are still a minority. Today, the recruitment of astronauts includes scientists and engineers to operate spacecraft and their experiments. Those selected to be career astronauts can expect to train for at least three and a half years before they are assigned a mission. Currently the major effort in human spaceflight is operating the ISS. A brief introduction to space operations on board the station is given to illustrate how the working day of a crew member is organized.

The book concludes with ESA astronaut Dr. Gerhard Thiele's personal account of a Space Shuttle mission, providing an insight into the experience of being assigned to a crew, of training, and of flying into space. He describes living in space and scientific aspects of the mission. Now that the Shuttle has been retired, it is opportune to evaluate its role as a reliable space transportation system. Two of the five orbiters were lost in flight, and a disproportionate amount of time was devoted to training for the launch phase in relation to the mission as a whole. Dr. Thiele reflects on the Shuttle program.

1.5 REFERENCES/RECOMMENDED LITERATURE

Avdeyev, A., Clervoy, J., Comtois, J., Doi, T., Hoffman, J., Mohri, M., and Thiele, G.: *Human Space Exploration: A Quest of Cultures around the World*, European Space Policy Institute (2010).

International Space Exploration Coordination Group (ISECG): *Annual Report* (2009).

Sverige i rymden: *Dædalus: Yearbook of the Museum of Technology*, No. 75, Museum of Technology (2007).

The Global Exploration Strategy: The Framework for Coordination, ASI (Italy), BNSC (United Kingdom), CNES (France), CNSA (China), CSA (Canada), CSIRO (Australia), DLR (Germany), ESA (European Space Agency), ISRO (India), JAXA (Japan), KARI (Republic of Korea), NASA (United States of America), NSAU (Ukraine), Roscosmos (Russia) (2007).

U.S. Standard Atmosphere, 1976, U.S. Government Printing Office (1976).

Websites

Fédération Aéronautique Internationale: *www.fai.org*
The Space Report: *www.thespacereport.org*
International Astronomical Union: *www.iau.org*
International Space Exploration Coordination Group: *www.globalspaceexploration.org*
Space agencies: *http://en.wikipedia.org/wiki/List_of_space_agencies*
NASA: *www.nasa.gov*
European Space Agency: *www.esa.int*
Roscosmos: *www.federalspace.ru*
CNES: *www.cnes.fr*
JAXA: *www.jaxa.jp*
DLR: *www.dlr.de*
ISRO: *www.isro.org*
ASI: *www.asi.it*

2

History of human spaceflight

Sven Grahn and Carol Norberg

ABSTRACT

Over a period of about two centuries humans learned how to travel farther and farther from the surface of the Earth. However, we still have limited experience of flight in space. Progress was fast during the Cold War, when the competitive spirit between the United States and the Soviet Union drove both nations' space engineers to work towards goals of ever increasing complexity.

In this chapter the reader is taken through the early successes of the Russians, and then the early space programs of the United States. A number of space stations were built and flown by both countries in order to investigate how the human body coped with being weightless for long periods, and to carry out experimental microgravity research. The most advanced station to date is the ISS. China has recently begun to develop its space program, with plans for its own space station. Meanwhile space, which has traditionally been tackled by national and international projects, is having increased interest from the commercial sector. We are potentially at the brink of a change in direction in space travel, and may soon see an increase in the number of private individuals purchasing tickets for trips that are literally out of this world.

2.1 INTRODUCTION

Almost a century before the first human spaceflight occurred, Jules Verne's fictional characters planned and set off on a trip to the Moon in his novel *De la Terre à la Lune*, published in 1865 (Fig. 2.1). The dog and the three-member crew in that story have analogues in the real-life characters Laika and the Apollo crews. Space travel has long been popular with science fiction writers, whose creative

Fig. 2.1. Passengers in Verne's space ship enjoying their first taste of weightlessness.

talents have been inspirational. For example, Jules Verne was followed by H.G. Wells, who published *The First Men in the Moon* in 1901.

Humans have been going ever higher since the first manned balloon flights in the 1700s. Pilatre de Roziers (the word "pilot" is derived from his name) and Marquis Arlandes made the first balloon flight from Paris in 1783, and it lasted about 20 minutes. Interest in balloons continued in the following decades. Gliders were developed in the 1800s, and towards the end of that century serious efforts were made to construct powered aircraft. In 1903 Orville and Wilbur Wright made the first successful flight in the "Wright Flyer" near Kitty Hawk, North Carolina, U.S.A. The aviation industry grew rapidly during the 1900s. High-altitude balloons were used in the United States during the period 1955–1958 for a project called Project Man High to study cosmic rays and human flight at high altitudes. Then, in Project Excelsior, parachute jumps were made from balloons to simulate parachuting from high-flying aircraft or space capsules.

Missile technology invented during the Second World War in Germany served as the basis for the rockets utilized to launch humans into space in the 1960s. Wernher von Braun (1912–1977) was a key figure in the production of the first guided ballistic missile, the V-2, during the war. After the war, he and some of his colleagues were taken to America to continue their work. Other German rocket scientists ended up in the Soviet Union. The leading player in the early Soviet space program was Sergei Korolev (1907–1966) from the Ukraine, who is often referred to as the Soviet Chief Designer.

In the beginning of what came to be known as the "space race", the Russians had the advantage because they had developed larger and more powerful rockets than the Americans. However, the Americans later caught up when Braun's team masterminded the production of the immense Saturn V rocket that launched the Apollo astronauts to the Moon.

After successful spaceflights with expendable rockets and space capsules, a move was made to more airplane-like vehicles in the form of the Space Shuttles within the scope of NASA's post-Apollo program in the United States and then, secretly, by the development of the somewhat similar Buran in the Soviet Union. But in 2011 NASA decommissioned its fleet of Shuttles without a clear replacement strategy. The Soviet Union decommissioned Buran after its unmanned maiden flight, and retained the use of rockets. This strategy has been very successful in transporting humans to and from space. In recent times China has developed its own transport system based on rocket technology, and has successfully sent a number of its citizens into space.

After the dissolution of the Soviet Union, Russia began to sell trips into space and their success stimulated interest in a commercial market for human space travel. We are now on the brink of a possible expansion of that market that could rapidly open up space to a much larger number of Earth's citizens than ever before. By 2011 just over 500 people had flown in space and astronauts had spent a total of about 90 years in orbit; about one human life-span. The total time humans have spent "spacewalking" is about 145 days, and the most experienced space traveler to date is cosmonaut Sergei Krikalev, who has accumulated 803 days in orbit during six spaceflights.

2.2 EARLY RUSSIAN SUCCESS

During the 1950s the Cold War between the East and the West was symbolized by the race to space. It was of national importance both in the United States and the Soviet Union to beat the competition. To conquer space was seen as a step towards dominance. The Russians had a series of successes beginning with the launch of the world's first artificial satellite, Sputnik, on October 4, 1957 using an intercontinental ballistic missile (R-7) inspired by the German V-2 rocket. The satellite had the form of a hollow steel sphere 58 cm in diameter, with four long radio antennas protruding from it. A replica is shown in Fig. 2.2. It contained a basic radio transmitter that sent out a signal that could be heard by amateur radio operators. The launch was linked to the International Geophysical Year, which lasted from July 1957 to December 1958. The satellite traveled around the Earth in about 96 minutes, and there were fears that in the near future satellites would carry weapons and launch them towards nations far away from the Soviet Union.

Less than a month after the surprise launch of Sputnik 1, the Soviets shocked the world by launching Sputnik 2 on November 3, 1957. It was much larger and

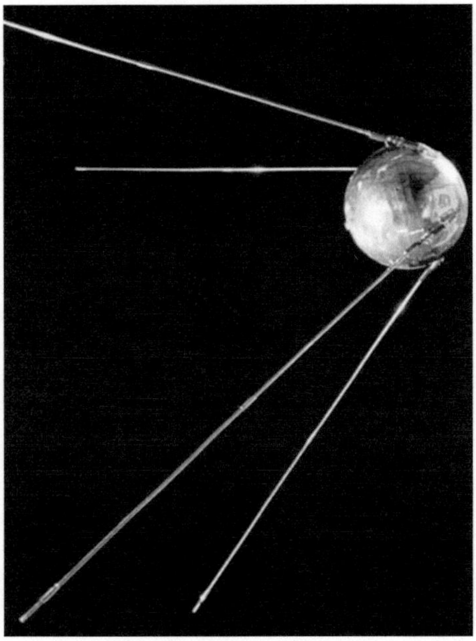

Fig. 2.2. The first artificial Earth satellite Sputnik 1 (source: RSC Energia).

heavier than its predecessor, with a mass of 508 kg. Most surprising of all, it carried a living passenger, the dog Laika. The spacecraft was built using parts from Sputnik 1 and high-altitude dog-carrying rockets. The life support system for Laika was designed to last for 7 days, but she perished 5 to 7 hours after launch due to overheating caused by a malfunction. The fact that Laika died so early in the flight only came to light recently. By illustrating that blood circulation was not disrupted by weightlessness, this flight was an important step towards launching a human into space. Sputnik 2 was followed by the scientific spacecraft Sputnik 3 on May 15, 1958. It had a mass of 1,327 kg and carried instrumentation to investigate the space environment near the Earth. Sputnik 4, with the then-amazing mass of 4,540 kg, was launched on May 15, 1960 and was the first launch of the Vostok series of spacecraft. It really made big headlines and was seen as a first step to manned spaceflight, even though the Soviets clearly stated that the craft would not be returned to Earth. The basic design could be adapted either to carry a passenger or automated reconnaissance apparatus, making it easier for politicians to allocate the substantial development funding than if it had only been intended for human use. Fig. 2.3 shows the design of the Vostok spacecraft with a spherical module accommodating the cosmonaut. After deorbit the craft split into two parts, and after reentry the cosmonaut ejected from his capsule to parachute separately to ground.

Sputnik 4, 5, 6, 9 and 10 (Table 2.1) were called Korabl-Sputnik 1 to 5 by the Soviets, and were test flights for the manned Vostok. In the last two, a

Fig. 2.3. Unmanned Vostok spacecraft, Korabl-Sputnik (source: RSC Energia).

Table 2.1. Vostok development flights

Mission	*Launch date*	*Details*	*Result*
Sputnik 4 (Korabl-Sputnik 1)	May 15, 1960	Unmanned Vostok spacecraft without a heat shield	Success
Sputnik 5 (Korabl-Sputnik 2)	August 19, 1960	Vostok spacecraft with dogs Belka and Strelka	17 orbits, dogs recovered
Sputnik 6 (Korabl-Sputnik 3)	December 1, 1960	Vostok spacecraft with dogs Pchelka and Mushka	Retro-rocket malfunction, the spacecraft was self-destructed, and the dogs not recovered
Sputnik 9 (Korabl-Sputnik 4)	March 9, 1961	Vostok spacecraft with a dummy cosmonaut Ivan Ivanovich and dog Chernushka	Success, dog recovered
Sputnik 10 (Korabl-Sputnik 5)	March 25, 1961	Vostok spacecraft with Ivan Ivanovich and dog Zvezdochka	One orbit, dog recovered

mannequin, nicknamed "Ivan Ivanovich", was flown dressed in a functional space suit. There was instrumentation inside the mannequin to gather flight data. It was ejected from the capsule during descent in order to test the escape system. Sputnik 7 and 8, which do not appear in the table, were unsuccessful attempts to send probes to Venus in early 1961.

In 1959 it was decided that only operational Air Force pilots would be considered for the first orbital space missions of the Soviet Union, since such pilots already had training that would be useful for preparing for a space mission (e.g., ejection seat training, and g-loading on various axes). Only men were considered, of course. Chief Designer Korolev specified the requirements for the candidates: they should be in the age range 25 to 30, be no taller than 1.70 m, and be no heavier than 72 kg. The physical requirements were necessary for a pressure-suited cosmonaut to fit into the capsule. At the end of the year, 20 men in the age range 23 to 34 were selected for training. Five of the group were outside the specified age limit but were accepted anyway, because of their performance during the selection process. A center for cosmonaut training was established in a Moscow suburb, about 30 km from the capital. It is now known as "Star City", and is still a cosmonaut training center. Six of the group were chosen for accelerated training. After some rearranging, the so-called "Vanguard Six" comprised Yuri Gagarin, Valerij Bykovskiy, Grigoriy Nelyubov, Andriyan Nikolayev, Pavel Popovich, and Gherman Titov. In January 1960 a ranking of these six trainees by an interdepartmental commission put Gagarin first and Titov second. They were hot favorites for the first flight, and the decision to send Gagarin was made only a few days prior to the launch. Titov served as backup, and was the obvious person for the next flight.

Human space travel became a reality on April 12, 1961, when Lieutenant Yuri Gagarin (Fig. 2.4) flew into space on Vostok 1. The launch was not announced in advance, only when under way. Following launch from the Baikonur Cosmodrome, Kazakhstan, Gagarin's flight lasted 108 minutes and reached a maximum altitude of about 328 km. Gagarin himself had no control over his spacecraft; he reported what he observed and how he felt throughout the flight. After automatic deorbiting, his capsule reentered the atmosphere. At an altitude of about 7,000 m, he was ejected to parachute to ground. This landmark event instantly made

Fig. 2.4. Yuri Gagarin on the way to the launch pad on April 12, 1961. His backup Gherman Titov is sitting behind him (source: Zvezda).

Table 2.2. Vostok manned flights.

Mission	*Launch date*	*Crew*	*Result*
Vostok 1	April 12, 1961	Yuri Gagarin	1 orbit, success
Vostok 2	August 6, 1961	Gherman Titov	17 orbits, success
Vostok 3	August 11, 1962	Andriyan Nikolayev	64 orbits, success
Vostok 4	August 12, 1962	Pavel Popovich	48 orbits, success
Vostok 5	June 14, 1963	Valerij Bykovskiy	81 orbits, success
Vostok 6	June 16, 1963	Valentina Tereshkova	48 orbits, success

Gagarin an international celebrity. Unfortunately the most famous space traveler of all time did not enjoy his glory for long. He died in a plane crash in 1968, aged 34, without making a second flight, and was interred in the Kremlin wall in Red Square in Moscow alongside other famous Russians. In celebration of the 50th anniversary of Gagarin's flight, a movie called *First Orbit* was made by matching audio recordings from his historic flight with movies of the Earth along his ground track photographed from the cupola on the ISS.

Some months after Gagarin's historic flight, his backup Gherman Titov, aged 25, made 17 orbits of the Earth on Vostok 2. Titov is still the youngest person to have flown in space. He took the first pictures of the Earth from space, ate meals, and slept in a weightless state for the first time. He also experienced symptoms of space sickness. The Vostok flights continued as detailed in Table 2.2. Vostok 3 and 4 were coplanar at the launch of Vostok 4 (the same inclination and right ascension of the ascending node, RAAN) and had the same orbital phase (angle along orbit from the node). The orbital planes of Vostok 5 and 6 differed 27.8° in RAAN but had the same initial orbital phasing. Valerij Bykovskiy orbited alone for 5 days, a record for spaceflight by an individual person that stands to this day. Another first was achieved when Valentina Tereshkova became the first woman to fly in space, on board the final Vostok mission in 1963. She had learned to parachute, which led to her applying to become a cosmonaut. Her flight lasted 3 days. It was 19 years before another woman flew into space. That was Svetlana Savitskaya, who flew twice and on her second flight became the first woman to make a spacewalk.

The logical next step was to develop a spacecraft with the capacity to carry at least two people. The Vostok was upgraded to create Voskhod, and on October 12, 1964 Voskhod 1 was launched with three cosmonauts. It was difficult to fit them into the craft. In particular, they did not have pressure suits or ejection seats since there was not room for these. After 16 orbits it returned to Earth. The cosmonauts landed inside the capsule. The Russians had once again beaten the Americans by succeeding in flying a crew which included the first medical doctor.

Fig. 2.5. The Soviet Vostok and Voskhod cosmonauts: Gagarin (Vostok 1), Titov (Vostok 2), Nikolayev (Vostok 3), Popovich (Vostok 4), Bykovskiy (Vostok 5), Tereshkova (Vostok 6), Feoktistov (Voskhod 1), Komarov (Voskhod 1), Egorov (Voskhod 1), Belyayev (Voskhod 2) and Leonov (Voskhod 2) (source: RSC Energia).

Voskhod 2, launched March 18, 1965, had two cosmonauts on board and made 17 orbits of the Earth. The highlight of the mission was the first ever spacewalk, carried out by Alexei Leonov. Yet another Soviet victory!

2.3 MERCURY AND GEMINI

The United States of America successfully launched a number of satellites into space in the years 1958 to 1960, and on April 1, 1959 NASA selected seven test pilots for its first manned spaceflight program, Project Mercury. Fig. 2.6 shows the "Mercury Seven" in space suits.

The main goal of the program was to orbit a manned spacecraft around Earth and recover the pilot in the capsule. It consisted of 20 unmanned launches, of which four successfully carried monkeys, and then two suborbital manned flights and four orbital ones. The Mercury spacecraft was built to carry one man. Fig. 2.7 illustrates the design. The astronaut had some control over the spacecraft during the flight, and the capsule was to parachute into the ocean for recovery. Three boosters with different launch capabilities were used during the program: Little Joe, Redstone, and Atlas. Table 2.3 summarizes the facts about the six manned missions.

The first manned Mercury flight, with Alan Shepard, was suborbital and occurred less than a month after Gagarin's historic orbital flight. In contrast to the furtiveness involved in the Soviet mission, Shepard's launch was watched by millions of people on television. America entered the space race with its first successful manned flight and, just as Gagarin became an instant hero in the Soviet Union, Shepard received a medal and was celebrated with ticker tape parades.

On May 25, three weeks after Shepard's flight, President John F. Kennedy made a bold speech to a joint session of Congress (Fig. 2.8) in which he set out a

Fig. 2.6. The Mercury Seven: from the left Walter (Wally) Schirra, Alan Shepard, Donald (Deke) Slayton, Virgil (Gus) Grissom, John Glenn, Gordon (Gordo) Cooper, and Scott Carpenter (source: NASA).

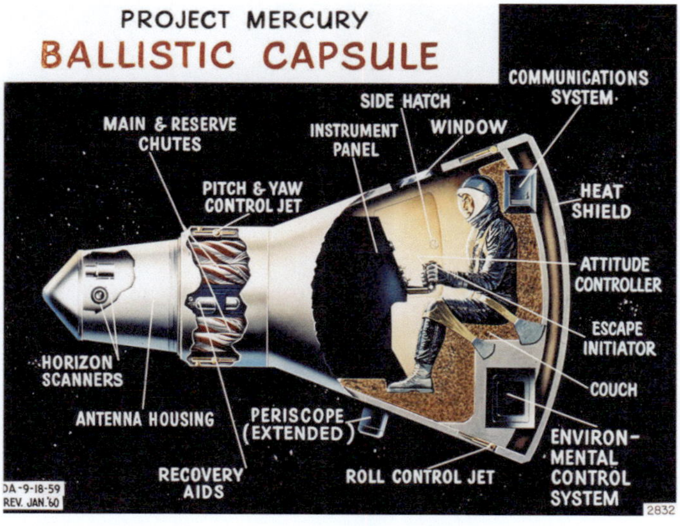

Fig. 2.7. The Mercury capsule, shown in a cutaway drawing made in January 1960 (source: NASA).

Table 2.3. Mercury manned flights

Mission	**Launch Date**	**Crew**	**Result**
Mercury-Redstone 3	May 5, 1961	Alan Shepard	First manned suborbital flight by an American
Mercury-Redstone 4	July 21, 1961	Gus Grissom	Second manned suborbital flight, the spacecraft sank, astronaut rescued
Mercury-Atlas 6	February 20, 1962	John Glenn	First American to orbit the Earth, 3 orbits
Mercury-Atlas 7	May 24, 1962	Scott Carpenter	Success, 3 orbits
Mercury-Atlas 8	October 3, 1962	Walter Schirra	Success, 6 orbits
Mercury-Atlas 9	May 15, 1963	Gordon Cooper	Success, 22 orbits

Fig. 2.8. President John F. Kennedy giving his historic speech to Congress, May 25, 1961. Shown in the background are (left) Vice President Lyndon Johnson and (right) Speaker of the House Sam T. Rayburn (source: NASA).

plan for space exploration with an ambitious schedule, and encouraged Congress to fund it. He said, in part

> ... I believe this nation should commit itself to achieving the goal, before this decade is out, of landing a man on the Moon and returning him safely to the

Table 2.4. NASA astronauts of group 2.

Name
Neil Armstrong
USAF Major Frank Borman
Navy Lt. Charles Conrad Jr.
Navy Lt. Cdr. James Lovell Jr.
USAF Capt. James McDivitt
Elliott M. See Jr.
USAF Capt. Thomas P. Stafford
USAF Capt. Ed. H. White II
Navy Lt. John W. Young

Earth. No single space project in this period will be more impressive to mankind, or more important for the long-range exploration of space; and none will be so difficult or expensive to accomplish.

The Mercury program in the United States was followed by the Gemini program. The Gemini spacecraft resembled its predecessor, but was bigger and could carry two people. It was also more maneuverable than the Mercury craft. The main goals of the new program were to demonstrate rendezvous and docking, to perfect landings, and to collect biomedical data over a period comparable with a lunar mission. The launch vehicle used was a two-stage Titan II rocket. In order to carry out the Gemini project and fly missions to the Moon, in September 1962 NASA selected a second group of astronauts, the members of which are listed in Table 2.4. As the program would have more technically complicated elements, the new astronaut group included engineers as well as test pilots. Table 2.5 summarizes the crew assignments and mission results of the Gemini program.

Gemini IV was of huge significance, because Ed White made the first U.S. spacewalk. Gemini VI-A and Gemini VII achieved a rendezvous in space. Fig. 2.9 shows Gemini VII photographed from Gemini VI-A. The 2-week flight of Gemini VII provided essential data on the effects of space travel on astronauts, in preparation for missions to the Moon. Gemini VIII docked with the Agena target vehicle, but when the combination started to spin the astronauts were obliged to undock and perform an emergency landing. The remaining flights undertook further docking training and a number of spacewalks.

Prior to the first Gemini launch, NASA's third group of astronauts was announced on October 18, 1963 (Fig. 2.10). In the beginning all American

Table 2.5. Gemini missions.

Mission	*Launch date*	*Crew*	*Result*
Gemini 1	April 8, 1964	None	First test flight, success
Gemini 2	January 19, 1965	None	Test flight, test of heat shield during reentry into the atmosphere
Gemini 3	March 23, 1965	Grissom & Young	3 orbits
Gemini IV	June 3, 1965	McDivitt & White	98-hour flight, 62 orbits, White performed the first U.S. spacewalk
Gemini V	August 21, 1965	Cooper & Conrad	120 orbits, 191-hour flight
Gemini VII	December 4, 1965	Borman & Lovell	220 orbits, 330-hour flight, first U.S. spacecraft rendezvous
Gemini VI-A	December 15, 1965	Schirra & Stafford	26-hour flight, first U.S. spacecraft rendezvous
Gemini VIII	March 16, 1966	Armstrong & Scott	First successful spacecraft docking, emergency landing after 10-hour 41-minute flight
Gemini IX-A	June 3, 1966	Stafford & Cernan	44 orbits, 72-hour flight, spacecraft rendezvous, spacewalk
Gemini X	July 18, 1966	Young & Collins	43 orbits, 71-hour flight, successful spacecraft rendezvous and docking, spacewalks
Gemini XI	September 12, 1966	Conrad & Gordon	71-hour flight, successful spacecraft rendezvous and docking, spacewalks, Gemini altitude record 1,189 km
Gemini XII	November 11, 1966	Lovell & Aldrin	94-hour flight, successful spacecraft rendezvous and docking, spacewalk

astronauts were male. In the early 1960s some efforts were made to initiate a "Girl Astronaut Program", and although female pilots took privately organized astronaut-like physical examinations they had little if any experience in high-performance aircraft and thereby fell short of NASA's basic requirements.

Fig. 2.9. Gemini VII spacecraft in flight, photographed from Gemini VI-A (source: NASA).

Fig. 2.10. NASA astronauts—group 3. They are (seated, left to right) Edwin E. Aldrin Jr., William A. Anders, Charles A. Bassett II, Alan L. Bean, Eugene A. Cernan, and Roger B. Chaffee. Standing (left to right) are Michael Collins, R. Walter Cunningham, Donn F. Eisele, Theodore C. Freeman, Richard F. Gordon Jr., Russell L. Schweickart, David R. Scott, and Clifton C. Williams Jr. (source: NASA).

2.4 SOYUZ

The development of the Soyuz spacecraft started in the early 1960s, initially to support a manned circumlunar mission. The first flight, Soyuz 1, was in April 1967. During the landing a malfunction led to the death of cosmonaut Vladimir Komarov. Soyuz 4 and 5 performed the first docking between two piloted spacecraft in January 1969. In an effort to upstage the American lunar missions, in October of that year Soyuz 6, 7, and 8 flew simultaneously in Earth orbit with a total of seven cosmonauts.

Hundreds of Soyuz spacecraft have flown in a number of forms, and it is still in use. The latest version is called Soyuz TMA-M (Fig. 2.11). It can carry up to three crew members and is the principal mode of transportation to and from the ISS. Launches are carried out from Baikonur using a three-stage Soyuz rocket (Fig. 2.12). The spacecraft has a three-module design, consisting of an orbital module for use by the crew in orbit, a module (in the middle) in which the crew ride for launch and landing, plus a service module. The orbital module and the service module are jettisoned when it is time to return to Earth, and left to burn up in the atmosphere. The rate of fall of the descent module is slowed using parachutes and engines in the base of the capsule. Each crew member has a custom-made seat lining and a shock-absorbing couch to enhance comfort during the impact of landing. The target area for returning Soyuz missions is in Kazakhstan.

Fig. 2.11. The Soyuz TMA-03M spacecraft (left) eases toward its docking with Mini-Research Module 1 (MRM-1) of the ISS, 2011 (source: NASA).

Fig. 2.12. Soyuz rocket with the Soyuz TM-14 capsule, MIR-92 (source: DLR).

A non-man-rated version called Progress can deliver up to 1,700 kg of cargo to the ISS. Prior to leaving the station it is filled with rubbish, which burns up together with the spacecraft when it reenters the atmosphere.

2.5 APOLLO

After Kennedy's speech NASA worked intensively towards the goal of putting a man on the Moon. The question was how to get there. Three options were

Fig. 2.13. John Houbolt at a blackboard, showing the lunar orbit rendezvous concept for lunar landings (source: NASA).

considered

- Direct ascent from the Earth to the Moon, requiring a very powerful rocket
- Earth orbit rendezvous in which the components of the spacecraft would be docked in an Earth-parking orbit prior to setting off for the Moon
- Lunar orbit rendezvous involving docking in a lunar-parking orbit.

The lunar orbit rendezvous approach was selected in June 1962. John Houbolt at NASA is seen lecturing on the concept in Fig. 2.13. The strategy was to launch the spacecraft into Earth orbit, transfer it to lunar orbit, separate the lunar module so that two of the three astronauts could land on the Moon while the third man remained in orbit in the command and service module with the propellant for the return journey. Once lunar surface activities were complete, the lower stage of the lander served as a launch pad for the upper stage, which returned to lunar orbit and docked with the command and service module. The ascent stage of the lunar module was discarded, and the main spacecraft returned to Earth to parachute into the ocean. The advantage of this strategy was that it did not require a huge rocket, but the disadvantage was the risk of something going amiss whilst attempting to dock two manned spacecraft in lunar orbit, far from the Earth. The next question was how to get the Apollo missions into space. The solution was the Saturn V, the largest rocket ever to be successfully flown. It had three stages and stood 111 m high. Fig. 2.14 illustrates the integration of the Apollo modules with the upper stage of the rocket. En route to the Moon, the command and service module separated from the rocket stage, rotated 180°, and docked with the lunar module, which was then released by the rocket. Only the command module returned to Earth, carrying the astronauts.

The Apollo program was operational from 1968 to 1972. Twenty-four astronauts left Earth orbit on Apollo spacecraft. Twelve astronauts walked on the Moon. No one has been back since. In total about 80 hours were spent working on the lunar surface. The crew assigned to the Apollo 1 mission were killed in a fire during a rehearsal on the pad a month before their scheduled launch.[1] Apollo 2 to 6 were unmanned tests in preparation for the manned flights, which are detailed in Table 2.6.

As the first manned spacecraft to orbit the Moon Apollo 8 was highly significant, especially to the Russians. The most famous flight of course is Apollo 11, which made the first lunar landing. Michael Collins remained in orbit around the Moon while Neil Armstrong and Buzz Aldrin descended to the surface in their lunar module (Fig. 2.15). The moonwalk was televised to a global audience, the blurry black-and-white images captivating millions and inspiring many children to become active in modern-day space activities. Returning to Earth, the command module landed in the ocean in the same manner as the Mercury and Gemini capsules had done. On their return to Earth the astronauts were held in quarantine for nearly three weeks in case they had contracted any extraterrestrial germs. They were transported back to America in the Mobile Quarantine Facility (Fig. 2.16 shows President Richard Nixon welcoming them home on the recovery ship) and transferred to the Lunar Receiving Laboratory in Houston. After their release they were treated like movie stars.

As the Apollo missions continued, the public's interest waned until the explosion that crippled Apollo 13 en route to the Moon. The uncertainty of whether the astronauts would return to Earth alive rekindled interest in the space program. Fortunately, they made it back. On Apollo 15, 16, and 17 a lunar rover (Fig. 2.17) was used to extend the area that the astronauts could explore on the surface. The final lunar crew took the picture in Fig. 2.18, which has entered popular culture as a symbol of the Earth. NASA had hoped for another three Apollo lunar landings, but these were canceled due to budgetary constraints.

The final Apollo mission in July 1975 docked with a Soyuz spacecraft in Earth orbit as a symbol of international cooperation in space activities, thereby drawing to a close the space race.

[1] The mission was originally named Apollo 204, but was officially assigned the name Apollo 1 following the accident.

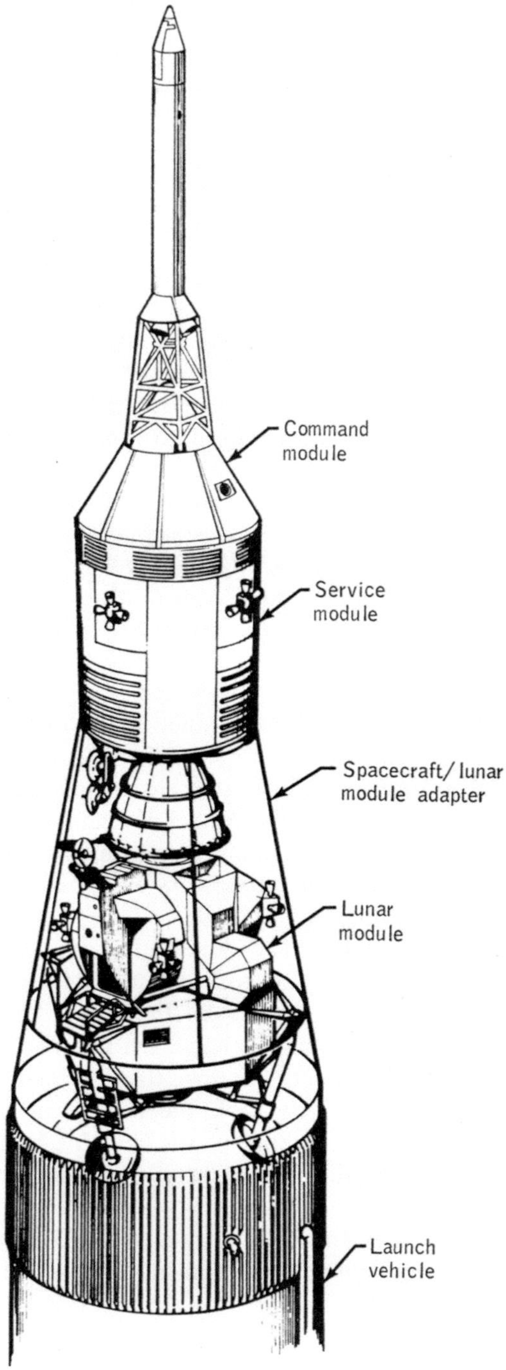

Fig. 2.14. Illustration of the stacking of the Apollo command module, service module, and lunar module at launch (source: NASA).

Table 2.6. Apollo manned flights.

Mission	*Launch date*	*Crew*	*Result*
Apollo 7	October 11, 1968	Schirra, Eisele, Cunningham	Command and service modules tested in Earth orbit, mission duration 260 hours
Apollo 8	December 21, 1968	Borman, Lovell, Anders	Lunar orbit
Apollo 9	March 3, 1969	McDivitt, Scott, Schweickart	Lunar module tested in Earth orbit
Apollo 10	May 18, 1969	Stafford, Young, Cernan	Full test of Apollo except the lunar landing
Apollo 11	July 16, 1969	Armstrong, Collins, Aldrin	First lunar landing, 21.6 hours spent on the Moon, 2 hours 14 minutes moonwalking
Apollo 12	November 14, 1969	Conrad, Gordon, Bean	Lunar landing, 31.5 hours spent on the Moon, 7 hours 50 minutes moonwalking
Apollo 13	April 11, 1970	Lovell, Swigert, Haise	An on-board explosion of an oxygen tank on the way to the Moon led to an aborted landing on this dramatic mission
Apollo 14	January 31, 1971	Shepard, Roosa, Mitchell	Lunar landing, 33.5 hours spent on the Moon, 9 hours 25 minutes moonwalking
Apollo 15	July 26, 1971	Scott, Worden, Irwin	Lunar landing, 66.9 hours spent on the Moon, 18 hours 20 minutes spent exploring the lunar surface, a lunar rover was used for the first time
Apollo 16	April 16, 1972	Young, Mattingly, Duke	Lunar landing, 71 hours spent on the Moon, 20 hours 14 minutes moonwalking
Apollo 17	December 7, 1972	Cernan, Evans, Schmitt	Lunar landing, 75 hours spent on the Moon, 22 hours 5 minutes moonwalking

Fig. 2.15. Buzz Aldrin during the Apollo 11 moonwalk, the lunar module, and experiments that had been deployed on the lunar surface. The photograph was taken by Neil Armstrong (source: NASA).

Fig. 2.16. President Richard Nixon visits the Apollo 11 crew in quarantine aboard the recovery ship (source: NASA).

Fig. 2.17. David R. Scott, Apollo 15 Commander, is seated in the Lunar Roving Vehicle (LRV) during Apollo 15's first lunar surface extravehicular activity. The photograph was taken by James B. Irwin (source: NASA).

Fig. 2.18. View of the Earth as seen by the Apollo 17 crew en route to the Moon (source: NASA).

2.6 EARLY SPACE STATIONS

2.6.1 Salyut and Mir

The Soviet Union also had plans to land men on the Moon, but when the Americans beat them to it they shifted the emphasis of their program to building a space station. The Salyut program led to the first ever space station in Earth orbit. It included both civilian (DOS[2]) and military (OPS[3]) stations, with the crews being ferried by Soyuz spacecraft. The military space station project, Almaz, was conceived in 1964 by Vladimir Chelomey's facility, currently known as NPO Mashinostroyenia,[4] in the Moscow suburb of Reutov. It was indeed a grandiose concept which impresses even today: a 20-ton station was to have been retained in orbit for several years and hosted a succession of spacecraft delivering cosmonauts and supplies for photographic and radar reconnaissance missions.

For a brief period in 1966–1967 a variant of the Soyuz was envisaged as the crew ferry. However, Chelomey proposed to let cosmonauts go into space with the station by adding a capsule for the ride into orbit and return to Earth. Each station would be discarded after the return of its crew. As the concept developed, Chelomey proposed a large spacecraft named the TKS (Transport Supply Ship) of similar mass to the station, to enable a series of crews to visit a station. As it turned out, delays in developing the TKS led to the decision in 1970 to use a version of the Soyuz to deliver crews to Almaz. This was seen as an interim measure until the TKS was ready.

The Almaz program was approved in 1967. Although the decree set the inaugural launch date for 1970 to celebrate Lenin's 100th birthday, Chelomey regarded this as unrealistic. In 1969, in response to the successful Apollo 11 lunar landing, the Soviet leadership ordered the Korolev design bureau—nowadays S.P. Korolev Rocket and Space Corporation (RSC Energia)—to launch a space station that would make use of elements of the Almaz design. Chelomey had to transfer the Almaz drawings to his rival. By 1970 Chelomey had manufactured eight prototypes and a pair of flight-worthy OPS hulls. The latter were transferred to the Korolev design bureau, which modified them to use some of the subsystems of the Soyuz spacecraft and fly as the DOS. In addition, Chelomey was directed to use Korolev's Soyuz instead of the TKS during the flight-testing phase of his OPS. The first DOS was launched in 1971 under the name Salyut. Although the arms race was being pursued vigorously, the Korolev bureau did not incorporate into its Salyut the reconnaissance capabilities of the Almaz concept, installing instead instruments for scientific research. The transfer of know-how to Korolev's bureau delayed the Almaz program for two years. Western observers were confounded by the amazing situation of two Soviet manned station programs running in parallel.

There were persistent reports that Almaz was equipped with weapons, creating

[2] Dolgovremennaya Orbitalnaya Stanziya; Long-duration Orbital Station.

[3] In English, the acronym stands for Orbital Piloted Station.

[4] In English, Machine Building Scientific Production Association.

a "fortress in space". The basic space station block was shaped like a cylinder with two different diameters, 4.15 m and 2.9 m. Its overall length was about 11 m, and it had a mass of about 15 tons. In the original version, the crew capsule, known as the VA,[5] would be mounted on the smaller diameter section of the station. An axial docking unit was included at the rear of the station, connected to an airlock for making spacewalks and for ejecting a small Information Return Capsule (KSI). Two solar panels at the rear spanned 23 m, with an area of 52 m^2 generating 3.1 kW. Also at the rear of the station were two orbit correction engines, each of which delivered a thrust of about 4,000 N.

The KSI could return film from the main reconnaissance camera, Agat 1. The capsule had a mass of about 360 kg and a diameter of 0.85 m. After ejection from the space station it was spun up by a small solid rocket motor for stability while the main retro-rocket fired, then another small solid rocket de-spun the capsule and the motor pack was jettisoned before reentry. Before landing, the drogue chute pulled the capsule out of the heat shield. An airbag was inflated to cushion the impact. The Agat 1, a colossal telescope, was approximately 1 m in diameter with a folded optical system having a focal length of 6.4 m. It was mounted in the floor of the station. The crew could use it to photograph targets such as airfields and missile complexes. An optical sight enabled them to "come to a standstill" above a target. Furthermore, there was a topographical camera and a device for taking pictures in the near-infrared range. The film frames were quite large, 50×50 cm, and according to U.S. estimates would have had a resolution of at least 50 cm assuming that the film was capable of resolving 100 line pairs per millimeter. Data from this reconnaissance payload could be returned not only using the KSI, but also through a radio link to the ground. It seems that reconnaissance film could be developed on board, scanned, and transmitted to the ground all within 30 minutes.

To be able to test Almaz without either the TKS or the VA being ready, the "base block" was launched alone and the cosmonauts sent up later on a Soyuz ferry. Even though white wedges on the smaller diameter section resemble antenna structures, this early version of the Almaz probably did not have the side-looking radar that was under development for the final station. This scaled-back configuration was launched as Salyut 2 on April 3, 1973, as Salyut 3 on June 24, 1974, and as Salyut 5 on June 22, 1976. Salyut 2 was crippled by an on-board explosion before a crew could be sent up. Salyut 3 had one crew, who spent 16 days in space. Salyut 5 hosted two crews, the first one making a 49-day flight and the second spending 18 days in space. Flight testing of the OPS was completed in 1978, but the Ministry of Defense decided not to accept it for operational service because by that time its capabilities were inferior to automatic reconnaissance satellites.

The story of how the DOS space station program was created is fascinating. One can easily imagine the frustration felt by the leadership of the Korolev design bureau when Apollo 11 landed on the Moon while their equivalent program

[5] Vozvrashaemiy Apparat; Return Apparatus.

(N1-L3) was in deep trouble having suffered two catastrophic launch failures, leaving no alternative role for their Soyuz spacecraft. Their rival, Vladimir Chelomey, was working on a high-profile military space station. At the end of 1968 the basic space station structures were being readied in the Khrunichev factory in Fili in the outskirts of Moscow. The OPS–Almaz design included the use of gyrodynes (electrically driven wheels similar to control moment gyroscopes) and other novel features such as a central computer called "Argon".

The first stirrings concerning the Korolev bureau's role in a space station program occurred in August 1969. Three leading engineers proposed a whirlwind program to overtake Almaz. The original idea was to equip a section of the hull of an R-7 rocket with Soyuz subsystems in order to form a space station. As the concept was studied subsystem by subsystem, it was realized that perhaps a mixture of Soyuz and Almaz hardware would be better. That is how a tight group of "conspirators" driven by the engineer-cosmonaut Konstantin Feoktistov devised the idea of outfitting Almaz hulls with Soyuz subsystems. After the semi-successful Soyuz "troika" mission of October 1969, this group, partly behind the back of their own managers, managed to convince government leaders to support the idea. The political leadership of the space program realized that because Almaz was a secret project it required a "cover", and making it visible in the manner proposed by the Korolev bureau would serve this purpose. The enthusiastic Korolev engineers explained that by combining the hulls of Almaz with Soyuz-derived subsystems, it would be possible to put a space station into orbit in a year's time! Almaz was not expected to fly for at least two years (three and a half as it turned out) so this was excellent news. Furthermore, Korolev engineers (probably erroneously) gave assurances that their plan could be implemented without adversely affecting the N1-L3 manned lunar-landing program. The engineers were told that the proposal would be submitted to the Military-Industrial Commission (VPK), and the decision would be made by the Central Committee in January 1970. Decrees "On creation of the DOS using Almaz as a basis" were issued on February 9 and 16, 1970.

Work began that month to modify the two Almaz hulls to the DOS configuration. In fact, DOS was simpler than Almaz with its very sophisticated payload. The main modifications were achieved by augmenting the basic Almaz structure

- An Exit Section with a passive Soyuz docking system and airlock was added to the front of the station
- An Engine Section modified from that of Soyuz was added to the rear of the station
- Soyuz solar panels were mounted to these two new sections.

This "kidnapping" of the space station program was brilliant! The much delayed, and secret, Almaz promised no immediate response to Apollo. The time for DOS was ripe. In just 16 months the Korolev bureau managed to adapt Chelomey's design to make extensive use of Soyuz subsystems. Early in January

1971 *Pravda* forecast the orbiting of a space station "in the very near future". The first DOS was shipped to Baikonur in February 1971, where work on it continued day and night. On April 15, newspapers reported rumors from Moscow about an imminent experiment involving a manned space station. Salyut 1, with a dry mass of 18,425 kg, was launched from Baikonur on April 19.[6]

Soyuz 10 was launched with three men on board on April 22. Because of a fault in the docking mechanism of the Soyuz no "hard dock" with Salyut could be achieved. Another three-man crew was launched on June 6, and successfully entered the space station. Their stay on board lasted until June 29, when the descent was initiated. Unfortunately, when the orbital module was jettisoned a pressure equalization valve in the capsule carrying the crew prematurely opened, reducing the cabin to vacuum and causing the deaths of Georgi Dobrovolski, Vladislav Volkov, and Viktor Patsayev. On October 11, having exhausted the fuel needed to counteract the effects of air drag, the station reentered the atmosphere. One result of the tragedy was that the crew of the Soyuz spacecraft was reduced from three to two to accommodate pressure suits in case of cabin depressurization during launch and reentry.

Attempts to launch further DOS space stations were made in 1972 and 1973. The first attempt on July 29, 1972 never reached orbit and the second attempt failed when the space station lost all orientation capability soon after achieving orbit on May 11, 1973. Its role was masked by giving it the name Cosmos 557; it burned up in the atmosphere after 11 days. Soviet space station operations continued with the launch of the first successful Almaz on June 24, 1974 and its subsequent occupation for 15 days by a crew of two. It was named Salyut 3 in an attempt to confuse the general public into believing that there was only one type of space station, but it was easy for observers in the West to spot the difference between OPS and DOS since they used different radio frequencies and transmitted different signals, were placed at different altitudes, and had different crew compositions with the OPS being manned only by military officers! On December 26, 1974 another DOS successfully reached orbit. Named Salyut 4, it was occupied by two crews for a total of 93 days. Salyut 4 had the same limitation as all stations launched up to this time—it could not be refueled, so when it ran out of orbital maneuvering capability the mission was essentially over. Atmospheric drag would lower the orbit so that the station reentered the atmosphere and burned up. Salyut 5, an Almaz launched on June 22, 1976, had the same design characteristics and two crews spent a total of only 65 days on board.

A major step forward in space station technology was achieved with the launch of Salyut 6 on September 29, 1977, a DOS with a docking port at each end. The port at the rear was equipped to transfer propellants. In early 1978 an unmanned version of the Soyuz named Progress was introduced for cargo and propellant delivery. These cargo ships could dock with the station while it was occupied by a crew. This was the first step towards creating a continuously and

[6] On April 28, Radio Sweden reported that informed sources in Moscow indicated that Salyut 1 was to have been launched March 23, but was delayed for technical reasons.

permanently manned station. By this means, Salyut 6 was occupied for a total of 676 days by five long-duration crews and eleven crews visiting for about a week each. In a major publicity initiative by the Soviets, the visiting crews included cosmonauts from other socialist countries. In 1981, after the final crew had departed, a TKS transport ship developed for the Almaz program was launched as Cosmos 1267 and docked with Salyut 6. A solo test flight of the TKS had been carried out in 1977 as Cosmos 929. Rumors in the trade press indicated that both TKS vehicles had released a reentry capsule about a month after launch. Soviet media reports said that the massive Cosmos 1267 was a prototype for a future space station module.

The decision to develop the permanently occupied modular space station that was to become Mir was taken in 1976 and its overall configuration was decided in 1979. The Almaz military space station had been canceled in 1978, and the Mir project was delayed, so in the late 1970s it was decided to use the backup for Salyut 6, and this was launched on April 19, 1982 as Salyut 7 (Fig. 2.19). Cosmonauts from friendly countries—including France—made short visits. With the cancellation of Almaz, a large proportion of the experiments conducted on Salyut 7 had military objectives. These were in turn supported by docking two further TKS vehicles in succession with the station: Cosmos 1443 in March 1983 and Cosmos 1686 in October 1985, the former with a VA which returned cargo to Earth. Salyut 7 was occupied for a total of 812 days by 10 crews.

Upon the departure of a crew in November 1985 after one cosmonaut fell ill, the space station inadvertently drained its batteries and drifted out of control, essentially dead. This caused a couple of problems. On the one hand, there was unfinished business aboard that station, in particular military experiments involving Cosmos 1686. On the other hand, there was political pressure to launch the base block of Mir and occupy it promptly. To reconcile these two conflicting demands, Mir was launched on February 19, 1986 and Soviet space planners

Table 2.7. Salyut space stations.

Station	*Type*	*Launch date*	
Salyut 1	DOS	1971	Manned, crew perished during the return journey
Salyut 2	Almaz	1973	Unmanned due to an accident
Salyut 3	Almaz	1974	Manned
Salyut 4	DOS	1974	Manned
Salyut 5	Almaz	1976	Manned
Salyut 6	DOS	1977	Manned, first Soviet space station with two docking ports
Salyut 7	DOS	1982	Manned

pulled off a really magnificent operation by sending a crew first to Mir, then to Salyut 7, where they stayed for 50 days to complete the military experiments, prior to returning to Mir. This was achieved by placing Mir into an orbit that had its orbital plane very close to that of Salyut 7 and then using the engines of Progress freighters to maneuver Mir to control the planes and phasing of the orbits of the two stations in order to drop off and later pick up the Soyuz carrying the crew.

The Mir base block was a modernized DOS. In the spring of 1987 modules started to arrive to build up the station. These were all based on the service module part of the TKS, often called the FGB.[7] The first to be added, in April 1987, was the Kvant 1 astrophysical observatory. After docking, the service module of the TKS, which had performed the rendezvous, detached and withdrew. Kvant 1 had been intended to be sent to Salyut 7, but was reconfigured for Mir—in particular, it required plumbing to route propellants to the base block from docked Progress ships. Having gyrodynes, it would also improve the station's attitude control capabilities. Its payload was a suite of instruments for X-ray and ultraviolet astronomy. Kvant 2 was added in December 1989. This was an "augmentation" module with a cargo compartment, an experiment compartment, and an airlock for spacewalks. It also had a system to regenerate water from urine, a shower, a water storage system, and additional gyrodynes. Its scientific apparatus included a high-resolution camera, spectrometers, X-ray sensors, and a fluid flow experiment. The third module, known as Kristall, was added in June 1990. It had two main sections. The first was used for materials processing, astronomical observations, and a biotechnology electrophoresis experiment. The second section was a compartment with two docking ports, initially intended for use with the canceled Soviet Buran space shuttle and actually used for docking the U.S. Space Shuttle. The docking compartment also contained a camera used for Earth resources experiments.

After Kristall had been docked with Mir, political events started to dominate the Mir program. Although the dissolution of the Soviet Union at the end of 1991 led to the cancellation of the Spektr module and the loss of funding for the Priroda module, in September 1993 the U.S. and Russia signed an agreement for a two-phase program to build the ISS. In Phase One (1994–1996, extended to 1998) NASA paid Russia $100 million per annum to keep a "guest astronaut" on Mir, with the U.S. Space Shuttle flying to the station. These funds enabled Russia to launch the final two modules. In total, NASA paid $472 million for nine Shuttle dockings. This phase of the project was primarily to enable NASA to learn from Russian experience of operating a space station, and to promote cooperation between the space agencies prior to moving on to the ISS. The Spektr module arrived in June 1995 and served as the living quarters for American astronauts. It was designed for remote sensing of the Earth's atmosphere and surface, but also included NASA experiments. Its four solar arrays provided approximately half of the station's power. Unfortunately, Spektr became unusable after a

[7] Funktsionalno Gruzovoy Blok; Functional Cargo Block.

Fig. 2.19. Salyut 7, the second Soviet space station equipped with two docking ports (source: RSC Energia).

collision with a Progress cargo ship in 1997. The final module, Priroda, arrived April 1996 with apparatus supplied by 12 countries to develop remote-sensing methods using microwave, visible, near-infrared, and infrared sensors. It was equipped with a large synthetic aperture radar antenna.

In its fully assembled state Mir was 19 m long and roughly 30 m wide, and had a mass of about 130,000 kg (Fig. 2.20). It was capable of supporting a resident crew of three and larger numbers for short-term visits. It hosted 125 spacefarers from 12 nations. They arrived on the 31 spacecraft that docked, including 9 flights by U.S. Space Shuttles. Seven NASA astronauts made long-duration visits, including one woman, Shannon Lucid. Their visits varied in length from 115 to 188 days, and in total American astronauts spent 975 days on Mir. This was very

Fig. 2.20. Russia's Mir space station is backdropped by Earth in this photo taken by a departing NASA crew on June 12, 1998 (source: NASA).

useful experience in preparation for the ISS. A total of 64 Progress ships delivered cargo and consumables. Having been occupied for a total of 12.5 years of its 15-year life, it set a record for the longest uninterrupted human presence in space of 3,644 days which was not surpassed by the ISS until 2010, and it still holds the record for the longest single human flight, when Valeri Polyakov spent 437 days 18 hours on board. Over the years, Mir served as a laboratory for some 23,000 scientific and medical experiments.

In 1998 the Shuttle flights to Mir ended and the construction of the ISS began, with the result that it became impossible to keep Mir going by selling visits to foreign customers, so a modified Progress ship deorbited the station on March 23, 2001, with the debris falling into the Pacific near Fiji.

2.6.2 Skylab

The first American space station, Skylab (Fig. 2.21) was launched by a Saturn V on May 14, 1973. At about 77 tons, it is the largest spacecraft ever to be launched by a single rocket. Unfortunately, an integrated thermal and micrometeoroid shield was torn off during the ascent into orbit and this led to the loss of one of

Fig. 2.21. Skylab with its improvised sunshield (source: NASA).

Fig. 2.22. A close-up view of astronaut Jack R. Lousma, Skylab 3 pilot, taking a hot shower in the crew quarters of Skylab (source: NASA).

the solar panels. The project was rescued by astronauts erecting an improvised "sunshade".

The enormous volume of the station enabled many bulky items to be carried. For example, a shower was fitted to maintain cleanliness during a long mission. Fig. 2.22 shows an astronaut testing this in space. In deploying the facility, the shower curtain was pulled up from the floor and attached to the ceiling. The water came through a push-button showerhead on a flexible hose and was drawn off by a vacuum system. It proved to be more trouble than it was worth because in the microgravity environment in space water does not behave as on Earth, forming droplets that drift rather than "fall down", and the astronauts reverted to washing themselves using cloths. There was an educational aspect to the program, with a national contest for secondary school pupils to suggest experiments. There were 25 winners, and 19 proposals were deemed suitable for flight, including one to determine whether weightlessness would cause spiders to spin abnormal webs.

There were three missions to Skylab in 1973 and 1974, lasting 28, 59, and 84 days, respectively. They proved the feasibility of living in space for prolonged periods—at the end of the program the longest Soviet flight was 24 days. Skylab reentered the atmosphere when its orbit decayed in 1979.

2.7 SPACE SHUTTLES

The successor to Apollo in the United States was the Space Shuttle. The basic idea behind the Shuttle was that reusability and frequent flights (initially it was expected to launch a Shuttle almost every week) would lead to cost effectiveness. The Phase A study for the Space Shuttle program began in 1969, and the initial design work in 1971. President Richard Nixon approved the program on January 5, 1972 and Rockwell was selected as the contractor for the Orbiter the same year. Political approval of the Shuttle was not driven by the scientific and technical achievements that could be achieved, but rather by preserving the image of the United States as a leader in space activities, by potential military applications of the Shuttle, and by creating employment in key states with respect to gaining votes in the upcoming presidential election.

The enormous financial input to NASA to achieve success with the early manned spaceflights came to an end during the Nixon presidency when he made it clear that space activities had to be funded with regard to other priorities of national importance. This meant tighter budget constraints. Riding on earlier successes NASA had a vision of continued space activities with regular flights, civil and military applications, and even terrestrial point-to-point transportation. There were many configurations considered for the new space transportation system. The initial plan was for a totally reusable system in which a large supersonic aircraft released the Orbiter at high altitude to complete the ascent to orbit, but budget constraints resulted in a partially reusable system. It was clear that liquid rocket boosters would have high development costs, but relatively lower operational costs which considering the planned launch frequency would have been an attractive option. Solid rocket boosters would cost more to operate but had a lower development cost which meant that selecting them deferred some of the cost issues to the future. This was a contributory factor in the final selection of solid rocket boosters. The trimmed-down Shuttle design selected in the end was a partially reusable system using a large external fuel tank and two solid rocket boosters (Fig. 2.23).

To protect the aluminum structure of the Shuttle in space and during reentry into the atmosphere, the Orbiter had to have a thermal protection system. Many thousands of ceramic tiles of various shapes were individually fixed onto the skin with adhesive. Unfortunately, being very fragile, the tiles were easily damaged and torn off. The Shuttle had a minimum crew capacity of two astronauts and a maximum of eight, the largest crew size on a space transportation vehicle to date. It had a spacious payload bay in which to operate apparatus or to carry satellites to be released. During the first 100 Shuttle launches the Shuttle deployed a total of 25 commercial communications satellites. In the 1990s a docking system was placed into the payload bay to enable the Shuttle to dock with Mir and later the ISS.

Exactly 20 years after Gagarin's spaceflight, the first Space Shuttle, Columbia, was launched from the Kennedy Space Center in Florida, piloted by John Young and Bob Crippen. The launch is shown in Fig. 2.24. The external tank was originally painted white, but after the first few missions it was realized the paint was an unnecessary mass addition and thereafter they were produced showing their orange–brown insulation. The Shuttle is officially named the Space Transportation System, and each mission is labeled with the acronym STS and a number, so the first mission was STS-1.

Five operational Orbiters were built (Table 2.8). There was an additional Orbiter named Enterprise, but it was used only for tests during the development phase of the program.

Europe obtained a role in the Space Shuttle program by contributing Spacelab, a laboratory module that fitted into the payload bay. Its first flight was STS-9, launched on November 28, 1984. Two Spacelab missions, D-1 in 1985 and D-2 in 1993, were carried out under German management, with two German astronauts on each crew. Spacelab was eventually superseded by a U.S.-made commercial module called Spacehab.

Challenger, Columbia, and their crews were both lost in accidents. Challenger exploded shortly after takeoff on January 28, 1986 as observers looked on in shock. The timing couldn't have been worse. Confidence in the ability to launch astronauts into space had increased to the extent that it had been decided to fly a teacher into space, Christa McAuliffe, making this mission of special interest to the public. The program was immediately put on hold. A presidential commission was established to determine the cause of the problem and recommend how best to resume operations. Among the illustrious members of the commission was Richard Feynman, a winner of the Nobel Prize in Physics. It was concluded that a rubber ring had been degraded by cold weather, failed to form a seal, and allowed a plume of flame to leak from one of the solid rocket boosters; this in

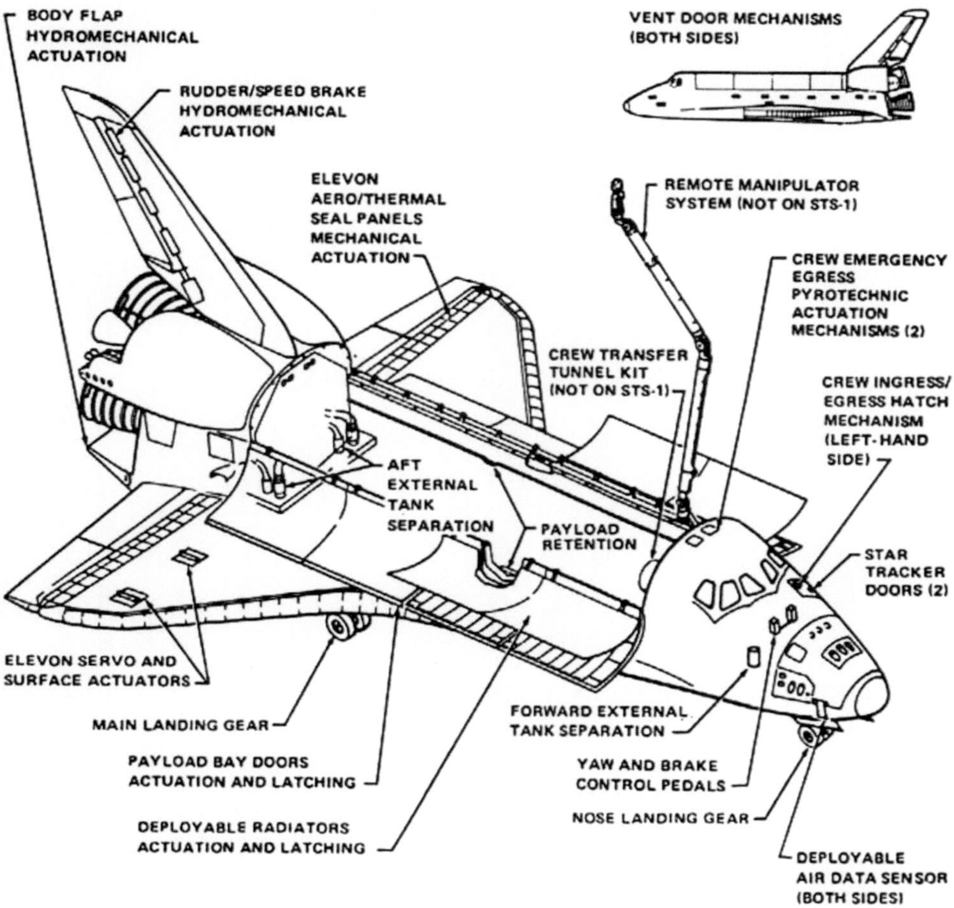

Fig. 2.23. Detail of the Space Shuttle Orbiter (source: NASA).

turn led to the loss of the vehicle. Conditions on that day were the coldest of any Shuttle launch to date. The issue was dramatically illustrated at a press conference by Feynman dropping a piece of rubber into a glass of chilled water. Problems in the management structure at NASA were also exposed. Feynman wrote in his personal observations regarding the Shuttle's reliability: "For a successful technology, reality must take precedence over public relations, for nature cannot be fooled." The program resumed on September 29, 1988, with the launch of Discovery as STS-26.

During the period 1982–1992 NASA launched 11 Shuttle flights with classified payloads for the U.S. Department of Defense. The military was not happy to rely on the Shuttle. Following the Challenger accident and launch delays they reverted to using conventional rockets to access space and recently have developed their own unmanned robotic Shuttle, X-37B.

Fig. 2.24. STS-1 Columbia launches from Kennedy Space Center, Florida, April 12, 1981 (source: NASA).

Table 2.8. U.S. Space Shuttle flights by Orbiter.

Orbiter	*Number of flights*	*Period of operation*
Columbia	28	1981–2003
Challenger	10	1983–1986
Discovery	39	1984–2011
Atlantis	33	1985–2011
Endeavour	25	1992–2011

Columbia disintegrated during atmospheric reentry on February 1, 2003, but its loss was traced back to an event during the ascent when a large piece of foam came loose from the external tank and damaged one wing of the Orbiter to the extent that it was unable to withstand the heat of reentry. The program was halted again for an investigation and necessary changes. Discovery resumed operations with STS-114 on July 26, 2005. One of the new procedures was to use a lengthened robotic arm to carry out an inspection of the tiles in orbit. When there proved to be tile damage during this mission, there was another pause for a year before the program resumed with Discovery lifting off on July 17, 2006 as STS-121 (which was being flown out of sequence). With the exception of a mission to service the Hubble Space Telescope, all remaining flights were devoted to completing the construction of the ISS. Atlantis launched on the final Space Shuttle mission, STS-135, on July 8, 2011. It carried a crew of four. The purpose of the mission was to deliver a large quantity of supplies to the ISS, essentially in anticipation of delays in establishing a reliable resupply of the station using future commercial carriers. This done, the Shuttle was retired and the surviving Orbiters are now on display in museums.

2.8 THE INTERNATIONAL SPACE STATION

To trace the ISS back to its origins is a major task because many factors contributed to its creation, so only a rough outline will be given here.

In the early 1980s the Cold War was still very much a reality and the space station efforts by the Soviet Union were seen by the U.S. as a "strategic challenge" in much the same way as the Soviets saw the U.S. Space Shuttle as a potential threat. In fact, the Shuttle was conceived as a tool for building a space station, and as its development wound down there was room in NASA's budget for a significant new program and a station was, as James Beggs, the agency's administrator, put it, "the next logical step". Visions of "a permanent presence in space" and the benefits this would deliver to the nation also abounded in the aerospace community.

Reagan's space station

As envisaged in the mid-1980s the U.S. space station marked a departure from previous major space programs. In March 1984 Robert Freitag, Deputy Director of Advanced Programs at NASA's Office of Space Flight, explained that the station would be a case of "design to capability" rather than, as previously, to "cost", "requirements", or "objectives". The use to which the station would ultimately be put was impossible to foresee. A more flexible approach was therefore needed. NASA also imagined that it would be possible to "get rid of the ten thousand men mission control army on the ground" by letting the station astronauts "control themselves". Another concept that was discussed as part of the design was extensive "use of in-hand technology" and standardization of the modules from which the station would be assembled, even to the extent that it would be possible to "buy it by the yard". The capabilities foreseen for the space station in the early 1980s were

- a laboratory for space research and space technology
- a permanent observatory of the sky above and the Earth below
- a transport node (refueling and cargo depot)

- a service facility and data relay node for free-flying laboratories
- a microgravity "factory" facility
- an assembly point for large structures (e.g., antennas)
- a storage facility for equipment spares, consumables, and backup subsatellites.

Freitag expected metal cutting for fabrication of the hardware to start in 1987 and stations operations to commence in 1993–1995; it was an optimistic schedule!

The space station program was announced by U.S. President Ronald Reagan in his State-of-the-Union address to Congress on January 25, 1984, shortly after the Shuttle had been declared operational: "We can follow our dreams to distant stars, living and working in space for peaceful economic and scientific gain."

The early design concepts for the station had fascinating names like the "Power Tower" and the "Dual Keel". An important part of all of the designs were framework trusses on which to mount solar panels, thermal radiators, and pressurized modules. After preliminary design work, the detailed design phase was initiated in December 1987.

As a "riposte" to the Soviets naming their modular space station "Mir" (Peace), in June 1988 Reagan gave the U.S. station the name "Freedom".

Europe's contribution

Europe enthusiastically joined the space station program. It felt that it had missed out on the Shuttle, being limited to providing Spacelab for carriage in the cargo bay, and so in 1985 it approved an ambitious program named Columbus.

The Columbus program was initially envisaged with three elements: an Attached Pressurized Module (APM) that would be integrated into the station; a Man-Tended Free-Flyer (MTFF) laboratory that would orbit nearby and be periodically visited by the Hermes mini-shuttle launched by the new Ariane 5; and a Polar Platform Facility (PPF). However, the costs escalated far beyond expectations. First, the French-led Hermes ran into problems of feasibility and costs, and was canceled. Thus, the MTFF was also canceled and the PPF was revised from an in-orbit serviced spacecraft to a remote-sensing satellite that was launched by Ariane 5 in 2002 as Envisat. The APM was renamed the Columbus Orbital Facility and later just Columbus. It was launched by Shuttle on February 7, 2008 and attached to the ISS. Columbus has research tools for biology, physiology, materials science, and fluid science. Slightly reminiscent of the MTFF is the Automated Transfer Vehicle (ATV) built by Europe.

The ATV is launched from Kourou by Ariane 5 and can carry up to 6.6 tons of cargo to the ISS. At the station it acts as an integrated module for up to 6 months and is used for orbit boosts. After that it is loaded with waste, detaches, and reenters the Earth's atmosphere where it is destroyed. ATV-1, Jules Verne was launched on March 9, 2008, followed by ATV-2, Johannes Kepler on February 16, 2011 and ATV-3, Edoardo Amaldi on March 23, 2012.

Japan's contribution

Japan's participation in the station program is as old as Europe's. Its contribution is the Japanese Experiment Module (JEM), also known as "Kibo" (Hope). At 14.8 tons, it is the largest single module of the ISS. It comprises a pressurized module, a facility exposed to space, an experiment logistics module, and a remote manipulator system. These were delivered on board three Shuttle missions. Japan also contributes its 16.5-ton H-II Transfer Vehicle (HTV), which is an unmanned spacecraft used to resupply the ISS in general and the Kibo module in particular. Design work on the HTV began in the early 1990s, and the first HTV was launched from the Tanegashima Space Center on September 10, 2009 using the Japanese H-IIB rocket.

Development kicks off

In April 1984 the Space Station Program Office at the Johnson Space Center in Houston issued a station configuration which was intended to serve as the basis for further planning. This concept, known as the "Power Tower", was based on a vertical truss or keel that had five modules attached at its Earth-facing end and solar arrays at the top end. The idea was that placing the heavy mass at the bottom end would act to stabilize the entire structure in the gravity gradient. Industry started definition studies and preliminary design work in April 1985.

In March 1986 the configuration was modified to the "Dual Keel" design, with the modules placed at the center of gravity to improve their microgravity environment. The truss was now two large keels, joined by end spars on which remotely controlled instruments could be mounted. A plan to base an Orbital Maneuvering Vehicle at the station was postponed.

Following the Challenger accident, a Critical Evaluation Task Force was set up to reassess the viability and safety of the station design. Meanwhile, post-Challenger safety concerns led to changes in the assembly plans. The Johnson Space Center had expressed misgivings about the amount of extravehicular work needed to assemble the station. The goal had been to limit this to 500 hours per year, but some estimates indicated a figure four times as high. In September 1986 the total development cost for the Dual Keel was estimated at $18.2 billion (in FY1989 dollars), with the first element due to be launched in January 1994—about when NASA had originally expected the station to be fully operational.

The Space Station Intergovernmental Agreement was signed in September 1988. Canada would receive 3% of the capability of the U.S. research modules in return for its contribution to the program—a robotic arm. Europe and Japan would keep 51% of their own modules. Six Americans and two other astronauts would be permanently based on the station. The plan published in 1987 said that each crew member would serve a 90-day tour, but this was soon increased to 120 days to reduce the number of Shuttle flights per year. In the spring of 1987 the project ran into financial problems in Congress, which demanded a less costly design. Furthermore, budget limitations more or less suspended the program pending the outcome of the presidential election in 1988. Nevertheless, NASA signed 10-year contracts for development of the station in September 1988, thereby finally advancing the project to the hardware phase.

Bush's space station

George Bush endorsed Space Station Freedom in August 1988 during his campaign for the presidency, and reaffirmed his support following his election. The design was revised again in late 1989—"Scrub 89"—after NASA's 1990 budget was reduced. The objective was to reduce the number of assembly flights. It was decided that the station would be permanently occupied by 1997, but the launch of the modules of the international partners would be delayed. The power available to station users would be cut by 15 kW to 30 kW. The new plan was also dependent on the introduction of a new solid rocket motor to correct the flaws that caused the Challenger accident, and increase the payload capacity of the Shuttle.

In a speech on the 20th anniversary of the first landing on the Moon, Bush announced the Space Exploration Initiative. The station would be one of the elements of this bold vision for returning humans to the Moon and onwards to Mars.

The Preliminary Design Review of the station at the element level in March 1990 showed it to be way over budget, 23% too heavy, too complicated to assemble, and providing too little power to users. Congress demanded another redesign. This was initiated in November 1990 and in March 1991 NASA revealed another design. This would reduce the number of Shuttle assembly flights to four per year. There would be no free-flying platforms, and the research emphasis would be life sciences and microgravity. The Preliminary Design Review at the system level was in November 1991.

Although the likely costs of the Space Exploration Initiative far exceeded what the President and Congress had envisaged, Bush continued to argue for it. Ironically, the space agency was skeptical, fearing that such a massive new program would threaten both the Shuttle and Space Station Freedom. Vice President Dan Quayle, chairing the National Space Council, became increasingly dissatisfied with NASA's commitment. This conflict led to the dismissal of NASA Administrator Richard Truly in February 1992. He was replaced in March 1992 by the energetic Daniel Goldin. During 1992 NASA tried to satisfy the White House by proposing a manned lunar outpost, even though it would have required the development of a Saturn V–class booster and would have cost $25 billion!

During President Bush's final year in office, NASA continued to struggle with the space station. Boeing was selected as the prime contractor in August 1992. Projects like the National Aerospace Plane and the heavy-lift booster had to be sacrificed to save the station. It had been intended that station astronauts would be delivered and retrieved only using Shuttles. After the loss of Challenger it was realized that if the fleet were again grounded this would strand the crew of the station in space. It was therefore decided that the station should have a "lifeboat", a Crew Emergency Return Vehicle. Lifting-body designs were studied, and the Soyuz was studied as an interim option.

Clinton's space station

In the election campaign of 1992, Bill Clinton, the Democratic candidate, endorsed Space Station Freedom but not the development of a heavy-lift launcher, and rejected the Space Exploration Initiative. On taking office Clinton directed NASA on March 9, 1993 to redesign the station in 90 days and report by June 7. Europe suggested that Space Station Freedom be merged with Russia's proposed Mir 2! The redesign report contained recommendations to join with the Russians and use the Soyuz as the Crew Emergency Return Vehicle, and because such collaboration would require changing the orbital inclination from 28° to 51.6°, elements of the station should be launched from Baikonur by heavy Russian rockets. Simultaneously, in June 1993, an effort in Congress to scrap the project was defeated by a single vote (215 to 216). Arguments against proceeding were that the station had lost its political role following the end of the Cold War, and in any case scientists were critical of its limited facilities.

In June Clinton had favored a small four-man U.S.-only "Alpha" station, but in May the White House had directed NASA to open discussions with the Russians and in September NASA suggested a larger and much more capable "Russian Alpha" with a crew of six. Clinton agreed to this.

In October 1993, with the gunfire of the battle between Russia's President Boris Yeltsin and the national Duma outside their windows, NASA negotiators in Moscow agreed to a merger of their plans. In November 1993 the discussions among the space station partners were complete, and Russia was formally invited to join the program. Freedom, Mir 2, and the European and Japanese modules were incorporated into a single station. Clinton also managed to strike a deal with Congress that established a fixed budget for the U.S. contribution to the station of $2.1 billion per annum. A final attempt to cancel the project was rejected in the House of Representatives in 1994.

The Russian Functional Cargo Block (FGB, derived from the TKS), later named Zarya, was introduced as the first element of the station. It would serve as an interim propulsion module providing attitude control for the passive U.S. elements. The new inclination and the use of heavy Russian rockets was confirmed in a November 1993 meeting, in which it was foreseen that Russia would supply the FGB, the Mir 2 base block (essentially a DOS equipped as a habitat), and Soyuz and Progress vehicles for crew and cargo transport. In June 1994 NASA decided to purchase the FGB from Russia in order to assist the cash flow crisis of the Russian space agency. Despite this purchase, it was estimated that Russian participation would save $2 billion. In March 1994 what was now being called "International Space Station Alpha" passed its System Design Review. Later the name "Alpha" was dropped, in part because the Russians objected to the implication that this would be the first space station.

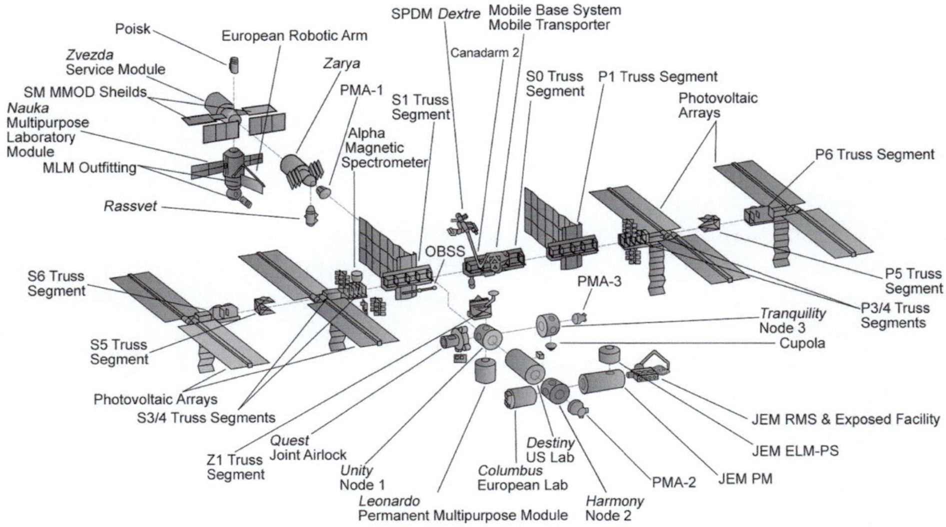

Fig. 2.25a. Configuration of the International Space Station (source: NASA/Wikipedia).

Fig. 2.25b. The International Space Station viewed from STS-134 Endeavour on May 29, 2011 (source: NASA).

Fig. 2.26. Space Shuttle Endeavour docked to the ISS on May 23, 2011, photographed from a Soyuz spacecraft (source: NASA).

On November 20, 1998 the 20-ton Zarya was launched by a Proton rocket as the first element of the ISS. On December 6 a Shuttle delivered the U.S. Unity node. The Zvezda base block docked on July 26, 2000, with the three modules in line. After a series of Shuttles added early elements of the U.S. part of the station, a Soyuz docked with William Shepherd, Sergei Krikalev and Yuri Gidzenko, the Expedition 1 crew who stayed on board the station for 136 days from November 2000 to March 2001. It was the beginning of an uninterrupted human presence on the station that still continues. By the end of 2010, the ISS had received 297 visitors (196 different people) and had a crew of up to 6.

Fig. 2.25 shows the configuration of the completed ISS and the ISS as of May 2011. Fig. 2.26 shows an iconic image of Endeavour docked with the station, as photographed from a departing Soyuz spacecraft. With the retirement of the Shuttle in July 2011, the only way to transport crews to the station is the venerable Soyuz spacecraft until a new spacecraft is produced for the purpose. The assembly of the ISS will be completed with the addition of Russia's science module, Nauka, at which time the total mass will be about 450 tons. According to current planning, the ISS will be operational through to 2020, with the prospect of further use.

Table 2.9. Shenzhou flights up to June 2012.

Mission	*Launch date*	*Details*
Shenzhou 1	November 19, 1999	Unmanned, 21 hours, 14 orbits
Shenzhou 2	January 9, 2001	Unmanned, 7 days
Shenzhou 3	March 25, 2002	Unmanned, 7 days
Shenzhou 4	December 29, 2002	Unmanned, 7 days
Shenzhou 5	October 15, 2003	Manned, 21 hours, Yang Liwei became the first Chinese astronaut
Shenzhou 6	October 12, 2005	Manned, 4 days, 21 hours, two astronauts, Fei Long and Nie Haisheng
Shenzhou 7	September 25, 2008	China's first three-person mission, 2 days 20 hours. First spacewalk by Zhai Zhigang, Liu Boming and Jing Haipeng remained inside the craft
Shenzhou 8	November 1, 2011	Unmanned, the spacecraft docked in low Earth Orbit with Tiangong 1, which was launched September 29, 2011 on a Long March rocket
Shenzhou 9	June 16, 2012	Liu Wang, Jing Haipeng, and Liu Yang (the first Chinese woman in space) spent 13 days in space. The mission included the first manned docking with Tiangong 1 and the crew also performed the first manual docking with the station

2.9 SHENZHOU

China became the third nation to send a human into space in 2003 when Yang Liwei, a military pilot, made 14 orbits of the Earth in Shenzhou 5 (Fig. 2.27).[8] The design of the Shenzhou spacecraft is similar to the Soyuz, with a capacity of three people. It is launched from the Jiuquan Satellite Launch Center in northern China using a Long March IIF rocket. Table 2.9 summarizes Shenzhou flights to date.

China's space industry is developing quickly. On September 29, 2011 China put an unmanned module, Tiangong 1,[9] into orbit at 350 km altitude. It is 10.5 m long, with a mass of 8.5 tons. On November 3 the unmanned Shenzhou 8 rendezvoused with Tiangong 1 and made that nation's first docking in space (Fig. 2.28). This step is reminiscent of the way that NASA built up its space engineering competence with the Gemini missions. In June 2012 astronauts visited Tiangong 1 as a preliminary to establishing a Chinese space station similar to Mir.

[8] Chinese astronauts are given the name "taikonauts".
[9] Translated into English, "Tiangong" means "Heavenly Palace".

Fig. 2.27. Yang Liwei, China's first astronaut (source: Xinhua News Agency).

Fig. 2.28. Schematic of Shenzhou 8 docking with the Tiangong 1 space laboratory module (source: Xinhua News Agency).

2.10 COMMERCIAL SPACE TRAVEL

The first humans to travel in space were almost exclusively test pilots with military backgrounds. After some time, engineers and scientists were included in some crews. In 1984 Charles Walker flew on the Shuttle as the first industrial payload specialist; McDonnell Douglas paid NASA a token sum for his training and flight. In 1986 the attempt to send a teacher into space ended disastrously when Christa McAuliffe died in the Challenger accident. Her backup, Barbara Morgan, flew as a mission specialist in August 2007. The late 1980s and early 1990s was a period of economic reform in Russia that influenced space activities. Toyohiro Akiyama, a Japanese journalist, became the world's first business traveler in space when his employer paid for him to make a 7-day trip to Mir in December 1990 as a member of a Soyuz crew, all in the pursuit of a television exclusive. He was not a typical candidate with the "right stuff", being a not particularly fit chain-smoker.

In the summer of 1989 an advertisement appeared in the British press. "Astronaut Wanted, No Experience Required". It began the process to select a British citizen for a visit to Mir. Named Juno, the project was a commercial venture. The catalyst was the Moscow Narodny Bank. The cost for the mission, estimated at £16 million, was to be raised by sponsorship. The project almost faltered due to difficulties raising the funding. Of some 13,000 people who applied, 150 candidates were called for medical and psychological tests. Those interviewed were issued written warnings stating: "If you are selected to be an Astronaut or the backup you must realise that you will be embarking upon a risky and potentially dangerous project." The successful candidate who flew to Mir in May 1991 was Helen Sharman, a chemist in the chocolate section of the research and development department at Mars Confectionary Ltd. (Fig. 2.29).

The American businessman Dennis Tito (Fig. 2.30) became the first fee-paying private space traveler in 2001 when he bought a ticket from a company called Space Adventures for a visit to the ISS. This unusual travel agency, founded in 1998, specializes in spaceflight experiences.

Tito flew up to the ISS as part of a Soyuz crew. His visit was not popular with the other ISS partners. Nevertheless, other "space tourists" followed; as detailed in Table 2.10. Most of these spacefarers are mature individuals, entrepreneurs who have acquired over time the economic resources to purchase an expensive travel ticket.[10]

The Ansari X PRIZE was modeled on the Orteig Prize.[11] It was founded in 1996 by the entrepreneur Peter H. Diamandis and administered by the X PRIZE Foundation. To win the $10 million prize, a non-governmental organization had to launch a reusable spacecraft capable of carrying three people to an altitude of 100 km twice within the period of 2 weeks. The challenge was taken up by 26

[10] The oldest person to have traveled to space is John Glenn, who made his second spaceflight at the age of 77, on STS-95 in 1998.

[11] The Orteig Prize was a sum of U.S.$25,000 offered by Raymond Orteig for a non-stop flight between New York and Paris. The prize was won by Charles Lindbergh in 1927.

Fig. 2.29. Helen Sharman with some of her experiments on board the space station Mir (source: Helen Sharman).

teams from seven countries, and the prize was won by SpaceShipOne, which was actually a high-altitude research aircraft designed by Burt Rutan of Scaled Composites in Mojave, California, funded by Microsoft co-founder Paul Allen.

SpaceShipOne has a carbon composite construction and a hybrid rocket motor that burns a liquid nitrous oxide and rubber fuel. A carrier aircraft, White Knight, carried SpaceShipOne for the first hour of its journey. Upon being released, the craft ignited its engine for a little more than a minute and climbed steeply. The craft then coasted on a parabolic trajectory to a peak altitude over 100 km, after which it fell back towards the Earth. The tail was canted up into a feathered configuration inspired by a shuttlecock to generate drag for stability as it reentered the atmosphere. The tail then reverted to its original aerodynamic

Fig. 2.30. Dennis Tito at the 40th Space Congress in 2003 at Cape Canaveral, Florida (source: NASA-KSC).

configuration, and the craft landed like a conventional glider. Fig. 2.31 shows White Knight and SpaceShipOne in flight.

Mike Melvill became the first commercial astronaut when he flew SpaceShipOne to just over 100 km on June 21, 2004. Melvill also flew the first

Table 2.10. Spaceflight participants who have paid their own fare.

Name	*Country of origin*	*Age during flight*	*Flight year*
Dennis Tito	U.S.A.	60	2001
Mark Shuttleworth	South Africa	28	2002
Greg Olsen	U.S.A.	60	2005
Anousheh Ansari	Iran	40	2006
Charles Simonyi	Hungary	58 & 60	2007 & 2009
Richard Garriott	U.K./U.S.A.	47	2008
Guy Laliberté	Canada	50	2009

Fig. 2.31. White Knight and SpaceShipOne in flight (source: Jim Koepnick and Virgin Galactic).

X PRIZE flight to 103 km on September 29, 2004.[12] Test pilot Brian Binnie flew SpaceShipOne to 112 km for the prize-winning flight on October 4, 2004, after which it was retired. In fact, this flight set a new altitude record by beating the 108 km achieved by U.S. Air Force pilot Joseph Walker flying an X-15 rocket plane on August 22, 1963. The X-15 was released by a NASA B-52 from a similar height to SpaceShipOne. Flights of the X-15 were made from 1959 to 1968, providing research results that were used in the development of the Mercury, Gemini, and Apollo spacecraft.

Scaled Composites is working with Virgin Galactic to build a fleet of spacecraft and carrier aircraft, appropriately named SpaceShipTwo and WhiteKnightTwo, as more advanced versions of the vehicles that won the X PRIZE. SpaceShipTwo is about double the size of its predecessor (Fig. 2.32), having a capacity of two pilots and six passengers. The entire flight will last approximately two and a half hours, with a few minutes of weightlessness at the top of the trajectory. The flight profile is illustrated in Fig. 2.33. To enhance the experience, each passenger will have a window beside his/her seat and another one overhead. An image captured by Brian Binnie during his famous flight illustrates the kind of view that passengers should expect (Fig. 2.34). At the altitude reached

[12] There were no passengers on the X PRIZE flights, instead there was an equivalent mass of two passengers on board.

Fig. 2.32. Illustration comparing SpaceShipOne with SpaceShipTwo (source: Virgin Galactic).

Fig. 2.33. Illustration depicting the journey of SpaceShipTwo during her Virgin Galactic voyage (source: Virgin Galactic).

Fig. 2.34. Photograph of Earth taken by pilot astronaut Brian Binnie on his X PRIZE winning flight in October 2004 (source: Brian Binnie and Virgin Galactic).

by SpaceShipTwo, the atmosphere is so thin that they will be able to see the blackness of space, as well as the curvature of the Earth and the thin blue layer of atmosphere above it.[13] It is expected that passengers will be able to move around the cabin during the weightless portion of the flight. It has been a hotly debated question whether or not these space tourists should wear some kind of pressure suit.

Other commercial companies also intend to take passengers on suborbital flights to the edge of space, and there is an interest in expanding space tourism with orbital flights. In particular, the American Entrepreneur Robert Bigelow is extending his real estate business into space. His company, Bigelow Aerospace, is specializing in the use of inflatable structures. A subscale test of this habitat technology was launched on July 12, 2006 as Genesis I (Fig. 2.35). A more advanced test followed on June 28, 2007 as Genesis II. The next step is the launch of a full-scale module, the Sundancer, with a capacity for three passengers on a long-term basis and six for short-term stays.

Space Exploration Technologies (SpaceX), which is funded by entrepreneur Elon Musk, is developing a family of rockets to launch its own spacecraft to undertake a variety of roles. On September 28, 2008 a Falcon 1 test rocket lifted off from Omelek Island in the Central Pacific and became the first privately devel-

[13] The atmosphere is not actually blue but appears that color due to the scattering of light by molecules in the air.

Fig. 2.35. Genesis I in orbit (source: Bigelow Aerospace, LLC).

Fig. 2.36. The SpaceX Dragon commercial cargo vehicle is grappled by the Canadarm2 robotic arm at the ISS. The arm was used to berth the spacecraft to the station on May 25, 2012 (source: NASA).

oped liquid rocket to achieve orbit. This was a precursor for the larger Falcon 9, which was first tested in a launch from Florida on June 4, 2010. The company has developed the Dragon capsule as a commercial vehicle to deliver crew and cargo to the ISS. On December 8, 2010 a Falcon 9 launched a test version of the Dragon, which flew two orbits and was then recovered from the Pacific, thereby making SpaceX the first commercial company to recover a satellite from orbit. Following this success the Dragon capsule was launched again on a Falcon 9 on May 22, 2012 and docked with the ISS on May 25 (Fig. 2.36), becoming the first commercial spacecraft to do so. It delivered supplies and was reloaded with return cargo that was recovered together with the Dragon capsule on its return to Earth.

2.11 MILESTONES IN THE DEVELOPMENT OF HUMAN SPACEFLIGHT

Table 2.11. Milestones in the development of human spaceflight.

Date	*Event*	*Nation*
April 12, 1961	Yuri Gagarin became the first human to orbit the Earth on Vostok 1	Russia
May 5, 1961	Alan Shepard made a suborbital flight to space on Freedom 7	U.S.A.
February 20, 1962	John Glenn orbited the Earth three times on Friendship 7	U.S.A.
June 16, 1963	Valentina Tereshkova became the first woman to go into space on Vostok 6	Russia
October 12, 1964	Vladimir Komarov, Konstantin Feoktistov, and Boris Yegorov became the first space crew to orbit the Earth on Voskhod 1	Russia
March 18, 1965	Alexei Leonov carried out the first spacewalk	Russia
June 3, 1965	Edward White became the first American to carry out a spacewalk	U.S.A.
December 24, 1968	Frank Borman, James Lovell, and William Anders orbited the Moon, Apollo 8	U.S.A.
July 20, 1969	First lunar landing by Neil Armstrong and Buzz Aldrin, Apollo 11	U.S.A.

(*continued*)

Table 2.11. Milestones in the development of human spaceflight (*cont.*).

Date	*Event*	*Nation*
March 2, 1978	Vladimir Remek traveled to Salyut 6 on board a Soyuz spacecraft. He was the first non-Soviet European in space	Czechoslovakia
June 24, 1982	Jean-Loup Chrétien traveled to Salyut 7 on a Soyuz spacecraft and became the first West European to travel to space	France
June 18, 1983	Sally Ride flew into space on the STS-7 Space Shuttle mission and became the first American woman to fly into space	U.S.A.
November 28, 1983	Ulf Merbold became the first ESA astronaut to fly to space and the first non-American to fly on the Space Shuttle	Germany
July 25, 1984	Svetlana Savitskaya (who had been the second woman to go into space) became the first woman to carry out a spacewalk	Russia
December 9, 1988	Jean-Loup Chrétien became the first European (and first non-American and non-Soviet) astronaut to carry out a spacewalk	France
October 20, 1995	Thomas Reiter performed the first spacewalk by an ESA astronaut	Europe
April 28, 2001	Dennis Tito became the first space tourist	Private
October 15, 2003	Yang Liwei became the first Chinese astronaut on Shenzhou 5	China
June 21, 2004	Mike Melvill made the first spaceflight in a privately built spacecraft, SpaceShipOne, built by Scaled Composites	Private
September 27, 2008	Zhai Zhigang performed the first spacewalk by a Chinese astronaut	China
June 16, 2012	Liu Yang launched on Shenzhou 9 and became the first Chinese woman to fly into space	China

2.12 REFERENCES/RECOMMENDED LITERATURE

Burrough, B.: *Dragonfly, NASA and the Crisis aboard Mir*, HarperCollins (1998).

Clark, P.: *The Soviet Manned Space Program: An Illustrated History of the Men, the Missions, and the Spacecraft*, Orion (1988).

Harford, J.: *Korolev*, Wiley (1997).
Linehan, D.: *SpaceShipOne: An Illustrated History*, Motorbooks International (2008).
Portree, D.S.F.: *Mir Hardware Heritage*, NASA RP 1357, NASA (1995).
Rogers, W.P., Armstrong, N.A., Acheson, D.C., Covert, E.E., Feynman, R.P., Hotz, R.B., Kutyna, D.J., Ride, S.K., Rummel, R.W., Sutter, J.F. *et al.*: *Report of the Presidential Commission on the Space Shuttle Challenger Accident*, U.S. Government Printing Office (1986).
7 Victories in Space and Other 42 Events in Russian Cosmonautics to Remember, Roscosmos (2011).
Ryan, C.: *The Pre-Astronauts: Manned Ballooning on the Threshold of Space*, Bluejacket Books, Naval Institute Press (1995).
Sharman, H. and Priest, C.: *Seize the Moment: The Autobiography of Britain's First Astronaut,* Victor Gollancz (1993).
Shayler, D.J. and Moule, I.: *Women in Space Following Valentina*, Springer/Praxis (2005).
Siddiqi, A.A.: The Almaz Space Station Complex: A History, 1964–1992, Part 1, *Journal of the British Interplanetary Society*, **54**(11/12), November/December (2001)
Siddiqi, A.A. and NASA History Office: *Challenge to Apollo: The Soviet Union and the Space Race, 1945–1974*, NASA History Series SP 2000-4408, NASA (2011).
Sparrow, G.: *Spaceflight: The Complete Story from Sputnik to Shuttle—and Beyond*, DK Books (2007).
Summerlin, L. (Ed.): *Skylab, Classroom in Space*, NASA (1977).

Websites

Bigelow Aerospace: *www.bigelowaerospace.com*
China National Space Administration: *www.cnsa.gov.cn*
European Space Agency: *www.esa.int*
NASA: *www.nasa.gov*
Russian Federal Space Agency: *www.federalspace.ru*
Scaled Composites: *www.scaled.com*
Space Adventures: *www.spaceadventures.com*
Space Exploration Technologies: *www.spacex.com*
Sven's Space Place: *www.svengrahn.pp.se*
Virgin Galactic: *www.virgingalactic.com*
X PRIZE Foundation: *www.xprize.org*

3

The space environment

Carol Norberg

ABSTRACT

The space environment is hostile to people and equipment. In this chapter the space environment is considered in terms of separate categories: gravity, neutral particles, vacuum, plasma, radiation, micrometeoroids, and space debris.

The microgravity environment available in space is an interesting laboratory for researchers in many disciplines but as a workplace it poses physiological challenges to humans. Weightlessness can be produced for short periods of time in ground-based experiments in drop towers, parabolic flights on board aircraft, and sounding rockets.

One of the most challenging aspects of a long-duration space mission for human exploration of the solar system is the susceptibility of both people and equipment to ionizing radiation consisting of relatively constant levels of galactic cosmic rays and sporadic high-energy solar particle events. Our current knowledge is insufficient to estimate with certainty the risk to astronauts from an extended period of exposure to this radiation.

Since the space age began, spacecraft have been launched with little concern for the resulting pollution. As a result the most popular Earth orbits are littered with the debris of space missions, and this puts new missions at risk of potentially damaging collisions.

3.1 INTRODUCTION

The space environment around the Earth is often discussed in terms of the properties of, and effects caused by, the various components of the environment; namely

- gravity
- vacuum
- neutral particles
- plasma
- micrometeoroids
- orbital debris.

We must understand the space environment in the vicinity of typical satellite orbits in order to design spacecraft appropriately and also to ensure high levels of safety in human spaceflight. Recently there has been a developing interest in space situational awareness encompassing space surveillance and space weather. Space surveillance is the monitoring of the environment around the Earth to detect natural and man-made objects that are a potential threat to satellites and the Earth itself. Space weather is the term used to describe the status of the Sun, the solar wind it emits, and the status of the Earth's magnetosphere. The monitoring of space weather enables potential threats to infrastructure such as electrical power grids to be identified, as well as to humans as a result of increased levels of radiation arriving from space. This area is addressed in sections dealing with plasma, radiation, micrometeoroids, and space debris. The phenomenon that makes space inaccessible to most of us is the gravitational force exerted by the Earth, and that is where this chapter begins.

3.2 GRAVITY

Ancient astronomers believed that the motions of all of the heavenly bodies that they observed by the naked eye were uniform circular motions. They also believed that the objects themselves were made of a perfect material that always had the same intrinsic properties, such as brightness. In addition, they were certain that the Earth was at the center of the universe. It was thought that the planets traveled in small circular paths, known as "epicycles", whose centers pursued larger circular paths that were centered on the Earth. Claudius Ptolemy recorded this understanding of planetary motions in the second century AD, and this model remained the state of the art for an admirable 1,400 years.

It was not until 1543 that the Polish astronomer Nicolaus Copernicus introduced a heliocentric frame for the planetary motion in our solar system.[1] It was still assumed, however, that the planets moved in circles. Towards the end of the 16th century the Danish astronomer Tycho Brahe gathered a large amount of observational data and when his German assistant Johannes Kepler analyzed these data in the first decade of the 17th century he found that the planets actually trace out elliptical orbits around the Sun. The three empirical laws that Kepler formulated in order to describe their motion are nowadays common knowledge to all

[1] "Helios" is Greek for "Sun" and hence heliocentric means Sun centered.

Table 3.1. Eccentricity of the orbits of solar system planets.

Planet	*Eccentricity*
Mercury	0.2056
Venus	0.0067
Earth	0.0167
Mars	0.0935
Jupiter	0.0489
Saturn	0.0565
Uranus	0.0457
Neptune	0.0113

physics students

1. The planets move in elliptical orbits, with the Sun at one focus.
2. The law of areas: the line joining the Sun and a planet sweeps out equal areas in equal times.
3. The squares of the period of revolution are proportional to the cubes of the average distances of the planets from the Sun

$$\frac{T^2}{R^3} = C$$

where T is the period of revolution, R is half of the major axis of the ellipse, and C is a constant that has the same value for all of the planets in our solar system (2.5×10^{19} km^3/d^2).

The shape of the orbit of a planet around the Sun is described by its orbital eccentricity, whereby an eccentricity of 0 represents a circular orbit and values from 0 to 1 represent increasingly elongated elliptical orbits. As shown in Table 3.1, with the exception of Mercury all of the planets have orbits that are close to circular.

Kepler's laws describe how the planets move, but not why they behave as they do. Since the planets describe closed orbits around the Sun and do not fly away, it can be inferred that there is an attraction between the Sun and the planets. As Isaac Newton realized later in the 17th century, this attraction is the gravitational force. The results of calculation and experiment have shown that the gravitational interaction is always attractive, depends on the mass of the interacting bodies, and varies inversely with the square of the distance between them. The gravitational force, F, between two bodies having masses M and m and separated by a distance r is

$$F = \frac{GMm}{r^2}$$

Table 3.2. Variation of acceleration due to gravity with altitude (with respect to the Earth).

Altitude above Earth's surface (km)	*Acceleration due to gravity* (m/s^2)
0	9.8
200	9.2
400 (typical ISS altitude)	8.7
35,786 (geostationary orbit)	0.22
380,000 (distance to the Moon)	0.0027

where G is the universal gravitational constant. This relation is known as Newton's universal law of gravitation. It applies equally to billiard balls on the Earth and to satellites orbiting planets.

In 2006 the Committee on Data for Science and Technology (CODATA) assigned to G the value of $(6.67428 \pm 0.001) \times 10^{-11}\ \mathrm{Nm^2/kg^2}$. It is actually one of the least accurately known physical constants.

Table 3.2 shows the acceleration due to gravity for various altitudes above the Earth's surface, assuming Earth to be a sphere with radius 6,371 km (in fact it is a little flat at the poles and bulges at the equator). As can be seen, the acceleration due to gravity at the orbit of the Space Shuttle is actually not much less than at the surface of the Earth. The assumption is often made that there is "no gravity" in space. What we call *weightlessness* on board a spacecraft arises because in orbit the spacecraft and everything in it are actually in a state of free fall.

As can be seen, the acceleration due to Earth's gravity at the Moon is very small. However, the Moon has its own gravitational field that causes an acceleration due to gravity on its surface that is about one sixth of that on the surface of the Earth. This is why the Apollo astronauts seemed to bounce around when they were on the Moon. Future visitors to Mars will experience an acceleration due to gravity on its surface of about one third of that on the surface of the Earth.

For a satellite in a circular orbit at altitude h above the Earth of radius R_E and mass M_E its period T is easily deduced from Newton's law to be

$$T = \frac{2\pi}{\sqrt{GM_E}}(R_E + h)^{3/2}$$

and its orbital speed, v, is

$$v = \sqrt{\frac{GM_E}{R_E + h}}$$

Table 3.3 shows satellite periods and orbital speeds for circular orbits at different altitudes above the Earth. At an altitude typical of the ISS the orbital period is about 90 minutes, so the astronauts on board see a sunrise and sunset about 16 times per Earth day. The speed of a satellite in a low circular orbit is

Table 3.3. Satellite periods as a function of altitude above the Earth's surface.

Altitude (km)	*Period*	*Orbital speed* (km/s)
200	1 hour 28 minutes	7.8
400	1 hour 32 minutes	7.7
35,786	23 hours 56 minutes	3.1

about 7.7 km/s (necessarily less than the 11.2 km/s required to escape the Earth's gravitational field).

A satellite in a geostationary orbit (GEO) has a period equivalent to the time it takes for the Earth to rotate around its axis (23 hours 56 minutes). As this orbit lies in the equatorial plane such a satellite appears to hover in position relative to the Earth's surface, making it useful for communications, weather forecasting, and television transmission. It is a very popular orbit.

Low Earth orbit (LEO) is a term used to describe orbits up to 2,000 km above the Earth's surface. The majority of spacecraft sent into space to date have been in LEO, and it is becoming congested. With the exception of the Apollo flights to the Moon, all orbital flights by humans have been carried out in LEO. The ISS orbit has varied between about 320 and 410 km and it is inclined at an angle of 51.6° to the equator. This inclination was selected to satisfy the requirement that it be accessible to the Soyuz and Progress vehicles launched from Baikonur and the Shuttle launched from Kennedy Space Center. It is also accessible from Kourou near the equator, and from Japan. A polar orbit passes over the polar regions and so has an inclination of approximately 90°. Since the Earth is rotating beneath the plane in which such a satellite is traveling, this offers the benefit of being able to observe the entire planet after a number of orbits.

As it is expensive and challenging to transport people and experiments into orbit to make use of the microgravity environment that can be achieved there, a number of complementary techniques are used nearer to home to create short-lived microgravity environments. These are drop towers, parabolic flights, and sounding rockets. As well as being useful facilities for microgravity research in their own right, they offer a relatively low-cost means of preparing and qualifying flight experiments, even for later use on satellites and the ISS. Ground-based facilities offer the advantage that experimenters can interact with their experiments right up until they occur, and then retrieve them shortly afterwards. The retirement of the Shuttle has limited the ability to return cargo from the ISS, so for the foreseeable future microgravity researchers may have to wait a long time to have their experiments returned to them from the station.

The duration and quality of the microgravity environment currently available in each category is presented in Table 3.4. The unit of quality is g, where 1 g is equivalent to the acceleration due to gravity at the surface of the Earth, 9.8 m/s^2. In addition, one must consider the vibratory accelerations that influence the

Table 3.4. Characteristics of microgravity environments

Microgravity environment	*Duration available*	*Quality* (g)
Drop tower	5–9 seconds	10^{-4}–10^{-6}
Parabolic flight	20–22 seconds	10^{-2}–10^{-3}
Rockets on a ballistic trajectory	3–12 minutes	10^{-4}–10^{-6}
Recoverable satellite	2 weeks–3 months	10^{-5}–10^{-6}
Space station	weeks–years	10^{-4}–10^{-6}

system and the experiment/payload in each case. It is interesting to do research in microgravity environments to investigate the importance of gravity in processes on the Earth. Each technique for carrying out microgravity research will be considered in turn in the following subsections.

3.2.1 Drop towers

There are a number of drop towers located around the world. They are specially built towers or underground shafts. Experiments inside special capsules experience microgravity while falling, with the period in this state essentially depending on the length of the drop. The towers are usually evacuated of air to reduce the effect of drag but need not be if a dual-capsule system is used. The outer capsule falls under normal conditions and the inner capsule falls in vacuum. One of the challenges is to ensure there is no contact between the inner and outer capsules during the fall. Drop towers provide relatively inexpensive access to high-quality microgravity with a short pre-launch development time and the ability to repeat experiments and modify them with a fast turnaround. They also provide a testbed for experiments before sending them on longer and more expensive space flights. These facilities can be used for experiments in fields such as fundamental physics, materials science, biology, and fluid physics.

In Table 3.5 the characteristics of some of the most frequently used drop towers are summarized. The NASA Glenn Research Center Zero Gravity Research Facility is 143 m long and provides up to 5.2 seconds of free fall. The National Microgravity Laboratory (NML) of the Chinese Academy of Sciences is 116 m high and provides about 3.5 seconds of microgravity in a dual-capsule system. The Japan Microgravity Center (JAMIC) has used a former mine shaft that is about 700 m deep to enable it to provide roughly 10 seconds of microgravity but it is not currently operational. (The operation of such a long underground shaft is challenging.) The drop tower in Toki City, Japan, is owned by the Microgravity Laboratory of Japan (MGLAB) and is also in a mineshaft.

The drop tower at the Center of Applied Space Technology and Microgravity of the University of Bremen (ZARM) is the only facility in Europe (Fig. 3.1a).

Table 3.5. Comparison of a selection of drop towers (adapted from a presentation by Dr. Peter Preu, DLR).

	NASA GRC (Cleveland, Ohio)	*NML (Beijing)*	*MGLAB (Toki City)*	*ZARM (Bremen)*
Type	Shaft	Tower	Shaft	Tower
Free fall distance	132 m	83 m	100 m	110 m
Duration of microgravity	5.2 s	3.5 s	4.5 s	4.7 s/9.3 s
Microgravity quality	$<10^{-5}$ g	10^{-5} g	10^{-5} g	10^{-6} g
Approximate vacuum level	1.3 Pa	<30 Pa between the inner and outer capsules	4 Pa	10 Pa
Upper limit on experiment mass	455 kg	Total mass of the capsule 600–630 kg	400 kg	274 kg

It was inaugurated in 1990 and has a shaft that is 146 m tall and can provide a microgravity environment of 10^{-6} g. Experiments to be dropped are first hoisted up to the top of the tower in the experiment capsule (Fig. 3.1b). The chamber is then evacuated to a residual pressure of 10 Pa, which takes about 90 minutes. Experiments released from the top of the tower experience 4.7 seconds of microgravity and reach a maximum speed of 170 km/h. Fig. 3.2a illustrates the residual acceleration in the drop capsule as a function of time, where zero seconds corresponds to the time of release. An 8 m high column of polystyrene pellets is used to decelerate the experiments at the bottom of the tower. Fig. 3.2b shows the deceleration of the capsule upon entering this column. The deceleration is such that even sensitive devices are protected. In 2004 a catapult system was introduced to extend the experiment time to 9.3 seconds. Now instead of simply dropping experiments, they can be made to perform an indoor parabolic flight. In this case the experiment capsule is accelerated upwards from rest to a speed of 170 km/h in 280 milliseconds, over a distance of 7 m. Masses of up to 500 kg can be catapulted upwards. The catapult system is then swiftly moved out of the path of the returning capsule.

3.2.2 Parabolic flights

Parabolic flights are made on a specially modified aircraft that can travel along a path that includes a series of parabolic trajectories. There are a number of

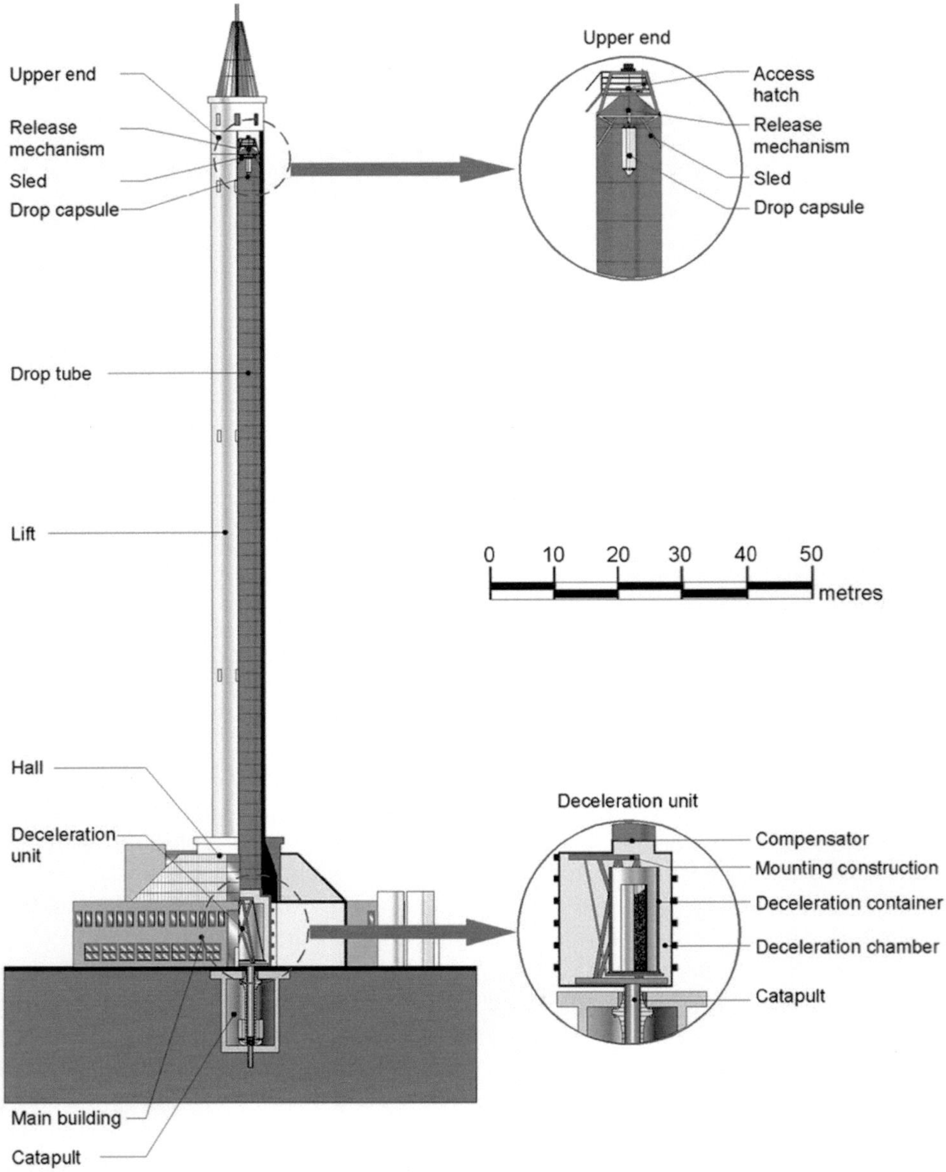

Fig. 3.1a. Schematic of the layout of the Bremen Drop Tower (source: ZARM, *www.zarm.uni-bremen.de*).

different aircraft available for this type of flight. In Europe, flights are usually carried out on an Airbus A300 "Zero-G" aircraft operated by Novespace. Fig. 3.3 illustrates its typical flight profile. It begins a parabola in horizontal flight at about 6,000 m altitude at a speed of 810 km/h. The pilot draws the nose upwards so that

Fig. 3.1b. Drop capsule suspended inside the Bremen Drop Tower (source: ZARM, *www.zarm.uni-bremen.de*).

the plane starts to climb. During this phase, which lasts for about 20 seconds, the acceleration inside the cabin is in the range 1.5–1.8 g (hypergravity). At an altitude of about 7,500 m and with the plane at an angle of about 47° to the horizontal the engine thrust is reduced and the aircraft follows a parabolic trajectory just like a ball thrown through the air. At the peak of the trajectory the plane's speed has reduced to about 370 km/h. When it has fallen back to 7,500 m the pilot pulls the plane back to a horizontal path and then the acceleration in the cabin increases again to the pre-parabolic levels. There is typically about 2 minutes of steady flight between each parabolic phase, with longer breaks after a number of parabolas to enable the experimenters to work on their experiments (Fig. 3.4). A campaign usually includes flights on several consecutive days, with the total flight time per day being typically 2 to 3 hours. Flight participants must have a medical certificate equivalent to that required of a private pilot in order to participate, and they have the option of taking medication to counteract nausea.

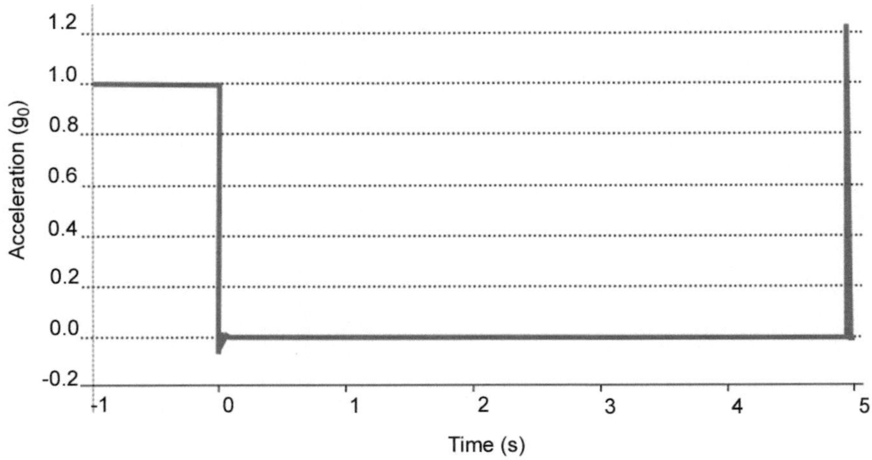

Fig. 3.2a. Graph illustrating the residual acceleration of the dropped capsule inside the Bremen Drop Tower (source: ZARM, *www.zarm.uni-bremen.de*).

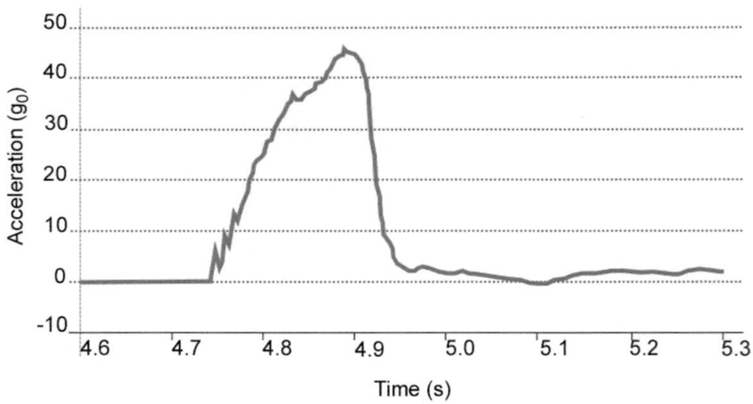

Fig. 3.2b. Graph illustrating the deceleration of the capsule inside the Bremen Drop Tower (source: ZARM, *www.zarm.uni-bremen.de*).

Similar kinds of scientific areas can be investigated on parabolic flights as in drop towers, but it is also possible to carry out physiology experiments for example. There is much more space available inside the cabin of an aircraft than in the capsule of a drop tower (Fig. 3.5). Experiments are fastened to rails on the floor of the cabin or allowed to float free during the weightless part of the flight. An important advantage of a parabolic flight is that the experimenter can interact directly with an experiment as it is being performed. The period of weightlessness is longer, but the quality of the microgravity environment (at typically 10^{-2} g) is not as good, and there are a limited number of campaigns per year.

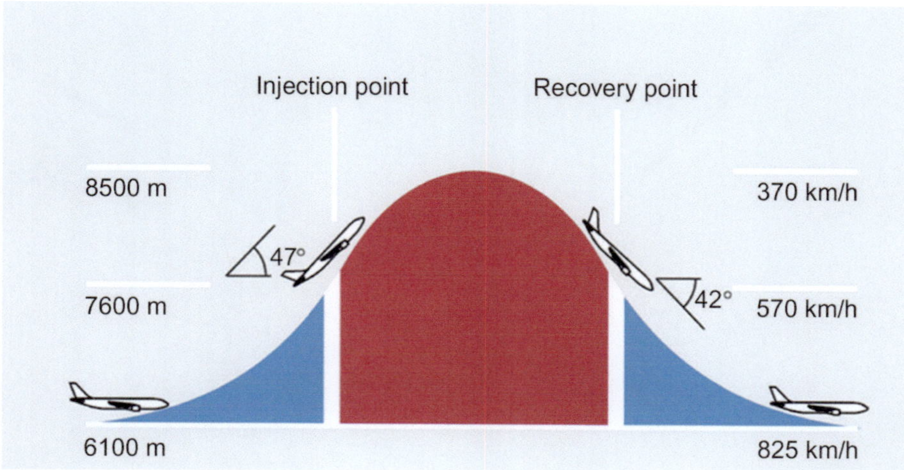

Fig. 3.3. Flight profile for a single parabola. Blue indicates hypergravity and red weightlessness (source: DLR).

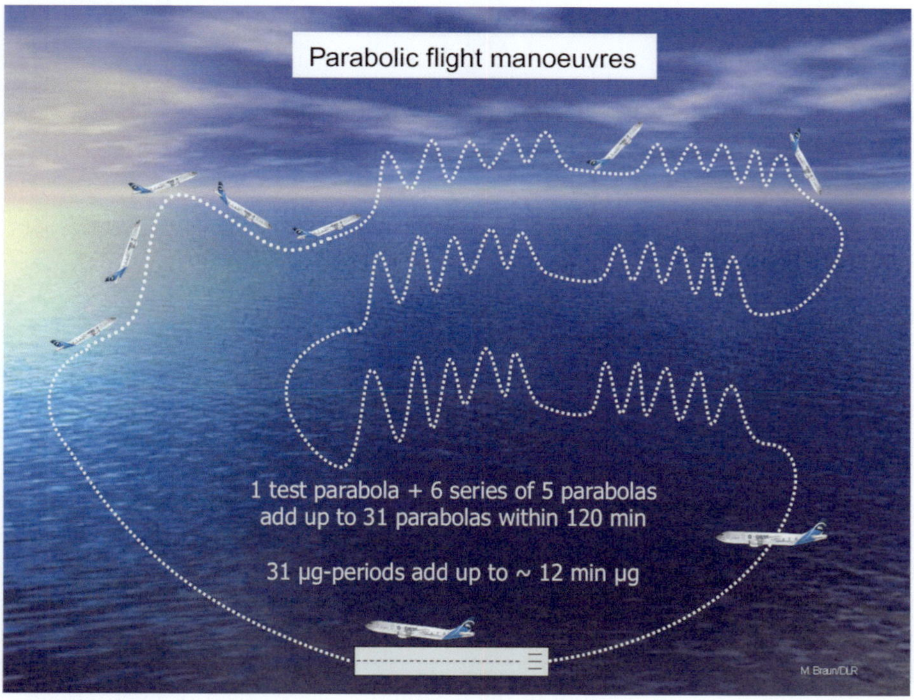

Fig. 3.4. Typical flight profile during a campaign (source: DLR/Markus Braun).

Fig. 3.5. Inside the cabin of the Airbus 300 during a period of weightlessness (source: DLR).

Parabolic flights were used by NASA to acclimatize the Mercury astronauts to weightlessness (Fig. 3.6) and are still used for astronaut training. Such flights can also simulate partial gravity, which can be useful in preparing for space missions to the Moon and Mars. Parabolic flights are not all work: there are companies that allow fee-paying private individuals to experience weightlessness for fun.

3.2.3 Sounding rockets

The first successful flight of Wernher von Braun's A4 ballistic rocket was conducted on October 3, 1942. It could carry one metric ton of payload on a ballistic trajectory with its apogee at an altitude of 190 km. In early 1944 Erich Regener at the Research Institute of Atmospheric Physics at Friedrichshafen, Germany, was invited to launch an atmospheric physics payload known as the "Regener Tonne" aboard the A4 rocket during its qualification flights. This landed via parachute for recovery and instrument readout. It began the routine use of rockets for research. They are particularly useful for upper-atmosphere studies at altitudes between those accessible to balloons, which have an upper limit of about 40 km, and the lower limit at which it is practical to use satellites at about 200 km. Rockets used in scientific research are often referred to as "sounding rockets", meaning that they make measurements. In the 1970s sounding rockets became the principal tool in the emerging research field of microgravity. A broad spectrum of research facilities, platforms, and tools have been developed

Fig. 3.6. Mercury astronauts in simulated weightlessness on board a C-131 aircraft (source: NASA).

over the years for research into materials sciences, fluid physics, metallurgy, physical and chemical sciences, biology, medicine, technology, etc. For example, scientists have advanced understanding of why shoots of plants grow upwards and their roots grow downwards by studying plants in the microgravity environment of rockets.[2] Rockets following suborbital trajectories provide a longer period of microgravity than drop towers and parabolic flights, but the payload volume is quite small and they are more expensive to operate.

Japan and Brazil are implementing national programs employing sounding rockets for microgravity research. However, the greatest interest in using sounding rockets is in Europe, involving ESA's Human Spaceflight, Microgravity and Space Operations Program, and the German National Life Science and Material Science Research and Technology Program. Launches are conducted from the Swedish Space Corporation's Esrange Space Center, close to the town of Kiruna in northern Sweden.[3] This location was chosen because it combines good communications and infrastructure with a low population density. The nominal landing point for payloads lies 75 km north of the launch site. The impact region has an area of 5,600 km^2. A prerequisite for launching is that the probability of personal injury or death to a member of the general public be less than 3×10^{-5}.

[2] Gravitropism is the term used to describe how plants orient themselves in relation to gravity.
[3] At 67°N, 21°E.

Table 3.6. Comparison of the characteristics of sounding rocket programs in Europe.

Program[a]	*Apogee* (km)	*Microgravity time* (min)	*Number of experiment modules*	*Scientific payload mass* (kg)
MAPHEUS	140	3–4	1–2	100
TEXUS	250	6	4	260
MASER	250	6	4	260
MAXUS	750	12.5	5	480

[a] MAPHEUS is a German acronym for Material Physics Experiments Under Microgravity; TEXUS is a German acronym for Technological Experiments in Weightlessness; MASER is an acronym for Materials Science Experiment Rocket; and the name MAXUS is derived from MAser-teXUS.

Sounding rockets currently employed in Europe can provide microgravity for 3–12 minutes. Table 3.6 summarizes the characteristics of the European sounding rocket programs. Launches of one or other of the first three rockets in the table occur several times per year, but the MAXUS rocket is launched only once every 2 or 3 years. A microgravity level of 10^{-6} g can be achieved while above an altitude of 100 km on the ascent and through apogee, until again reaching 100 km on the way down. Below this altitude there is too much atmospheric drag to maintain such a level. In flight, the payload can be stabilized by cold-gas systems and inertia platforms to maintain the microgravity for the science objectives. Starting in 2000 it became standard to provide experimenters with a command capability directly to the flying experiment or a transmission capability for telemetry data and video via the Internet to their home research laboratory. Fig. 3.7 shows a TEXUS payload and nose cone prior to launch at Esrange, and Fig. 3.8 records the launch of this rocket on November 27, 2011. Parachutes are used to ensure that the payload returns to Earth undamaged. It is then quickly returned to the launch site by a helicopter. An advantage of carrying out launches in the north of Sweden in winter is that the lakes are frozen and the ground is usually covered in snow, which means that there is no risk of the rocket parts sinking in water and they are relatively easy to spot from the air.

3.2.4 Recoverable satellites

Recoverable satellites offer long-term high-quality microgravity for experiments that do not require very much volume and can be run without the presence of an operator. Russia, China, and Japan have all launched such satellites.

The Russian Foton and Bion reentry satellites are similar spacecraft based on the designs of the Vostok spacecraft; each has a launch mass of about 6,500 kg. They are launched from Baikonur using a Soyuz launch vehicle. Fig. 3.9 shows the layout of a Foton spacecraft, which has three sections: from left to right the

Fig. 3.7. TEXUS 48 rocket payload and nose cone (source: Carol Norberg).

service module, the spherical descent module, and the power supply module. The descent module contains the experiments and returns to Earth gently with the aid of a parachute and a retro-rocket that fires immediately prior to contact with the ground. Foton satellites are used for research in physical and materials sciences and even life sciences, while Bion satellites are mainly used for research in the life sciences. The first Bion flight was in 1973 and the first Foton was in 1985. There are fewer than one flight of each type per year. The latest versions, Bion-M and Foton-M, orbit at altitudes in the range 400 to 600 km at an inclination of 64.9°. The active flight time for Bion-M is 30 days and for Foton-M from 60 to 90 days. This is longer than was originally possible, and was achieved by replacing the batteries by a solar power system. These satellites provide a high-quality micro-gravity environment. They can accommodate 650 kg of scientific payload inside

Fig. 3.8. TEXUS 48 rocket launch from the Swedish Space Corporation Esrange Space Center, Kiruna, Sweden, November 27, 2011 (source: EuroLaunch-SSC/ESRANGE-DLR MORABA).

and 250 kg outside the spacecraft to investigate the space environment, including the reentry environment.

3.2.5 International Space Station

The ISS offers the opportunity to undertake research in a microgravity environment over a period of about 6 months, with levels of 10^{-6} g for at least 30 consecutive days at a time. The microgravity environment aboard the ISS varies in quality depending on the position inside the station relative to the station's center of mass, with the best levels nearest the center of mass. Disturbances in the onboard environment, drag on the station, and even the movements of the astronauts can impair the quality of the microgravity. On the ISS there are laboratories owned by ESA, JAXA, and NASA. Two NASA accelerometers inside the centrally located U.S. laboratory module Destiny monitor the environment and their data are made available to experimenters to assist in interpreting the data from their experiments. The Microgravity Acceleration Measurement System measures small quasi-steady accelerations at frequencies below 1 Hertz caused by the interaction of the station with the ambient environment, while the Space Acceleration Measurement System measures accelerations in the range 0.01 to 400 Hertz caused by the astronauts, the equipment, or the station itself.

Staffing for microgravity experiments is available in the form of the crew of the ISS, who must be trained on the ground on how to run the experiments in

Fig. 3.9. Artist's impression of the Foton-M3 spacecraft (source: ESA).

space. The longest running microgravity experiment is a plasma crystal experiment. Plasma is an ionized gas with equal amounts of positive and negative charge. It is the fourth state of matter after solid, liquid, and gas. In fact 99% of the universe is in the plasma state. The Sun is a ball of plasma. Terrestrial examples include candle flames and the gases in fluorescent lights. If small dust particles are added to plasma under suitable conditions the dust particles form a stable lattice structure similar to that in solid crystals. These plasma crystals, as they are called, were discovered in 1994. Only thin crystals can be made in ground-based laboratories due to the presence of gravity but under microgravity conditions it is possible to create large three-dimensional plasma crystals. The Max Planck Institute for Extraterrestrial Physics in Germany is leading the plasma crystal research in microgravity. The initial research was carried out on parabolic flights and TEXUS rocket campaigns. Next, two devices, PK-1 and PK-2, were used to produce plasma crystals on board the Mir space station. In 2001 the plasma crystal experiment PKE[4]-Nefedov was the first scientific experiment to be set up on the ISS (Fig. 3.10). Its successor PK-3-Plus was delivered to the station in 2005. The next generation of the experiment will be PK-4.

[4] PKE is a German acronym for Plasma Crystal Experiment.

Fig. 3.10. Russian cosmonaut Oleg Kotov works with the PK-3-Plus experiment in the Zvezda Service Module of the ISS (source: NASA).

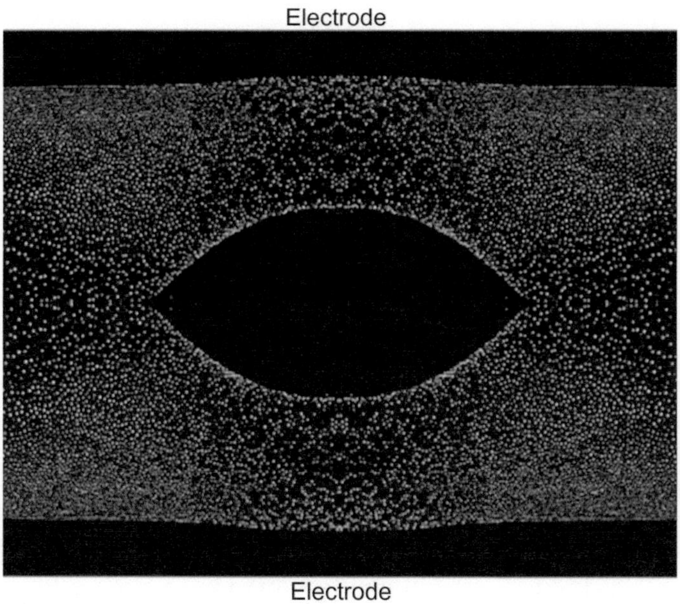

Fig. 3.11. Typical appearance of a dusty plasma in the PKE-Nefedov experiment (source: Thomas *et al.*, 2005).

To produce the plasma crystals, micron-sized polymer spheres are injected into a plasma discharge tube containing argon (neon can also be used). They accumulate negative charge and interact with the surrounding plasma and with one another via the Coulomb force. It can take some minutes for a stable structure to form. In this respect, the long periods of undisturbed microgravity on the ISS are a clear advantage over the more limited periods available in ground-based facilities. Plasma inside the PKE-Nefedov experiment is shown in Fig. 3.11. The plasma is 3–4 cm wide and the white dots are the polymer spheres. The particle-free zone in the middle is a common feature of the plasmas formed. These projects support fundamental physics research.

3.2.6 Space travel and the human body

Human beings have evolved in an environment with practically constant gravitational force acting upon us. When traveling into space the human body is first exposed to high accelerations during the ascent into orbit, then the weightlessness of space, and finally the deceleration of returning home. The launch and landing are relatively brief compared with the periods spent in weightlessness, during which the body must adapt to a new equilibrium state with many changes that can pose problems on returning to an environment with stronger gravitational forces.

Launch and landing

To prepare their bodies for the launch and landing of their spacecraft, all classes of astronauts participate in centrifuge tests. There are a number of facilities around the world that can be used for this work. Fig. 3.12 illustrates how tolerant humans are to acceleration. It is possible to tolerate higher levels of acceleration lying horizontally with respect to the direction of acceleration than it is when sitting upright because in the upright position the acceleration in the upward direction tends to cause blood to be displaced towards the lower part of the body, and the decrease in the blood volume in the head deprives the brain of its normal oxygen supply and increases the risk of passing out. In the horizontal position the blood is more evenly distributed along the length of the body. Keeping the legs above the head is the most desirable position and so this is the preferred position for astronauts during launch. The Mercury astronauts experienced a peak acceleration of 8 g during launches, Gemini astronauts 7.25 g, the Apollo astronauts 4 g, the Space Shuttle peaked at about 3 g and the Soyuz peaks at 3.5 g. The reentry forces in the Soyuz capsule normally go up to about 3.8 g chest to back, in the Space Shuttle they went up to about 1.5 g head to toe. The training flight profile that has been used during centrifuge runs for fee-paying astronauts preparing for a flight on SpaceShipTwo includes a peak of 4 g from head to toe during launch and a peak of 6 g from the chest backwards during reentry. In centrifuge tests people can be trained to increase to some extent their tolerance to acceleration.

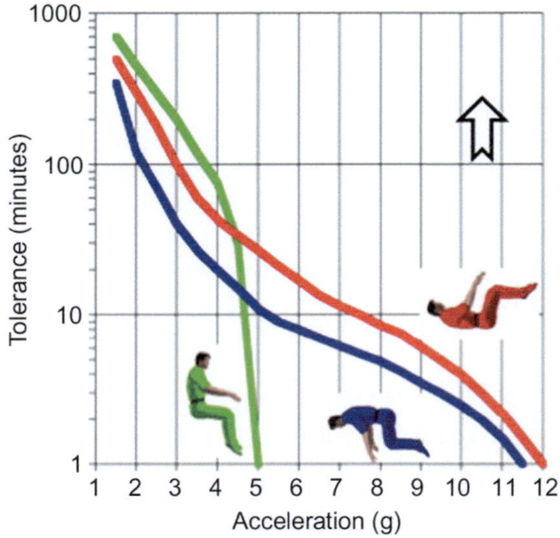

Fig. 3.12. Tolerance to acceleration measured as the elapsed time prior to fainting during a run in a centrifuge in various body positions. The arrow indicates the direction of the acceleration relative to the subject's body position (reproduced from *Fundamentals of Space Medicine*; courtesy of Gilles Clément).

Weightless environment

While weightless on board a spacecraft, the body must adapt to the new environment, and this involves a number of changes. Despite the first space traveler reaching space half a century ago there is actually very little data available on the adaptability of the human body to weightlessness, since the data we have cover in total little more than the equivalent of a human life-span. The majority of those data are for individuals who were weightless for between 1 and 2 weeks and who in addition were very healthy to start with as a result of the careful medical screening of astronaut selection. There are a small number of individuals who have spent much longer periods in space, and by studying them prior to, during, and after their flights researchers have obtained some insight into how the human body reacts to prolonged periods in weightlessness.

More than half of those individuals who have traveled into space have experienced space motion sickness or space adaptation syndrome with symptoms similar to those of Earth-bound travel sickness and most particularly dizziness, nausea, and vomiting. There were no reports during the first space missions, but when people started to fly in spacecraft large enough to unstrap and move around in the problem appeared. The symptoms of space motion sickness vary from mild to severe, and in the severe cases medication can be taken to overcome the symptoms. For those who are afflicted, the symptoms develop soon after reaching orbit and subside after several days. It is not possible to predict in advance who will be affected. Travel sickness on the ground is not a predictor of space sickness. For most of those who suffer, there is no apparent improvement during later flights. To avoid the problem arising during a spacewalk, extravehicular activities are not scheduled for the very beginning of a space mission.

It is not the microgravity environment itself that makes astronauts feel sick but rather the movements that they perform in it. Sudden head movements can provoke an attack, so astronauts are advised to move cautiously upon arriving in space. One explanation that has been proposed to explain this syndrome is that it is due to the brain receiving conflicting signals during movements in microgravity. There are a number of signals to the brain to tell it where we are in relation to our environment. The most obvious is from our eyes, showing where we are irrespective of whether we are on the Earth or in space. Signals from the skin, muscles, tendons, and joints can help us to sense our relationship to our surroundings; this is called proprioception and it seems not to function as well in space as it does on the Earth. Then there is the vestibular system that helps us to balance and move about smoothly on the Earth. Its sensors are found in the ears, and consist of three fluid-filled semicircular canals that are sensitive to angular accelerations and two otolith organs (the utricle and saccule) that register linear accelerations, including gravity (Figs. 3.13 and 3.14). The otoliths are crystals of calcium carbonate that are embedded in a gelatinous substance, which sits on top of small hairs, the cilia. When the head is tipped on the Earth, the otoliths are drawn down by gravity and this causes the cilia to move and generate a signal that is integrated with other inputs to inform the body that the head is being tipped. The problem in space is that when the head is tipped there is no gravitational force to move the otoliths, and although the eyes and neck muscles register the head motion there is no signal from the vestibular system to confirm it. This conflict may lead to space sickness. After several days the body learns to reinterpret the signals it receives in the weightless environment. The neurovestibular system quickly settles into a new equilibrium state. Astronauts returning to Earth from long missions show symptoms similar to space adaptation syndrome. For example, they can experience the feeling of linear motion when they tip their heads because whilst they were in space the brain has learned to interpret otolith movements as linear motion and

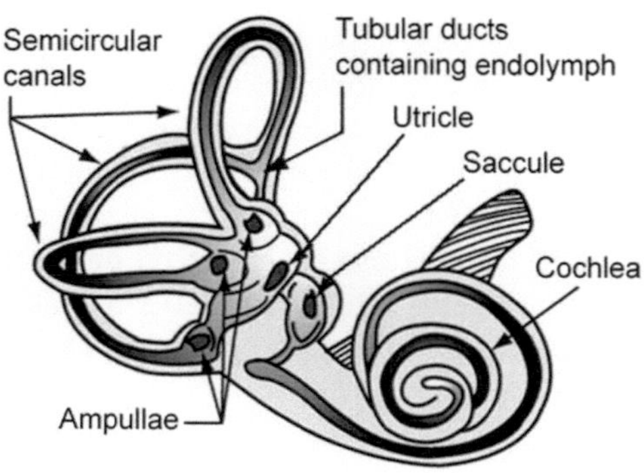

Fig. 3.13. The semicircular canals and the otolith organs in the ear (source: NASA).

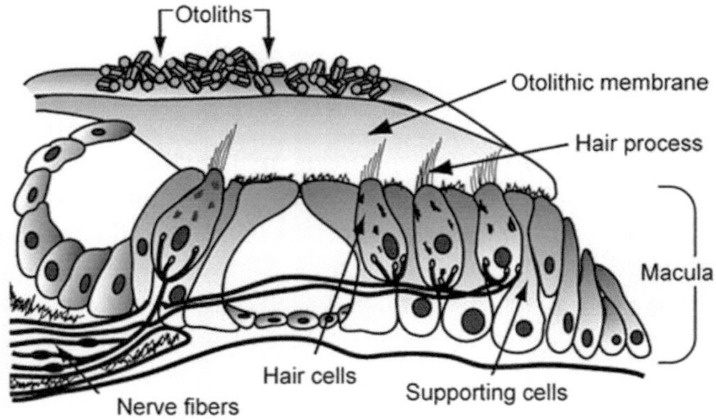

Fig. 3.14. The structure of the otolith organs (source: NASA).

excluded the possibility of gravity being the cause. In a process of reverse adaptation, their bodies must relearn to live in 1 g.

Some of the effects of weightlessness on the human body are evident to the naked eye, notably the so-called "puffy face" and "chicken legs" that astronauts develop as a result of the redistribution of body fluids in space. As we walk on the Earth, gravity acts downwards on our body and we have evolved to adjust the distribution of body fluids to deal with this, since otherwise fluids would tend to pool in our legs. Upon becoming weightless the body initially works as it would on Earth even though there is next to no net force acting on it, with the result that fluids tend to gather in the head and produce a feeling of fullness in the head, headaches, nasal congestion, distended upper veins, and puffy faces in comparison with when on Earth. This can be seen in Fig. 3.15, which shows ESA astronaut Christer Fuglesang in preparations for his flight on STS-128 and looking chubbier in the face during the flight itself. In contrast, astronauts' legs become thinner during spaceflight since there is less fluid in their legs than on Earth. This can be readily observed by measuring the circumference of the legs in each environment. The body adapts to the microgravity environment after a time. The kidneys decrease the volume of fluid in the body, which reduces the fluid pressure in the upper part of the body. As a result of having less blood to pump around the body, the heart diminishes in size. A lower body negative pressure suit that basically sucks the body fluids into the lower part of the body can be used as a countermeasure by simulating the effect of gravity on body fluids. As astronauts return to Earth the gravitational forces on their bodies increase and fluid is once again drawn towards the feet, and this can lead to a deficit in the head and possibly to fainting. To mitigate against this, astronauts are advised to drink a large quantity of fluid immediately prior to returning home. The shifts in body fluid are illustrated in Fig. 3.16.

On Earth the force of gravity leads to compression of the soft intervertebral disks in the spine. The body is usually horizontal when sleeping and the force of gravity is acting equally along the whole length of the back, which stretches a little

Fig. 3.15. ESA astronaut Christer Fuglesang pre flight on the left, and during flight STS-128 on the right (images: NASA).

compared with when the back is usually vertical. This effect can be detected by measuring the height of a person as soon as he gets out of bed in the morning and just before going to bed at night. The height in the morning is usually found to be slightly greater than at night. In space the effect is the same as if the person never got out of bed at all. After a few days an astronaut's height can increase by several centimeters, and the stretched spine can cause back pain.

Space travel is detrimental for musculature. As is well known, exercising builds muscle strength and in microgravity the low workload on the body's muscles quickly leads to muscle atrophy, especially of those muscles that we normally use to maintain our posture. Once astronauts were able to move around freely in spacecraft, it was discovered that the natural relaxed body position in weightlessness is with the back stooped rather than straight, the legs slightly bent and the arms bent at the elbow with the lower arms projecting almost horizontally in front of the body; this is called the neutral body position. Posture was studied during the Skylab missions and the results of analyzing 12 pictures of the three male astronauts of the final crew are shown in Fig. 3.17. These data have been used as a human integration standard by NASA. The study illustrates for example how the line of sight changes in space. This is of significance because astronauts must spend a good deal of time in space looking at screens. It is important to take

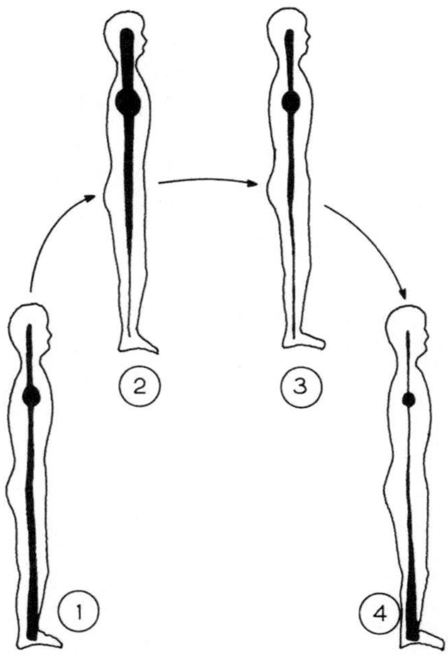

Fig. 3.16. Schematic representation of fluid homeostasis as a function of status: (1) normal conditions on Earth pre flight; (2) immediately after entry into weightlessness; (3) after adaptation to the weightless environment; (4) immediately after return to Earth (courtesy: Helmut Hinghofer-Szalkay).

the neutral body position into account in designing spacecraft and space habitats. It should be noted however that the standard illustrated in the figure is based on three male astronauts, and a later study of four male and two female astronauts on board the Space Shuttle in 1993 suggested a broader variation in the neutral body posture between individuals than first indicated.

In the neutral body position the muscles of the legs do not have to support the weight of the body as they do in walking around on the Earth, so they soon lose mass and strength. Muscle atrophy is not a significant problem for astronauts spending a few weeks in space, but it is for astronauts on longer missions. On return to the Earth muscle mass can be regained by exercising.

The bone in the human body is being formed and resorbed on a continuing basis. A contributing factor to bone formation is physical stress. Floating in a microgravity environment does not stress the skeleton in the same way as living on the Earth. Bone formation decreases but bone resorption remains unchanged, and this results in a net decrease in bone density that progresses with time. Bone loss is not evenly spread throughout the body, and in fact bone density in the head increases slightly. The amount of bone loss is highly individual, as is the rate and extent of recovery after a flight. Bone loss could be a serious health concern for astronauts spending very long periods of time in space, such as on a journey to Mars. They might return with the skeleton of a significantly older person.

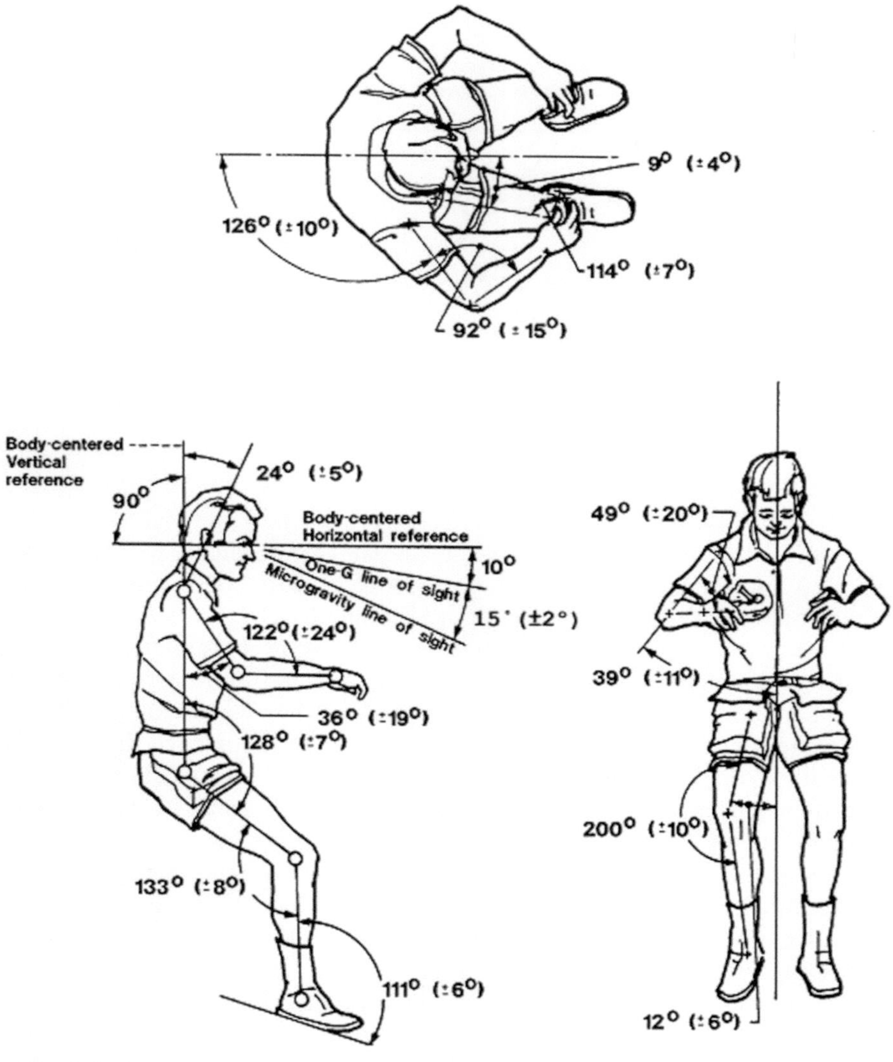

Fig. 3.17. Neutral body posture as determined by Skylab studies (source: NASA).

To minimize muscle atrophy and bone loss, astronauts on the ISS spend 2 hours per day exercising by riding a stationary bicycle, running on a treadmill, training on a rowing machine, and operating resistive training apparatus.

Other side effects of space travel include changes in the perception of smell and taste, an increased risk of developing kidney stones, and the risk of illness because of exposure to high radiation levels. Fig. 3.18 summarizes how the human body adapts to living in a microgravity environment. The horizontal line labeled "clinical horizon" represents the level where signs of illness appear. After about 6 weeks it can be said that the body has adapted to the new environment in an

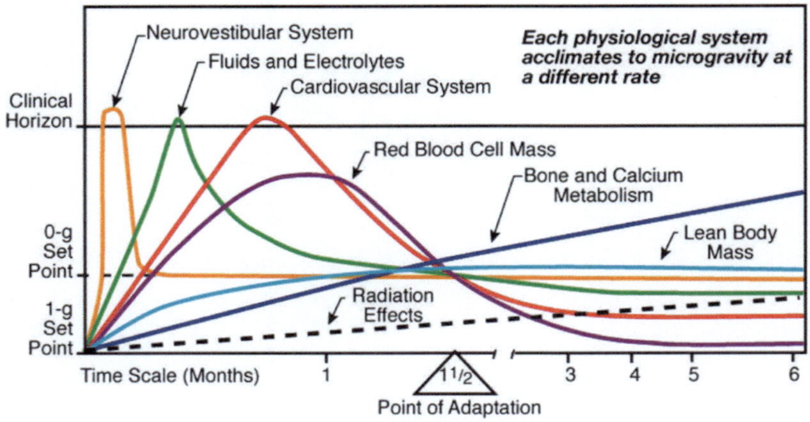

Fig. 3.18. Illustration of the physiological adaptation of the human body to weightlessness as a function of time (courtesy: Arnauld Nicogossian).

equilibrium state different from that at home. However, bone and calcium metabolism and the effects of radiation exposure increase continuously with time. We shall come back to radiation effects in Section 3.6.5.

Having traveled for about a year using today's propulsion technology and adapted to the weightless environment on their spacecraft, an interesting question is how the bodies of the first astronauts to travel to Mars would cope upon arrival at that planet? The gravitational force on the surface is about one third of terrestrial gravity. Would the astronauts have the strength to get out and walk on the surface after their journey? The measures required to ensure that they will be able to function in that environment are the subject of current research. It is generally accepted that the countermeasures to the effects of space travel that are available today are not effective enough. Current countermeasures address one physiological system at a time. An integrated stimulus nearer to that provided by gravity is needed. One countermeasure under consideration is a short-armed centrifuge for use inside a spacecraft. This would have the advantage of exerting forces on the entire body. A spinning spacecraft could provide a force to simulate a gravitational force en route but this still belongs to the realm of science fiction movies.

Analogue studies on the ground

To simulate the effects of weightlessness under controlled conditions on the Earth, bed rest studies are carried out. Participants in the studies spend weeks to months lying in a bed that is tipped at 6° to the horizontal so that their feet are slightly more elevated than their heads, a position which has been found to be the most effective for reproducing the effects of weightlessness. The studies can be useful to do research into the effects of weightlessness over prolonged periods of time without actually sending the subjects into space. Countermeasures can be tested on the ground prior to being put into use on the ISS, for example. The most extreme example of a bed rest study is probably that carried out in Moscow in 1986 when 11 men lived in this position for 370 days. Since fewer women than men have undertaken long-duration spaceflights, there are only limited data available on the long-term effects of weightlessness on the female body. Including women in bed rest studies helps to reveal gender differences in the reaction to microgravity environments. Some studies have included only women.

Another way to simulate reduced gravity is to suspend a person from above so that there is an upward force to act against gravity. The NASA Active Response Gravity Offload System works on this basis, continuously offsetting part of the weight of a human or robot (Fig. 3.19). However, the suspension rig

Fig. 3.19. The Active Response Gravity Offload System (source: NASA).

constrains the movements of the subject. The system is used to simulate microgravity, lunar gravity, and Martian gravity.

3.3 NEUTRAL PARTICLE ENVIRONMENT

The neutral particle environment is the dominant space environment for spacecraft in LEO, giving rise to drag, spacecraft glow, and surface erosion and degradation. Spacecraft can contribute to the neutral particle environment themselves through outgassing. The composition of neutrals varies with altitude, with predominantly nitrogen molecules up to about 100 km and atomic oxygen from about 100 km to 700 km.

3.3.1 Drag

The neutral particle density falls off with altitude above the surface of the Earth from 10^{25} particles/m^3 at the surface to 10^{15} particles/m^3 at about 400 km altitude. True densities vary with the solar cycle. In low orbits the neutral particle population is still sufficient to provide friction to slow down an orbiting spacecraft. Air drag, F_D, can be approximated as

$$F_D = \frac{1}{2}\rho A C_D v^2$$

where ρ is air density, A is the cross-sectional area perpendicular to the direction of motion, C_D is the drag coefficient, and v is the speed of the spacecraft. As a result of drag, spacecraft orbits deteriorate, making reboosts necessary to maintain altitude. Without reboosts, spacecraft in low orbits ultimately reenter the atmosphere and are destroyed. The large ISS descends by about 100 m per day. During periods of increased solar activity the atmosphere expands, increasing drag. Fig. 3.20 shows how the altitude of the ISS has varied over its lifetime. The decreases in altitude are due to atmospheric drag and the increases are the result of reboosts. After the construction of the station was completed and the Space Shuttle was retired in 2011 the station was boosted to a higher altitude than it has had in recent years. It is convenient to have the ISS higher now, since there is no longer a need to access it with the Space Shuttle and the greater the altitude the lower the air drag.

3.3.2 Spacecraft glow

A phenomenon that has been attributed to the neutral particle environment is a glow associated with the surfaces of spacecraft, and extending out to about 10 cm from the surface. The effect was noted in 1983 on one of the early flights of the Space Shuttle, and "Shuttle glow" experiments were carried out on several later missions to study the emission spectrum. An example of glow observed on the Space Shuttle is shown in Fig. 3.21. Scientists have considered various mechanisms

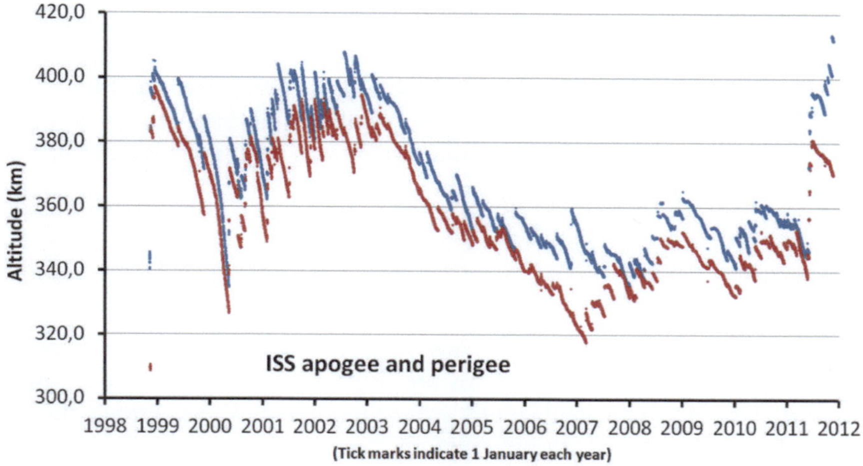

Fig. 3.20. Altitude of the apogee and perigee of the orbit of the ISS as a function of time since the start of construction in 1998 through to 2011. Blue and red represent the apogee and perigee respectively (data courtesy: *www.space-track.org*; image: Sven Grahn).

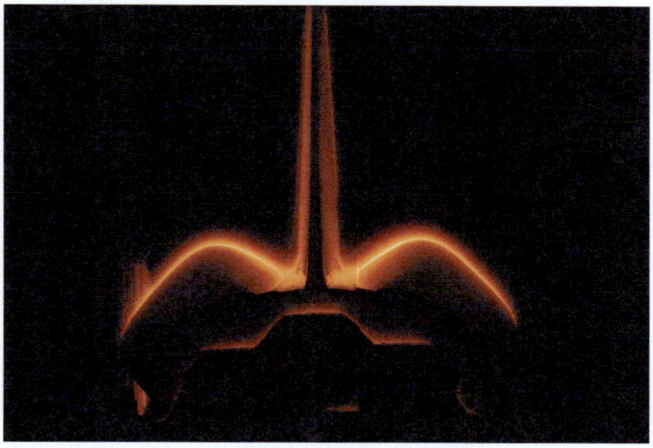

Fig. 3.21. Shuttle glow accompanying the Space Shuttle Columbia during STS-62 in 1997 (source: NASA).

to explain the origin of the glow. The surfaces on which most of the glow is observed are facing the direction of travel and the neutral particles are striking the ram surface at around 7.5 km/s. The source of the visible glow is most likely the recombination of fast oxygen atoms with nitrogen oxide, NO, absorbed on the spacecraft surface to produce excited nitrogen dioxide, NO_2^*, which relaxes by releasing a photon with frequency f

$$O_{fast} + NO_{ads} \rightarrow NO_2^*$$

$$NO_2^* \rightarrow NO_2 + hf$$

where h is Planck's constant. During STS-39 an experiment was carried out in which nitrogen oxide was released from the payload bay of Space Shuttle Discovery. The ensuing intensification of the Shuttle glow confirmed that nitrogen oxide could indeed give rise to glow. The source of the nitrogen oxide that produces glow is probably the ambient atmosphere around spacecraft and/or exhaust gases from spacecraft operations.

3.3.3 Surface erosion and degradation

Neutral particles interacting mechanically with spacecraft can lead to sputtering in which chemical bonds of surface atoms are broken, allowing them to escape from the surface. The effect can significantly reduce the lifetimes of materials on vehicles that are intended to spend very long periods in space, such as the ISS.

In LEO it is atomic oxygen that is the dominant component of the neutral particle population. It reacts with surface materials in a number of ways, producing oxides, causing erosion, and weakening spacecraft surfaces. Mitigation techniques are mainly focused on protective coatings.

3.4 VACUUM ENVIRONMENT

The atmospheric pressure at 100 km (officially defined as the edge of space) is about six orders of magnitude less than the pressure at sea level. At the altitudes of the ISS the pressure is about 10 orders of magnitude less. Hence it is reasonable to talk about the vacuum in space. Spacecraft must be hermetically sealed to prevent the breathable atmosphere inside from escaping. There are large pressure differentials between the inside and outside of crewed spacecraft. Anything that creates a hole in the structure could potentially result in sudden depressurization that would be fatal for the crew. In the case of automatic satellites, vents must be installed to allow atmospheric gases to escape after launch. There is outgassing from many materials when they are in a low-pressure environment; they lose mass and the outgassed material can contaminate the surface of a spacecraft. The highest levels of outgassing can be expected to occur in the first days that a spacecraft is in space but the process will continue throughout a mission. Observations have been made of water and organic material outgassing from the Rosetta spacecraft, forming a thin gas cloud around the vehicle that persisted for many years. In the vacuum of the space environment the intense ultraviolet radiation from the Sun can also degrade materials.

Spacecraft designers can mitigate against the problems that arise in vacuum by carefully selecting materials and by using design strategies to minimize the effects of degradation.

3.5 PLASMA ENVIRONMENT

The nature of plasma was introduced in Section 3.2.5 in describing the plasma crystal experiment on the ISS. The solar wind is the dilute plasma of electrons, protons, a few alpha particles, and some other minor constituents that streams out from the Sun in all directions. At the orbit of the Earth the density of the solar wind is about 5 particles/cm^3, and the speed of solar wind particles varies in the range 300 to 800 km/s with an average of 400 km/s. The solar wind carries with it the interplanetary magnetic field, which has a value of around 5 nT at Earth. The solar wind confines the Earth's magnetic field to a bubble, inside which the motions of charged particles are determined by the planetary magnetic field. This region of space is called the Earth's magnetosphere (Fig. 3.22). On the sunward side of the Earth the solar wind compresses the Earth's magnetic field and on the night side drags it out to form a long tail. The interaction between the fast-flowing solar wind and the Earth gives rise to a bow shock at a distance of six to ten Earth radii upstream on the sunward side. There are similarities to the bow shock in front of a boat traveling through water.

In reality the Earth's magnetosphere is highly dynamic, reacting to changes in the solar wind, which in turn is controlled by processes on the Sun. Solar activity

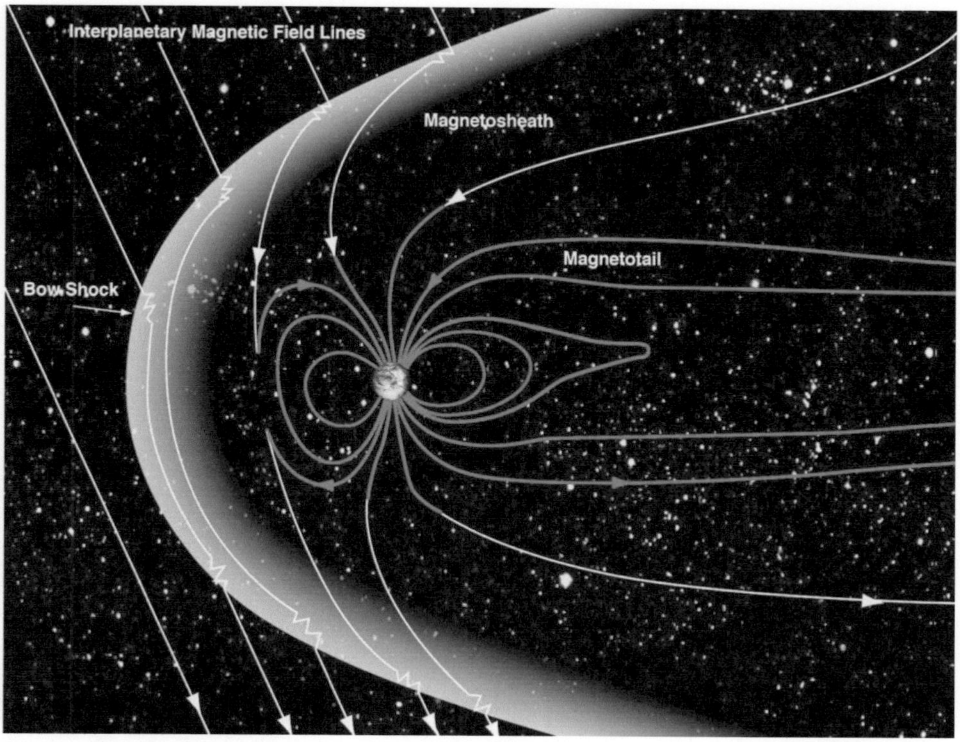

Fig. 3.22. The Earth's magnetosphere (source: NASA/Goddard/Aaron Kaase).

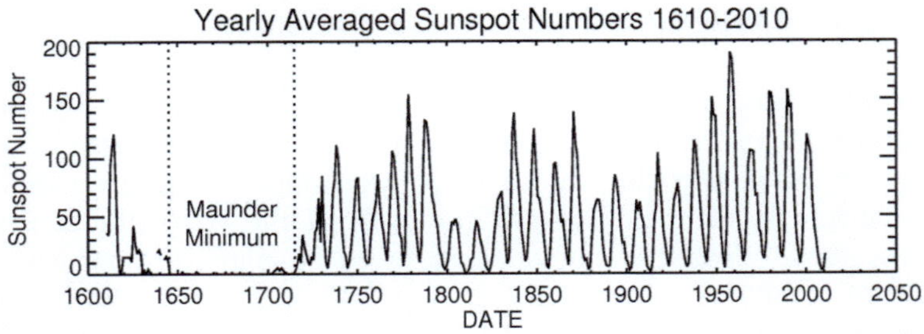

Fig. 3.23. Variation of sunspot number with time, including the Maunder Minimum (source: NASA).

has been observed to follow roughly an 11-year cycle. An indicator of solar activity is the number of sunspots. These are regions on the surface of the Sun that are cooler than the surrounding area, and thus appear darker. They are also associated with increased magnetic activity, solar flares, and coronal mass ejections. Fig. 3.23 shows a record of the sunspot number since the year 1600, when observations began. It is evident that there is a variation in activity between cycles. There was a period from 1645 to 1710, called the Maunder Minimum, when the Sun was inactive. Fig. 3.24 shows more detailed data for this century. The last cycle ended with a long deep minimum in activity, and the present cycle is expected to peak at a lower level of activity than the previous one. The length of a cycle is a weak predictor of the amplitude of the following cycle. Once a cycle

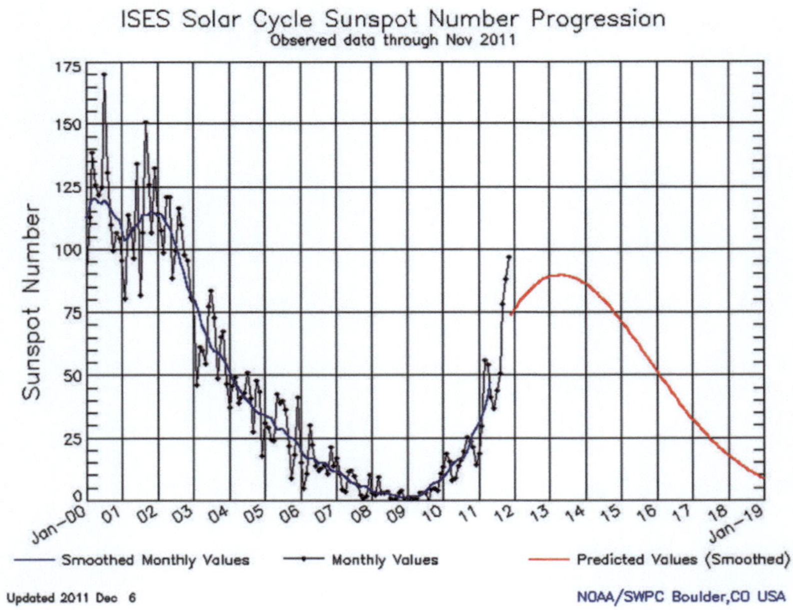

Fig. 3.24. Sunspot number progression January 2000 to November 2011.

begins, a better estimation of its amplitude can be obtained from the rate of increase of sunspots.

Auroral activity is a signature of solar activity that can be seen on Earth and some other planets. The Earth's magnetic field approaches the surface in the northern and southern polar regions (Fig. 3.22) thereby creating openings for charged particles to approach the surface. During magnetic substorms energetic charged particles, mostly electrons with energies in the range 1–15 keV, spiral along the magnetic field lines and collide with the atoms, molecules, and ions in the atmosphere. When an electron has a collision it loses a little energy and the particle it interacts with is excited. After a time the excited particle releases the excess energy as a photon of visible light. The electron continues to collide with atmospheric constituents, the number of collisions increasing as the atmospheric density increases, until it has yielded all of its energy, which typically occurs around the 100 km altitude. Most auroral activity occurs around this altitude, although the lower boundary can be as low as 80 km. An auroral display can extend upwards to several hundred kilometers. It mostly occurs at polar latitudes; in the north being called the "aurora borealis" and in the south the "aurora australis". Fig. 3.25 shows a picture of the aurora australis taken from the ISS. Two colors are seen in the aurora in the image: green and red. Green light with wavelength 558 nm is the most commonly observed color in aurora and is due to the relaxation of excited oxygen atoms. The red aurora present above the green aurora in the image also originates from oxygen atoms. The lower edge of the aurora, at around 90 km, is often seen to have a reddish color that originates from

Fig. 3.25. Aurora australis observed from the ISS on September 17, 2011 (source: Earth Sciences and Image Analysis Laboratory, NASA Johnson Space Center).

nitrogen molecules. At times of increased solar activity there is an increase in the number of energetic particles that interact with the atmosphere and the compression of the magnetosphere enables particles to reach lower latitudes, making aurora visible well away from the polar regions, but this occurs no more than once or twice per year.

During severe geomagnetic substorms there are systems of electric current in the magnetosphere that can induce currents at the Earth's surface and disrupt infrastructure such as electricity grids. One storm in March 1989 caused major power outages in Quebec and the northeastern United States. Storms can also impair communications with aircraft and satellites. One problem for spacecraft due to the plasma environment is known as spacecraft charging; this is mainly a problem in GEO, but can occur in LEO. It can disturb telemetry, electronic instruments, and scientific measurements on spacecraft. Charging is also an important issue with regard to docking of spacecraft and astronauts carrying out spacewalks, since there is a risk of arcing if potential differences are too large. Engineers employ a range of mitigation techniques to tackle the problem.

3.6 RADIATION ENVIRONMENT

In this section we begin by defining the terms and units used to describe radiation. Then we consider the three categories of radiation in the Earth's space environment: the Earth's radiation belts, galactic cosmic radiation, and solar particle events. Most of the space missions to date have been carried out within this environment. The space radiation environment poses a hazard to both equipment and the human body. Apart from the technological and psychological challenges of long missions in space, the most challenging hurdle in sending humans far from the Earth is the danger posed by the radiation environment consisting of galactic cosmic radiation at a fairly constant level and sporadic and unpredictable solar particle events. This has been described as a possible "showstopper" when it comes to human trips to Mars. The status of current research on this subject is summarized in this section.

3.6.1 Radiation, definition and units

Radiation is the term used to describe energy traveling through space in the form of either waves or particles. There are two types of radiation: ionizing and non-ionizing. Ionizing radiation consists of waves or particles with sufficient energy to ionize the atoms or molecules with which it interacts. It is the more penetrating of the two types and can cause the most damage to materials and human tissue. The term radiation is commonly used to mean ionizing radiation. The ionizing radiation in space is mostly in the form of alpha particles, beta particles, or gamma rays released with energies of hundreds of keV to MeV by the decay of radioactive nuclei. In addition, there are protons, neutrons, heavy ions, and X-rays.

Table 3.7. Units of radioactivity and radiation dosage.

Quantity	*Description*	*Unit*	*Symbol*	*In terms of other units*
Radioactivity	The activity of a radioactive material producing one disintegration per second	Becquerel	Bq	s^{-1}
Radioactivity	Radioactivity equivalent to the activity of 1 gram of the radium isotope ^{226}R	Curie	Ci	3.7×10^{10} Bq
Absorbed dose, D	The absorbed energy per unit mass	Gray (SI unit)	Gy	J/kg
Radiation absorbed dose	10 mJ/kg delivered by ionizing radiation	Roentgen absorbed dose	rad	0.01 Gy
Equivalent dose, H	Absorbed dose (in Gy) multiplied by a quality factor characteristic of the radiation, the radiation weighting factor, W_R ($H = W_R \times D$)	Sievert (SI unit)	Sv	J/kg
Roentgen equivalent dose	Absorbed dose (in rad) multiplied by W_R	Roentgen equivalent man	rem	0.01 Sv
Radiation exposure	The amount of gamma or X-rays needed to produce ions carrying one unit of electric charge in one cubic centimeter of dry air under standard conditions	Roentgen	R	2.58×10^{-4} C/kg

A number of units are used to describe radiation and its effects. The most common are presented in Table 3.7. Radiation weighting factors, W_R, are found in Table 3.8. The term "relative biological effectiveness" is used to describe this factor in medical applications. The reference form of radiation in assigning the values of the factors is 200 keV X-rays. Linear energy transfer, LET, is the average local energy deposition per unit length traveled by radiation in a material.

Table 3.8. Radiation weighting factors.

Radiation	W_R
Photons (e.g., X-rays, gamma rays)	1
Beta particles	1
Neutrons	
<10 keV	5
>10 keV–100 keV	10
>100 keV–2 Mev	20
>2 MeV–20 MeV	10
>20 MeV	5
Protons >2 MeV	5
Alpha particles, heavy nuclei	20

It has units of keV/μm. Typically, the higher the LET of the radiation the greater the ionization it can cause.

3.6.2 Earth's radiation belts

Discovery

America's first Earth satellite, Explorer 1 (Fig. 3.26), was launched on January 31, 1958, just a few months after Sputnik. James Van Allen's group at the University of Iowa had installed a Geiger counter capable of registering high-energy electrons and protons with the intention of making measurements of cosmic rays. The satellite was placed in an elliptical orbit. At low altitudes the Geiger counter data were as expected but at high altitudes no measurements were recorded, which was puzzling. Explorer 3 was launched later that year and found that the lack of a signal from its predecessor at high altitudes was actually due to the presence of such highly energetic particles that its counter was unable to register them. The source of the measurements was a belt of energetic particles linking the north and south poles of the Earth and encompassing it. Two Pioneer spacecraft launched on missions to the Moon at the end of 1958 and in early 1959 carried similar instruments and detected another belt of radiation at higher altitudes. The radiation belts are often referred to as the Van Allen belts (Fig. 3.27).

Belt characteristics

The radiation belts are located within the Earth's magnetosphere. The inner belt is in a region of space from about 500 km to about 2 Earth radii above the surface and contains protons in the energy range 0.1–100 MeV and electrons in the energy

Fig. 3.26. A full-size model of Explorer 1 being held aloft by (from left to right) William Pickering, James Van Allen, and Wernher von Braun, at a press conference on February 1, 1958 (source: NASA).

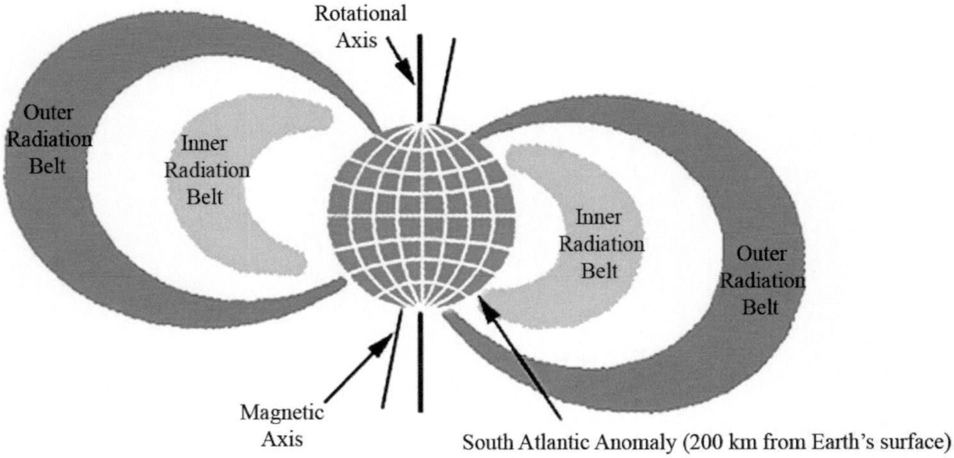

Fig. 3.27. Schematic of the Van Allen radiation belts and the South Atlantic Anomaly (source: NASA).

range 1–10 MeV. The outer belt is located about 3–9 Earth radii from the surface and consists predominantly of energetic electrons (10–100 MeV) and a mixture of ions, mostly energetic protons. The source of the inner belt is thought to be cosmic ray interactions with the Earth's atmosphere and the outer belt is due to the injection of particles from the Earth's plasma sheet. The inner belt is stable while the outer one is variable.

The topology of the belts is controlled by the Earth's magnetic field. The particles are trapped in it, and bounce back and forth between the two magnetic poles. It takes just a few seconds for typical belt electrons to bounce between the poles, and tens of seconds to several minutes for heavier protons. Precisely why the energetic particles become trapped by the Earth's magnetic field can be understood by considering how a single charged particle moves along a magnetic field line. A charged particle spirals around the field line in a direction dependent upon the charge of the particle. The greater the field strength the tighter the spiral becomes, and the component of the velocity parallel to the magnetic field decreases to conserve energy. The force parallel to the magnetic field acts in the opposite direction to the increasing magnetic field gradient, and the spiraling particle is reversed and sent in the opposite direction; this is called magnetic mirroring.

South Atlantic Anomaly

Apart from the Apollo lunar missions, human spaceflights have all taken place at altitudes significantly below the radiation belts. However, part of the inner belt dips down to about 200 km above a region off the Atlantic coast of Brazil. This distortion is known as the South Atlantic Anomaly. It occurs because the Earth's magnetic axis is offset from its rotational axis by about 11° (Fig. 3.27). The increased radiation in this region poses a threat not only to spacecraft carrying a human crew but also to those with delicate solid-state electronics.

3.6.3 Galactic cosmic rays

Galactic cosmic rays originate from outside our solar system. They consist mainly of protons (85%), helium nuclei, and ionized heavy ions such as iron. All of them are fully ionized, and because they are charged their motion is affected by the magnetic fields that they encounter in their passage through space. Owing to their having been accelerated by distant supernova explosions they are the most energetic component of the space radiation environment, with energies in the range 10 MeV to tens of GeV per nucleon. The highest fluxes have energies around 1 GeV per nucleon. The Sun's magnetic field helps to protect us from this class of radiation. The number of cosmic rays reaching the Earth is anti-correlated with the solar cycle because the more active the Sun the better the shielding effect. The Earth's own magnetic field and atmosphere also provide shielding from cosmic rays. The magnetic barrier is most efficient around the equator. In polar regions, where magnetic field lines approach the surface, it is easier for cosmic rays to enter the Earth's magnetosphere. The thicker the atmosphere the better able it is to absorb cosmic radiation, hence cosmic radiation increases with altitude above the surface up to a maximum at about 20 km. When a cosmic ray hits the atmosphere, it generates a shower of secondary particles. By studying the neutrons generated in the atmosphere it is possible to measure cosmic ray flux. There is a global network of about 50 neutron-monitoring stations at a variety of locations and altitudes; ideally the stations should be as high as possible. Fig. 3.28 shows cosmic ray variations measured by the Moscow Neutron Monitoring Station between 1958 and 2011.

Cosmic rays are also monitored from space. Fig. 3.29 shows measurements of iron nuclei in cosmic rays provided by NASA's ACE (Advanced Composition

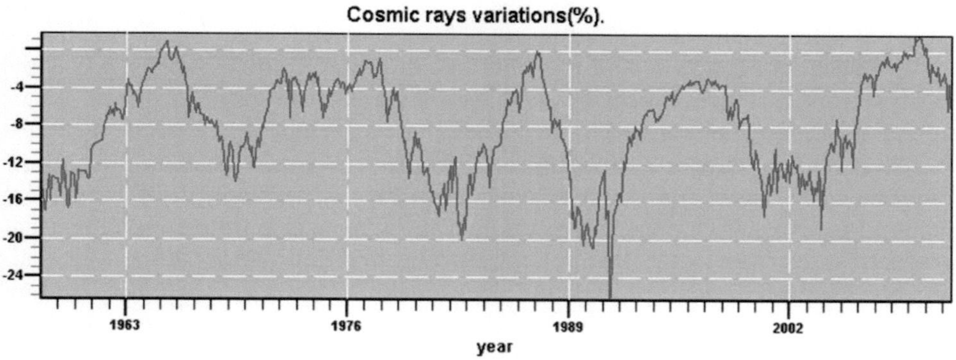

Fig. 3.28. Cosmic ray variations deduced from neutron measurements (source: Moscow Neutron Monitoring Station).

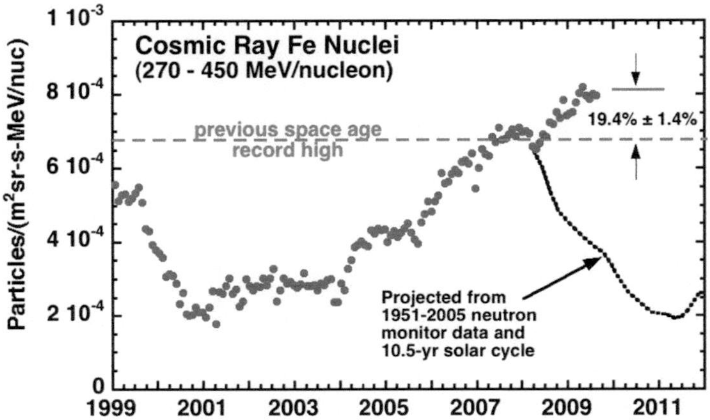

Fig. 3.29. The cosmic ray high in 2009 (source: NASA).

Explorer) satellite. The data show that cosmic ray intensity had reached a space age peak that correlated with the unusually deep solar minimum experienced at the end of the last solar cycle (see Fig. 3.24).

Recent research has found that there are also ultra-high energy cosmic rays with energies exceeding 10^{20} eV reaching Earth. These particles may originate outside our galaxy.

3.6.4 Solar particle events

Unpredictable outbursts on the Sun driven by solar flares and coronal mass ejections lead to the emission of protons with energies in the range of a few MeV to a few hundred MeV, electrons with energies up to a few MeV, and some alpha particles and heavier ions. These can produce disturbances in the radiation environment around the Earth lasting for hours to several days. Although the occurrence of individual events is unpredictable, their frequency follows the solar cycle, being more frequent during solar maximum and less frequent or indeed absent at solar minimum.

Strangely enough, solar storms can actually lead to an improvement in the Earth's radiation environment because the magnetic fields associated with major storms can deflect cosmic rays away from the Earth, giving rise to a "Forbush decrease" in the radiation environment. (The phenomenon is named for Scott Forbush, the American physicist who discovered it.) An example of such an event occurred in 2005 when solar flares resulted in high-energy proton emissions that produced a decrease in the cosmic rays reaching the Earth. Radiation levels temporarily dropped inside the ISS and the Moscow Neutron Monitoring Station noted a dip in the number of neutrons over a period of several days (Fig. 3.30).

A number of major solar particle events are presented graphically in Fig. 3.31. The event in August 1972 occurred between the Apollo 16 and 17 lunar landings and is one of the largest events ever detected. If there had been an Apollo mission

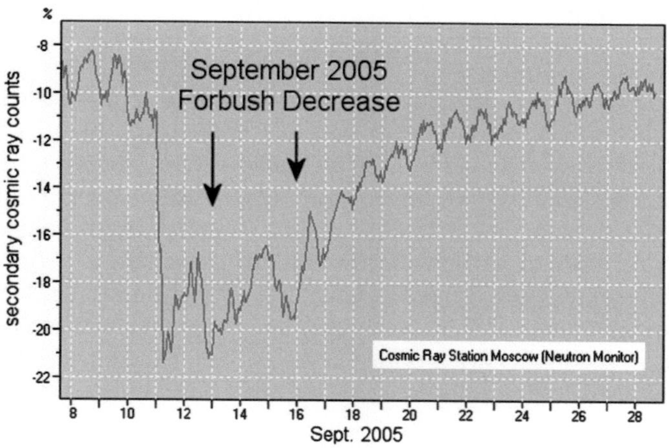

Fig. 3.30. The cosmic ray minimum in 2005 (source: NASA).

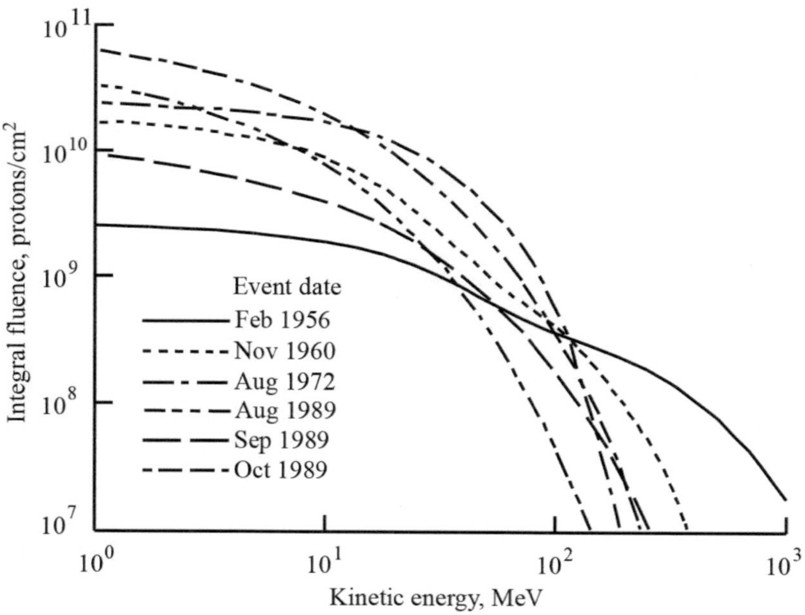

Fig. 3.31. Large solar proton event spectra at 1 AU (source: NASA).

in progress at that time, the astronauts would have been exposed to a dangerously high radiation dose. This is often cited as the worst case example in making predictions of radiation exposure for long-duration missions to the Moon and Mars.

The largest solar particle event on record was caused by a solar flare known as the Carrington flare of 1859, named for the British astronomer Richard Carrington who observed it with his telescope. Information about it has been obtained more recently by studying the concentration of nitrates in polar ice.

3.6.5 Radiation effects

The natural barriers of the Earth's magnetic field and atmosphere protect us from the worst space radiation. Humans are accustomed to living with low levels of radiation on the surface of the planet and our bodies are able to repair most of the damage that occurs. The amount of radiation we are exposed to depends upon where we live, the buildings we live in, our health care (e.g., how many medical X-rays we receive), and our activities. Radon is the primary component of the background radiation at ground level. Radon gas emerges from the ground (especially from some types of rock) and can penetrate homes. In addition, it can be found in building materials. It is dangerous because the low-energy alpha particles emitted by radon gas are a major cause of lung cancer. A simple measuring device can be used to check the level in buildings, and action taken to cut excessive concentrations. The average annual exposure to radiation of a person living on Earth is about 3.5 mSv.

When we travel on aircraft there is a thinner layer of atmosphere above us than at sea level and we are less protected from the radiation coming from space. In addition, in the polar regions the magnetic barrier is weaker than it is above the equator and the radiation exposure is greater. Air crew must be monitored to ensure that they are not exposed to unacceptably high radiation levels, and flights are sometimes re-routed to avoid high doses in the vicinity of the poles, which can occur, for example, if there is a solar particle event.

Table 3.9 gives approximate radiation doses to which humans are exposed. The radiation doses on commercial flights are estimated using the on-line version

Table 3.9. Human radiation doses.

Event	*Dose*
Average annual radiation exposure on Earth	3.5 mSv
Radiation exposure on a commercial flight from London to Paris, cruising altitude 8,500 m	3 μSv
Radiation exposure on a commercial flight from London to New York, cruising altitude 11,000 m	50 μSv
Average skin doses, Apollo crews	0.16–1.14 cGy
10-day Space Shuttle mission	7 mSv
6-month stay on the ISS at solar maximum	80 mSv
6-month stay on the ISS at solar minimum	160 mSv
Current career dose limit for astronauts (there is some variation from country to country and for males and females)	1–4 Sv

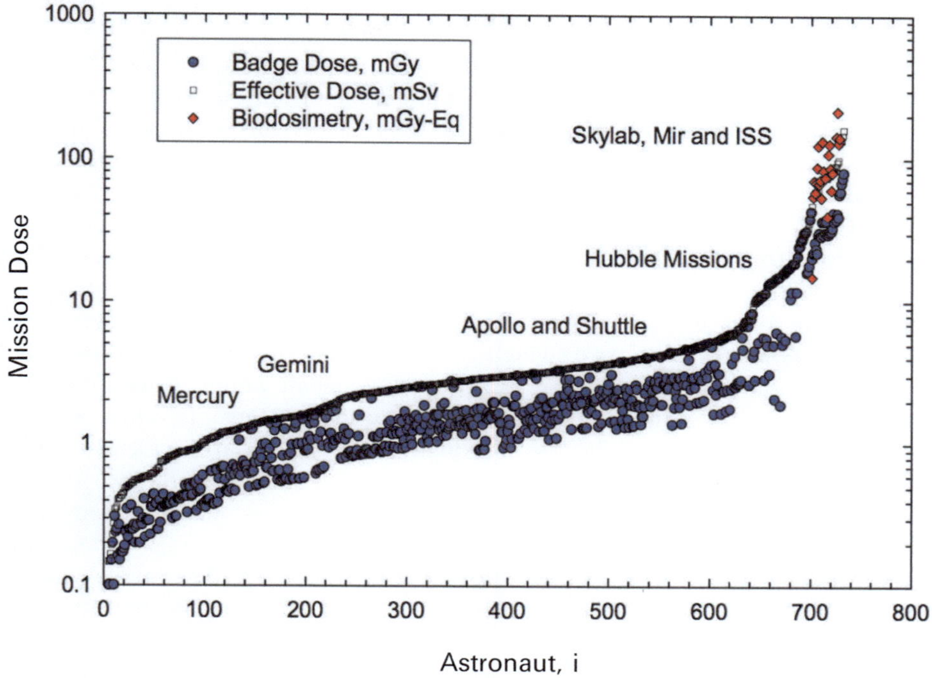

Fig. 3.32. Badge doses and effective doses from astronauts on NASA space missions up to 2007 for Mercury, Gemini, Apollo, Skylab, Apollo–Soyuz, Shuttle, Mir, and ISS (source: NASA publication by Cucinotta and Durante, 2009).

of the European Program Package for the Calculation of Aviation Route Doses (EPCARD). The maximum dose obtained during the Apollo program occurred when Apollo 14 passed through an intense part of the Van Allen belts. The different values for the ISS at solar maximum and solar minimum illustrate the shielding of cosmic rays provided by solar activity.

The early astronauts spent little time in space, but as missions became longer and ventured away from the local space environment the radiation exposure increased as illustrated in Fig. 3.32.[5] The Apollo astronauts had to pass through the radiation belts to travel to and from the Moon, but the duration of the exposure was brief. The curve is expected to increase sharply upwards for long-duration space exploration missions. Whole body doses of 1–2 mSv per day are expected in interplanetary space, which is equivalent to the dose on the Earth over a period of months. A mission to Mars with current technology is expected to expose astronauts to radiation levels comparable with or in excess of the current career dose limits.

[5] Biodosimetry involves using physiological, chemical, or biological markers to evaluate the exposure to ionizing radiation, instead of making direct measurements of radiation itself.

Light flashes

Astronauts can detect the radiation environment in space with their eyes because it causes them to see occasional flashes of white light of various shapes and sometimes moving. The flashes are similar to those experienced by some migraine sufferers on the Earth. In space they are most often seen just before the astronauts go to sleep. The first reports of such light flashes came from the Apollo astronauts. The frequency was about one every 3 minutes; however, the period varied from person to person and was different going to the Moon and coming back. During studies carried out on the Mir space station, one flash was noted every 7 minutes on average. There is an indication that astronauts observe fewer flashes the longer they spend in space, which might explain why the Apollo astronauts saw fewer on the way home. There has been some debate about the cause of these light flashes. Research indicates that high-energy ions hitting the retina probably cause the majority of the flashes, although some others are evidently due to Cherenkov radiation. Astronauts passing through the South Atlantic Anomaly tend to observe an increase in the number of light flashes, which is only to be expected due to the elevated levels of particle radiation in that region. On board the ISS, the ALTEA (Anomalous Long-Term Effects in Astronauts' Central Nervous System) project includes a helmet with inbuilt particle detectors that astronauts wear for limited periods in order to correlate particle measurements with flash observations and brain activity. It is hoped that this study will give a deeper understanding of the observed flashes.

Health risks

Excessive radiation exposure gives rise to acute radiation syndrome, the symptoms of which can include changes in the blood, nausea and vomiting, bleeding diarrhea, and death. At lower levels of exposure there are long-term effects such as cataracts, sterility, and damage to cellular DNA leading to cancer. There are difficulties in accurately estimating the risk of cancer and other diseases due to ionizing radiation during long-duration missions beyond the Earth's magnetosphere. Astronauts will be exposed to the full spectrum of galactic cosmic rays and unpredictable solar particle events. In LEO the Earth's magnetic field affords a degree of protection from galactic cosmic rays, and indeed the Earth itself serves as a shield to radiation. Current dose limits are calculated such that the risk of exposure-induced death of an astronaut from cancer shall not exceed 3%. Estimates have been made based on studies of radiation workers, patients exposed to radiation treatments, victims of nuclear accidents, and of the fallout from atomic bombs, but the radiation to which they have been exposed is mainly low-LET radiation that places a low amount of energy in materials and tissue. In interplanetary space the ionizing radiation is of a different character, being high-LET radiation mainly composed of

- high-energy protons
- highly charged and energetic heavy ions (HZE ions)

- secondary radiation, including neutrons and recoil nuclei resulting from primary radiation interacting with spacecraft surfaces or tissue.

This radiation places a large amount of energy in material and tissue. Not enough is known about the effects of this kind of radiation on the human body to determine reasonable exposure limits for space crews on interplanetary missions. For example, there is uncertainty regarding the accuracy of using the weighting factors presented in Table 3.8 to estimate the effects of HZE particles in comparison with the effects of a similar dose of low-energy radiation. Experiments on animals show HZE particles to be more carcinogenic than low-energy radiation, but the growth of tumors caused by HZE particles is still poorly understood.

Radiation damage to spacecraft

Radiation can damage solar cells on spacecraft, leading to decreased power output with implications for the lifetime of a satellite. Spacecraft must be built with an initial excess of power-generating capacity in order that they will be able to operate efficiently towards the end of the mission.

A single high-energy particle can cause unwanted ionization in electronics. This can lead to a common problem called a single-event upset in which an electronic bit is flipped from 0 to 1 or vice versa, corrupting the data. This problem has grown as electronic devices have become ever smaller and the operating voltages within them have decreased. Radiation can also cause devices to hang or break.

Mitigation techniques

Traditional mitigation techniques against unwanted radiation exposure are to increase the distance from the source, shorten the exposure time, and use shielding. To date the focus has been on identifying an effective shielding material for spacecraft intended for long space missions. Hydrogen has properties that make it attractive as a potential shielding material. Materials like polyethylene, which is a plastic with high hydrogen content, are of interest. The thicker the shielding the greater the protection, but there comes a point that continuing to thicken the shield provides little additional benefit. Shields 5–7 cm thick can stop about one third of the radiation. However, thick shields are cumbersome and expensive to launch. Building faster spacecraft can shorten the exposure time and the likely exposure can be minimized by selecting an appropriate travel time in relation to the solar cycle.

Astronaut selection strategies could be tuned to identify individuals whose genetics imply a low susceptibility to developing cancer. This could be detrimental to the selection of females, since in comparison with men they have the additional risk of developing breast or ovarian cancer. In addition, there is a risk to post-flight pregnancies because women are born with their life supply of eggs in their ovaries. Research indicates that the risk of cancer decreases with increasing age, although the magnitude of this effect is uncertain. For long missions in deep

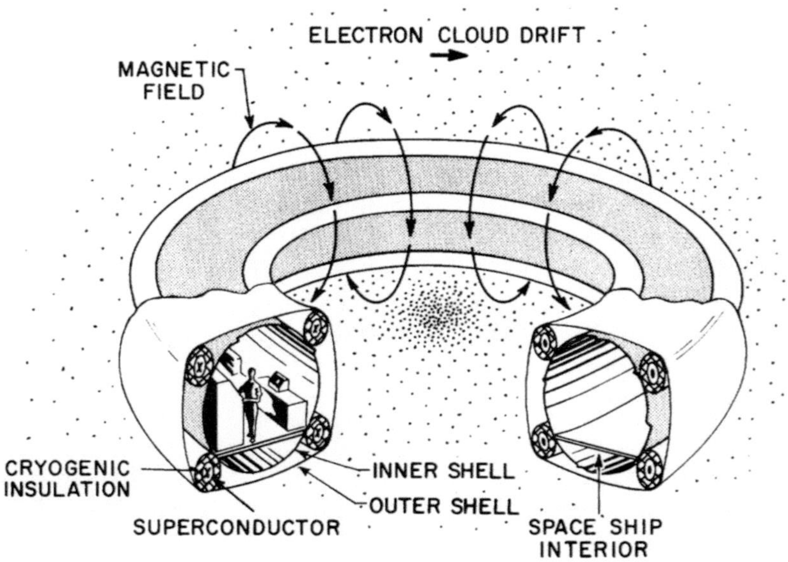

Fig. 3.33. Plasma radiation shield concept (source: NASA publication by French and Levy, 1967).

space, this makes older candidates somewhat more attractive than younger ones. In fact, it could be argued that older candidates are risking fewer years of their life.

Medication and a good diet can help to keep astronauts healthy. A substance called "amifostine" is often administered to patients undergoing radiation treatments since it has proven effective in protecting against ionizing radiation. It would not be suitable for astronauts because the associated side effects include vomiting, but if a similar medication without such side effects is developed it would be appropriate. A diet rich in antioxidants (e.g., in fruit, vegetables, cereal, and nuts) can protect against cancer, but in space these might help to rescue damaged cells and promote cancerous tumors.

It may be necessary to develop more innovative solutions to the radiation problem. Fig. 3.33 shows a plasma radiation shield that was proposed in the 1960s to protect against radiation from solar proton events. In recent years researchers have carried out experiments and computer simulations to investigate how an artificially generated "mini-magnetosphere" around a spacecraft could give it protection similar to that provided naturally for us on Earth by the magnetosphere around the planet (Fig. 3.34).

3.7 MICROMETEOROID AND SPACE DEBRIS ENVIRONMENT

Meteoroids are naturally occurring fragments of matter resulting from the breakup of comets and asteroids. They are larger than an atom or molecule, but

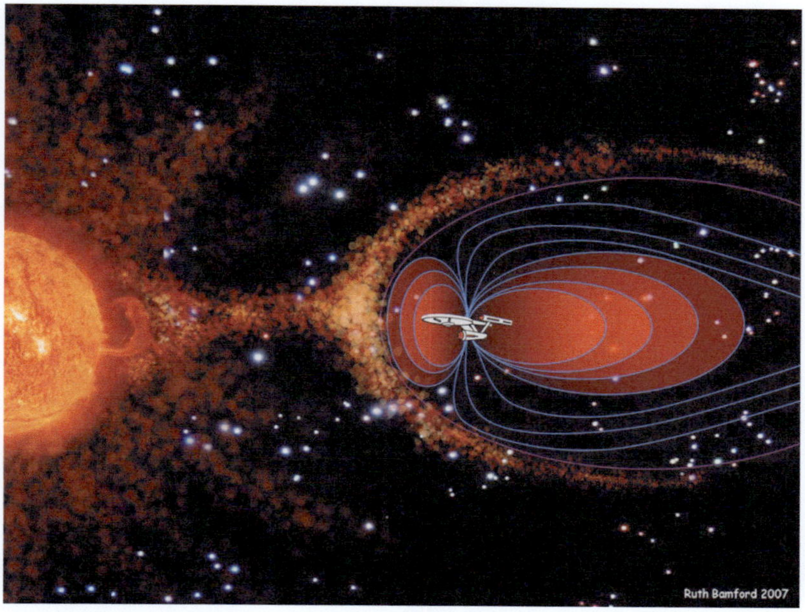

Fig. 3.34. Mini-magnetosphere radiation shield (source: Ruth Bamford).

smaller than an asteroid. The visible path of a meteoroid as it burns up in the Earth's atmosphere is called a meteor. These "shooting stars" are often visible in the altitude range 70–140 km. Fortunately, most of the meteoroids entering the Earth's atmosphere burn up before they reach the ground. If a meteoroid does reach the ground its remnants are called meteorites and any resulting impact crater, a meteor crater. Micrometeoroids are small meteoroids. They are usually less than a few millimeters in diameter and have a mass of less than 1 gram. The velocities of meteoroids lie in the range 11–72 km/s, with an average velocity of about 20 km/s upon entering the Earth's atmosphere. As they pass through LEO they have the potential to seriously damage anything they collide with. The incident meteoroid flux is not constant. At recurring periods of the year showers of meteors occur when the Earth passes through the orbit of an asteroid or comet. The showers are named after the star constellations from which they *appear* to emanate. The heaviest of the showers in the year 2011 are detailed in Table 3.10. The zenith hourly rate (ZHR) is the approximate maximum number of meteors per hour that an observer would see in perfectly clear skies with the radiant of the shower positioned directly overhead. The period of activity can vary from year to year, as can the ZHR, which increases with proximity to the source of the shower. The European Space Agency's Olympus telecommunication satellite was lost during the Perseid meteor shower of 1993. Although it has not been proven that a meteoroid strike caused the loss, this is thought likely.

Operations in the local space environment have led to an artificial population of debris orbiting the Earth, the main constituents of which are non-operative

Table 3.10. Common meteor showers observed on Earth (source of data: International Meteor Organization).

Meteor shower	*Period of activity*	*ZHR*
Quadrantids	December 28–January 12	120
April Lyrids	April 16–25	18
Eta Aquarids	April 19–May 28	70
Delta Aquarids	July 12–August 23	16
Perseids	July 17–August 24	100
Orionids	October 2–November 7	25
Leonids	November 15–19	20
Geminids	December 7–17	120
Ursids	December 17–26	10

spacecraft, expired stages from launch vehicles, particles from solid rocket motors, remnants of explosions in orbit, and paint flecks. These objects have dimensions of approximately micrometers up to meters. Fig. 3.35 illustrates how the objects that can be tracked are distributed in space. There is a high density close to the Earth, in LEO, and also in an equatorial ring around the Earth that coincides with the popular geostationary orbit. In LEO the amount of orbital debris exceeds that of micrometeoroids, but for orbits above 2,000 km micrometeoroids are in the majority. Currently the priority is on avoiding collisions as much as possible, and giving spacecraft appropriate protection from micrometeoroids and space debris particles that will bombard them while in space. Studies are also made to ensure that there is very little risk of casualties on the ground due to falling space debris.

Most pieces of orbital debris have dimensions less than 1 cm. As there is very little atmospheric drag on such tiny objects they are able to remain in orbit for a long time, only slowly migrating to lower altitudes. Orbital debris removal has been viewed as impractical due to the technical challenges of undertaking it, and the potentially high cost, but the increasing problem may prompt an effort to carry out active removal of the largest objects. When geostationary satellites reach the end of their lives they are boosted into a higher "graveyard orbit" out of the way of operational satellites, to prevent that valuable orbit becoming congested with unused spacecraft. This is a temporary solution, as the debris will ultimately break up and spread out.

Data on the populations of meteoroids and space debris are obtained using ground-based optical cameras and radar, by satellite measurements, and by studying surfaces that have been exposed to the space environment and returned to the Earth. The U.S. Strategic Command employs the global network of radar and

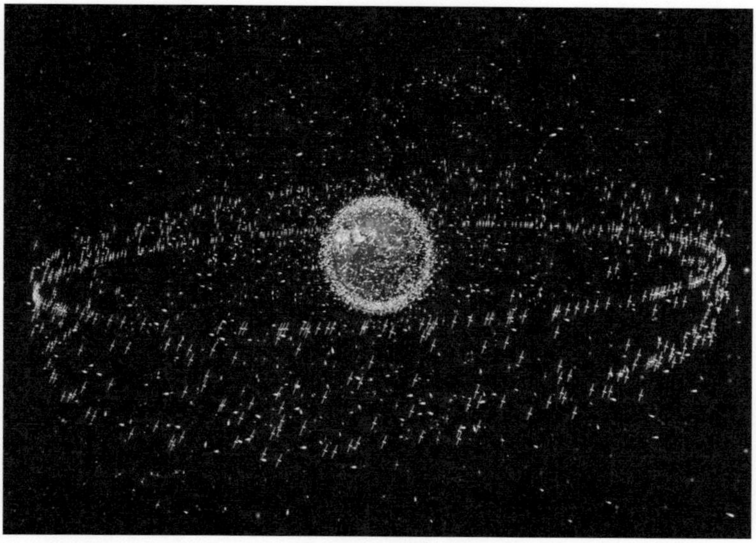

Fig. 3.35. Trackable objects in orbit around Earth. The debris objects shown in the image are an artist's impression based on actual density data. However, the debris objects are shown at an exaggerated size to make them visible at the scale shown (source: ESA).

optical telescopes of the Space Surveillance Network and the Space Based Space Surveillance (SBSS) satellite to track orbital material with a diameter of 10 cm or greater. Only about 5% of the 22,000 objects that it is currently tracking are functioning payloads and satellites. Fig. 3.36 shows how the number of catalogued objects in Earth orbit has increased over the years. The sharp increase in fragments in early 2007 was due to a Chinese anti-satellite test that used a polar-orbiting weather satellite at an altitude of 865 km as the target; it produced over 3,000 trackable objects.[6] The sharp increase seen in February 2009 was due to an accidental collision between the non-operational Russian military satellite Cosmos 2251 and the privately owned communication satellite Iridium 33 at an altitude of 776 km. The Russian Federation also maintains a catalogue of space debris.

The surfaces of objects returning from space have impact craters that testify to the presence of a large number of very small particles in orbit. Fig. 3.37a shows a hole made in a Space Shuttle and Fig. 3.37b is an example of damage to the ISS. Chemical analysis can help to determine whether these objects were hit by meteoroids or space debris. Ground-based hypervelocity impact tests have been used to obtain insight into the size of the impactors. In 1996 an ESA-funded impact detector was launched into GEO aboard a Russian satellite. During 5 years

[6] Strategic satellite destruction tests were also performed by the U.S.A. and the U.S.S.R. in the 1960s and 1970s—see Fig. 3.36 with several step increases (e.g., 1965, 1969, 1971, and 1974). The number of pieces of debris were smaller (200–500) due to smaller satellites, dedicated target satellites, and the fact that tracking capabilities were less advanced at that time.

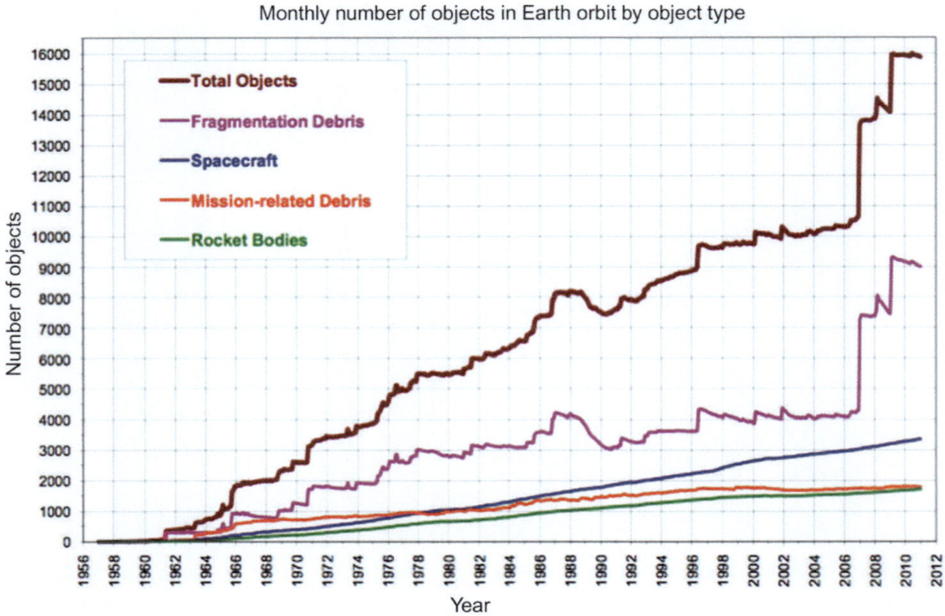

Fig. 3.36. A summary of all objects in Earth orbit officially catalogued by the U.S. Space Surveillance Network. The category "fragmentation debris" includes remnants of satellite breakup and anomalous event debris, while "mission-related debris" includes all objects dispensed, separated, or released as part of a planned mission (source: NASA).

of operation it reported an average of 2.4 impacts per day, with peak counts of 50 per day.

In 1947, a decade before the start of the space age, the American astronomer Fred Whipple considered the threat of a meteoroid penetrating a pressurized spacecraft. He predicted the probability to be low, and to be related to the thickness of the spacecraft's hull. He proposed equipping a spacecraft with a "meteor bumper". Often referred to as a "Whipple shield", a bumper consists of a thin layer of material, usually aluminum, offset at a small distance from the surface of the spacecraft. When the meteoroid hits the bumper it explodes dissipating its energy and momentum over a wide area of the inner shell. Fig. 3.38 illustrates the design of three types of shield, including a Whipple shield; in each case a particle is shown approaching the shield from above. These shields protect against impacts from particles up to about 1 cm in size. Whipple shields are used on the ISS to protect sections of the Russian Zvezda module and NASA's Unity Node and Destiny Laboratory, and typically have a multi-layer insulating (MLI) blanket between the bumper and the rear wall for increased efficiency. The Whipple shield can be stuffed with layers of Nextel and Kevlar, and an MLI blanket can be added to further improve protection compared with a standard Whipple shield that has the same spacing between the bumper and rear wall. Such shields are used to protect parts of the NASA, ESA, and JAXA modules, although each

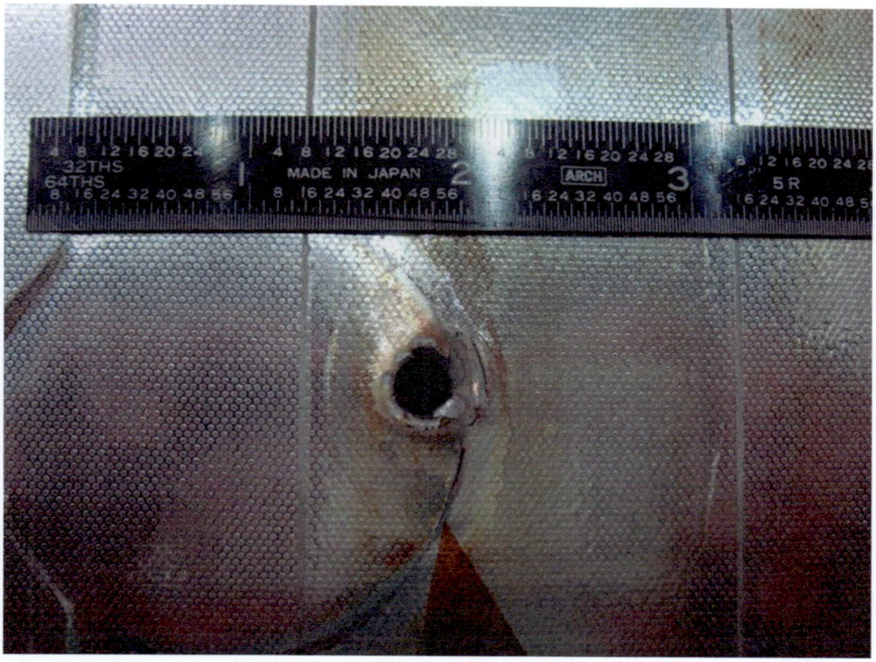

Fig. 3.37a. Entry hole damage (5.5 mm in diameter) to Space Shuttle Endeavour's left-side aftmost radiator panel observed during post-flight inspection, STS-118 (source: NASA).

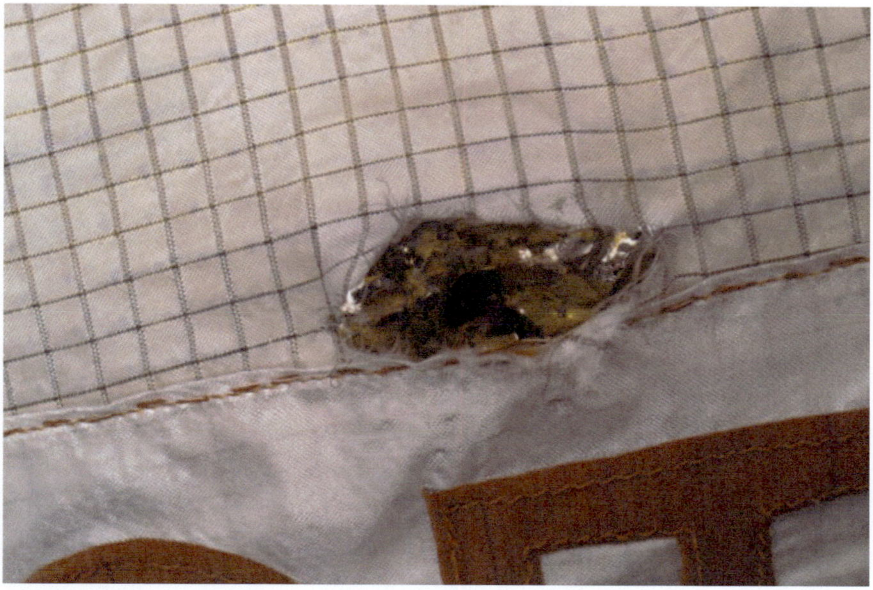

Fig. 3.37b. During a spacewalk, ISS Expedition 16 commander Fyodor Yurchikhin noticed damage to a multi-layer insulation protective blanket on the Zarya module (source: NASA).

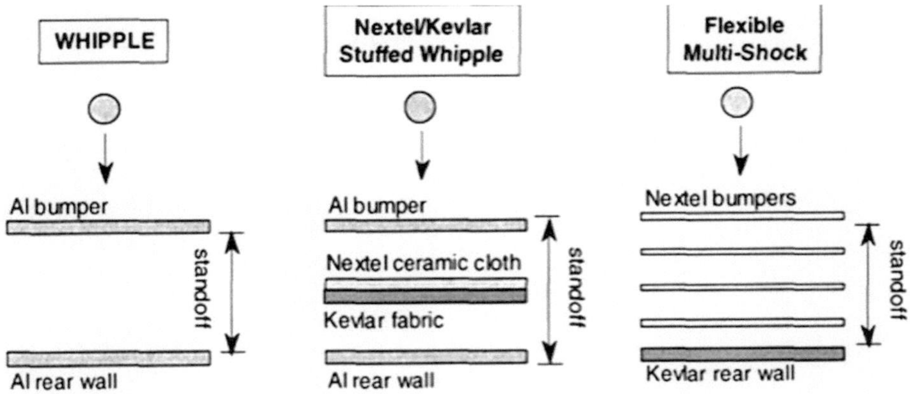

Fig. 3.38. Designs for meteoroid/space debris shields (source: NASA).

organization uses its own variation on the design. Multi-shock shields employ layers of ceramics such as Nextel to increase shield performance, and tend to be set at a larger standoff distance.

Another development is mesh double-bumper shields that have four components: a mesh bumper, a second continuous bumper, a high-strength fabric intermediate layer and a rear wall. A modified mesh double-bumper shield is used to protect parts of the ISS (e.g., steel mesh protects the Russian Zarya module). The modular design of the ISS provides the possibility to isolate a module should it be punctured by an impacting particle. If the position of a small hole in the structure can be located, there is the possibility of repairing it. The goal is to construct effective low-weight shields to protect spacecraft and crews in space. This requires knowledge of the expected debris environment and laboratory tests on the ground to simulate as authentically as possible the kinds of impacts expected in space, augmented with simulations to predict the behavior of shields with various characteristics under the full scope of conditions they may encounter. The analysis is used to determine the optimum shield to employ for a particular position on a specific spacecraft to counter the threat that it will face during its mission. Early spacecraft for human use were shielded against meteoroids only (e.g., the original shielding requirement for the Space Shuttle was for the meteoroid environment only). Now shielding is needed both for meteoroids and space debris, and the focus of concern for impacts by small particles is now space debris. In LEO the relative impact velocity for space debris is 1–15 km/s. Whenever possible, the Space Shuttle flew "backwards" so that if it were struck by debris the impact would be on the engines rather than on the windows at the front where the crew were to be found. When the ISS is threatened by a piece of orbital debris a collision avoidance maneuver is made. If there is insufficient time to perform such a maneuver, the crew must retreat to the Soyuz "lifeboats" in readiness for the possibility of an emergency evacuation. Astronauts carrying out spacewalks only have their space suits to protect them from the ambient environment, and these provide shielding against objects up to about 0.1 mm in size.[7] When possible, spacewalks are carried out in the shadow of the astronaut's spacecraft with regard to space debris.

According to a report from the U.S. National Research Council in September 2011 the situation regarding debris in the space environment around the Earth has already reached a "tipping point" as a result of having passed the threshold at which there will be continual collisions between items of space debris. This means that there will be a continual increase in the number of items in orbit, and in turn an increase in the number of collisions with operational spacecraft. At present there are no international laws governing the disposal of spacecraft in Earth orbit. The Inter-Agency Space Debris Coordination Committee consists of 12 space agencies collaborating to tackle the global problem posed by man-made and natural debris in space.[8] The committee formulated a set of guidelines based on the existing practices of a number of organizations. In response, the United Nations Office for Outer Space Affairs has issued space debris mitigation guidelines for mission planning, as well as the design, manufacture, and operation of spacecraft and launch vehicle orbital stages

1. Limit debris released during normal operations.
2. Minimize the potential for breakups during operational phases.
3. Limit the probability of accidental collision in orbit.
4. Avoid intentional destruction and harmful activities.
5. Minimize the potential for post-mission breakup resulting from stored energy.
6. Limit the long-term presence of spacecraft and launch vehicle orbital stages in the LEO region after the end of their mission.
7. Limit the long-term interference of spacecraft and launch vehicle orbital stages with the GEO region after the end of their mission.

3.8 REFERENCES/RECOMMENDED LITERATURE

Adams Jr., J.H., Hathaway, D.H., Grugel, R.N, Watts, J.W., Parnell, T.A. Gregory J.C., and Winglee, R.M.: *Revolutionary Concepts of Radiation Shielding for Human Exploration of Space*, NASA TM-2005-213688, NASA.

Avdeev, S.: Eye light flashes on the Mir Space Station, *Acta Astronautica,* **50** (2002).

Bothmer, V. and Daglis, I.A.: *Space Weather Physics and Effects*, Springer/Praxis (2007).

[7] A common test value for the space suit thermal micrometeoroid protection garment is to withstand the hit of a 0.5 mm aluminum ball at 7 km/s.

[8] The members of the IADC are ASI (Italian Space Agency), CNES (French Space Agency), CNSA (China National Space Administration), CSA (Canadian Space Agency), DLR (German Aerospace Center), ESA (European Space Agency), ISRO (Indian Space Research Organisation), JAXA (Japan Aerospace Exploration Agency), NASA (National Aeronautics and Space Administration), NSAU (National Space Agency of Ukraine), Roscosmos (Russian Federal Space Agency), and UKSA (U.K. Space Agency).

Ceglia, E.: *European User's Guide to Low Gravity Platforms*, ESA UIC-ESA-UM-0001 Issue 2, Revision 0, ESA (2005).

Christiansen, E.: *Meteoroid/Debris Shielding,* NASA TP-2003-210788, NASA (2003).

Churchill, S.E.: *Fundamentals of Space Life Sciences, Vol. 1*, Krieger (1997).

Clément, G.: *Fundamentals of Space Medicine*, Springer-Verlag (2011).

Cour-Palais, B.G.: *Meteoroid Environment Model—1969 (Near Earth to Lunar Surface)*, NASA SP-8013, NASA (1969).

Cucinotta F.A. and Durante M.: Cancer risk from exposure to galactic cosmic rays: Implications for space exploration by human beings, *Lancet Oncol.*, **7**, 431–435 (2006).

Cucinotta F.A. and Durante M.: *Risk of Radiation Carcinogenesis*, Chapter 4 in J.C. McPhee and J.B. Charles (Eds.), *Human Health and Performance Risks of Space Exploration Missions: Evidence Reviewed by the NASA Human Research Program,* NASA SP-2009-3405, NASA (2009).

Durante M. and Cucinotta F.A.: Heavy ion carcinogenesis and human space exploration, *Nat. Rev. Canc.*, **8**(6), 465–472 (2008).

French, F.W. and Levy, R.H.: *Plasma Radiation Shield Concept and Applications to Space Vehicles*, NASA-CR-84420, Avco-Everett Res. Rept.-258, NASA (1967).

Fuglesang, C., Narici, L., Picozza, P. and Sannita, W.G.: Phosphenes in low Earth orbit: Survey responses from 59 astronauts, *Aviation, Space and Environmental Medicine*, **77**(4), 449–452 (2006).

Hastings. D. and Garrett, H.: *Spacecraft Environment Interactions*, Cambridge University Press (2004).

Hathaway, D.H., Wilson, R.M., and Reichmann, E.J.: The shape of the sunspot cycle, *Solar Physics*, **151**(1), 177–190 (1994).

Jackson, J., Bond, R., and Gundersen, R.: Neutral body posture in zero-G, *Skylab Experience Bulletin*, **17** (1975).

Mount, F.E. Whitmore, M., and Stealey, S.L.: *Evaluation of Neutral Body Posture on Shuttle Mission STS-57 (SPACEHAB-1)*, NASA TM-2003-104805, NASA (2003).

Murad, E.: The Shuttle glow phenomenon, *Annu Rev. Phys. Chem.*, **49**, 73–98 (1998).

Narici, L., Bidoli, V., Casolino, M., De Pascale, M.P., Furano, G., Modena, I., Morselli, A., Picozza, P., Reali, E., Sparvoli, R. *et al.*, The Altea facility on the International Space Station, *First International Workshop on Space Radiation Research and Physica Medica—Vol. XVII, Supplement 1, 2001 11th Annual NASA Space Radiation Health Investigators' Workshop, Arona (Italy), May 27–31*, pp. 255–257 (2000).

National Research Council, Committee on the Evaluation of Radiation Shielding for Space Exploration: *Managing Space Radiation Risk in the New Era of Space Exploration*, National Academies Press (2008).

National Research Council, Committee for the Assessment of NASA's Orbital Debris Programs: *Limiting Future Collision Risk to Spacecraft: An Assessment of NASA's Meteoroid and Orbital Debris*, National Academies Press (2011).

Nicogossian, A.E. (Ed.), Huntoon, C.L., and Pool, S.L.: *Space Physiology and Medicine*, Second Edition, Lea & Febiger (1989).

Schläppi, B, Altwegg, K., Balsiger, H., Hässig, M., Jäckel, A., Wurz, P., Fiethe, B., Rubin, M., Fuselier, S.A., Berthelier, J.J.. *et al.*: Influence of spacecraft outgassing on the exploration of tenuous atmospheres with in situ mass spectrometry, *Journal of Geophysical Research*, **115**, A12313, doi: 10.1029/2010JA015734 (2010).

Seibert, G.: *The History of Sounding Rockets and Their Contribution to European Space Research*, ESA HSR-38, ESA (2006)

Thomas, H., Morfill, G.E., Demmel, V., Goree, J., Feuerbacher, B., and Möhlmann, D.: Plasma crystal: Coulomb crystallization in a dusty plasma, *Phys. Rev. Lett.*, **73**, 652–655 (1994).

Thomas, H., Morfill, G.E, Ivlev, A.V., Nefedov, A.P., Fortov, V.E., Rothermel, H., Rubin-Zuzic, M., Lipaev, A.M., Molotkov, V.I., and Petrov, O.F.: PKE-Nefedov: Complex plasma research on the International Space Station, *Microgravity Sci. Technol.*, **XVI**(1) (2005).

Tribble, A.C.: *The Space Environment, Implications for Spacecraft Design*, Princeton University Press (2003).

Viereck, R.A., Murad, E., Green, B.D., Joshi, P., Pike, C.P., Hieb, R., and Harbaugh, G.: Origin of the Shuttle glow, *Nature*, **354**, 48–50 (1991).

Whipple, F.L.: Meteorites and space travels, *Astronomical Journal*, **1161**, 131 (1947).

Wilson J.W., Cucinotta, F.A., Jones, T.D., and Chang, C.K.: *Astronaut Protection from Solar Event of August 4, 1972*, NASA Technical Paper 3643, NASA (1997).

Wilson J.W., Cucinotta, F.A., Tai, H., Simonsen, L.C., Shinn, J.L., Thibeault, S.A., and Kim, M.Y.: *Galactic and Solar Cosmic Ray Shielding in Deep Space*, NASA Technical Paper 3682, NASA (1997).

Zhang, X., Yuan, L., Wu, E., Tian, L., and Yao, K.: Some key technics of drop tower experiment device of National Microgravity Laboratory (China) (NMLC), *Science in China Ser. E Engineering and Materials Science*, **48**(3), 305–316 (2005).

The DLR Parabolic Flights, German Aerospace Center D-GB_05/06, DLR.

Human Integration Design Handbook, NASA/SP-2010-3407, NASA (2010).

Technical Report on Space Debris, U.N. (1999).

Space Debris Mitigation Guidelines of the Committee on the Peaceful Uses of Outer Space, U.N. Office for Outer Space Affairs, U.N. (2010)

Websites

Brazilian Space Agency: *www.aeb.br*

German Aerospace Center: *www.dlr.de*

German Research Center for Environmental Health, EPCARD portal: *http://www.helmholtz-muenchen.de/en/epcard-portal/epcard-home/index.html*

International Bureau of Weights and Measures: *www.bipm.org*

International Meteor Organization: *www.imo.net*

Inter-Agency Space Debris Coordination Committee: *www.iadc-online.org*

International Space Station Operations: *http://microgravity.grc.nasa.gov/MSD/MSD_htmls/issops.html#*

Japan Aerospace Exploration Agency: *www.jaxa.jp*

Max Planck Institute for Extraterrestrial Physics, Plasma Crystal project: *www.mpe.mpg.de/34063/PKE*

Mini-magnetosphere project, Science and Technology Facilities Council, U.K.: *www.minimagnetosphere.rl.ac.uk*

NASA Zero Gravity (Zero-G) Research Facility: *http://facilities.grc.nasa.gov/drop/*

United Nations Office for Outer Space Affairs: *www.oosa.unvienna.org*

ZARM Center of Applied Space Technology and Microgravity, Bremen: *www.zarm.uni-bremen.de/*

4

Space exploration

Carol Norberg

ABSTRACT

Even before the first satellites were launched into Earth orbit scientists and engineers became interested in carrying out missions to the Moon and to the planets Venus and Mars. There have been many setbacks in their efforts, but also many successes. This chapter summarizes our exploration of the solar system using unmanned spacecraft, plus the science undertaken by the Apollo missions. The preparations for a renewal of human exploration beyond low Earth orbit must take into account both the physical and psychological challenges which human beings will face in making such journeys. Projects addressing the psychological challenges of isolation are described, such as the recently concluded Mars 500 simulation of a mission to Mars.

4.1 SOLAR SYSTEM EXPLORATION

At its simplest, exploration is a quest to discover the unknown. Although we can send spacecraft and robots out to explore space for us, space exploration is at its most satisfying when carried out directly by a human being. A barrier to action in planning human exploration is the possibility of injury or even death. In this respect, robotic exploration has a distinct advantage. The benefit of sending a human being exists in the intelligence, spontaneity, and emotion of physical presence. Although Yuri Gagarin was Russian and the first Moon landing was carried out by Americans, these events had meaning for the world's population as a whole. Space exploration is in reality a global endeavor, as is evident in the United Nations Outer Space Treaty of 1967 "on principles governing the activities of states in the exploration and use of outer space, including the Moon and other celestial bodies." It stated

> The exploration and use of outer space, including the Moon and other celestial bodies, shall be carried out for the benefit and in the interests of all countries, irrespective of their degree of economic or scientific development, and shall be the province of all mankind.

The treaty was opened for signature in Moscow, London, and Washington on January 27, 1967. To date it has been ratified by 100 states including the space-faring ones, and signed by a further 26. The United Nations Committee on the Peaceful Uses of Outer Space produced four further treaties which, in chronological order, are

- The Agreement on the Rescue of Astronauts, the Return of Astronauts and the Return of Objects Launched into Outer Space ("Rescue Agreement"), opened for signature on April 22, 1968 and entered into force on December 3, 1968
- The Convention on International Liability for Damage Caused by Space Objects ("Liability Convention"), opened for signature on March 29, 1972 and entered into force on September 1, 1972
- The Convention on Registration of Objects Launched into Outer Space, ("Registration Convention") opened for signature on January 14, 1975 and entered into force on September 15, 1976
- The Agreement Governing the Activities of States on the Moon and Other Celestial Bodies ("Moon Agreement"), opened for signature on December 18, 1979 and entered into force on July 11, 1984.

Fewer and fewer states ratified each successive treaty; the Moon Agreement which addresses, for example, the exploitation of resources on the Moon and other celestial bodies, is ratified by just 13 countries[1] and signed by another 4. Notably, it has not been ratified by the United States, the Russian Federation, or China.

Space exploration has tended to focus upon the Moon, Venus, and Mars, so these are the main subjects of this chapter. Despite the large number of setbacks in trying to send spacecraft to these bodies, scientists, engineers and the governments that have funded them have persevered to learn more about the history of the solar system and thereby understand the early history of the Earth. Missions to the more distant bodies of the solar system have been fewer in number, but they have had a high success rate. The Apollo lunar landings remain the highlight of space exploration to date. In the years since then, the next logical step has appeared to be a human mission to Mars but this venture faces major hurdles.

Before the dawning of the space age our knowledge of the solar system was based on observations made from the ground, first by the naked eye, then with telescopes of increasing complexity. Optical and radar observations are the ground-based methods used today. The birth of the satellite made it possible to

[1] Australia, Austria, Belgium, Chile, Kazakhstan, Lebanon, Mexico, Morocco, The Netherlands, Pakistan, Peru, Philippines, and Uruguay.

place detectors above the Earth's atmosphere in order to gain a clearer view of celestial bodies across a broader range of wavelengths. The advent of spacecraft for the deep-space environment enabled us to examine our celestial neighbors in close proximity and even land on them. At first, the focus was on taking close-up pictures of the Moon and planets, finding out whether they had intrinsic magnetic fields, and making radiation measurements. Nowadays, instruments are more sophisticated but they still have a limited range of capabilities, and part of a space scientist's job is to have a good idea of what to expect to find, and then design instruments that will enable theories to be tested. The process of building an instrument involves calibrating it and testing it to ensure that it will survive the violent vibrations of launch and go on to function properly on board the spacecraft in the extreme conditions of the space environment.

Table 4.1 provides an overview of the most successful planetary missions to date, apart from Mars which is discussed in Section 4.3. Venus, which has been the subject of many missions, is slightly smaller than Earth, lacks an intrinsic magnetic field, has a thick layer of cloud, and an atmosphere composed almost entirely of carbon dioxide that provides an average temperature of about 460°C and a pressure of 92 bar at the surface. Launched by the U.S.S.R. in 1961, Venera 1 was the first spacecraft dispatched to Venus, but contact was lost en route. The next attempt, Mariner 1 launched by the U.S. in 1962, had to be destroyed shortly after launch when its rocket went off course. Mariner 2, launched later the same year, made a successful flyby of Venus collecting data that significantly increased our knowledge of the planet. Venera 2 also fell silent en route; in 1966 Venera 3 became the first probe to reach the surface of another planet, but contact had already been lost; Venera 4 in 1967 successfully returned data to Earth.

The Space Task Group appointed by President Richard Nixon to draw up recommendations for how the U.S. ought to move forward after the end of the Apollo program reported in 1969 and, regarding unmanned planetary exploration, called for

> ... missions continuing throughout the [coming] decade, both for science returns and, in the case of Mars and Venus, as precursors to later manned missions. The program should include progressively more sophisticated missions to the near planets as well as multiple-planet flyby missions to the outer planets taking advantage of the favorable relative positions of the outer planets in the late 1970s.

While the U.S.S.R. continued to focus on Mars and Venus in the 1970s the U.S., in line with the recommendations of the Space Task Group, set out to investigate other planets. The primary mission of the two Voyager spacecraft launched in 1977 was to explore Jupiter and Saturn. After encountering those planets, Voyager 1 began its trip out of the solar system while Voyager 2 continued on to Uranus and Neptune before also heading out of the solar system. These spacecraft are searching for the boundary where the region of space dominated by the Sun yields to the interstellar environment. As of 2012, 35 years after they were

Table 4.1. Spacecraft missions to solar system planets.

Planet	*Average distance from the Sun* (AU)	*Mission, launch year, country*
Mercury	0.39	Mariner 10, 1973, U.S.A. MESSENGER, 2004, U.S.A.
Venus	0.72	Mariner 2, 1962, U.S.A. Venera 4, 1967, U.S.S.R. Mariner 5, 1967, U.S.A. Venera 5 and 6, 1969, U.S.S.R. Venera 7, 1970, U.S.S.R. Venera 8, 1972, U.S.S.R. Mariner 10, 1973, U.S.A. Venera 9 and 10, 1975, U.S.S.R. Pioneer Venus 1 and 2, 1978, U.S.A. Venera 11 and 12, 1978, U.S.S.R. Venera 13 and 14, 1981, U.S.S.R. Venera 15 and 16, 1978, U.S.S.R. Vega 1 and 2, 1984, U.S.S.R. Magellan, 1989, U.S.A. Galileo, 1989, U.S.A. Cassini, 1997, U.S.A./Europe MESSENGER, 2004, U.S.A. Venus Express, 2005, Europe Akatsuki, 2010, Japan[a]
Mars	1.52	See Section 4.3.2
Jupiter	5.20	Pioneer 10, 1973, U.S.A. Pioneer 11, 1974, U.S.A. Voyager 1, 1977, U.S.A. Voyager 2, 1977, U.S.A. Galileo, 1989, U.S.A. Ulysses, 1990, U.S.A./Europe Cassini, 1997, U.S.A./Europe New Horizons, 2006, U.S.A.
Saturn	9.58	Pioneer 11, 1974, U.S.A. Voyager 1, 1977, U.S.A. Voyager 2, 1977, U.S.A. Cassini, 1997, U.S.A./Europe
Uranus	19.2	Voyager 2, 1977, U.S.A.
Neptune	30.0	Voyager 2, 1977, U.S.A.

[a] The Japanese mission Akatsuki, originally named Planet-C and also referred to as the Venus Climate Orbiter, inadvertently flew by Venus instead of entering orbit around the planet.

launched, both continue to return data. Voyager 1 has traveled the farthest of the pair, being about 121 AU from the Sun.[2]

In the 1990s other countries began to undertake space exploration. So far Europe, Japan, China, and India have entered the field. It should be noted that the scientists and engineers who supply the instruments for such missions and analyze the data are typically from many nations, making space exploration a truly international effort.

Pluto is not included in Table 4.1 because it was demoted to the category of dwarf planet by the International Astronomical Union in 2006, following the discovery that there are other similarly sized objects at the edge of the solar system. It is at an average distance of 39.5 AU from the Sun. The NASA New Horizons spacecraft is en route to Pluto and the Kuiper Belt, which lies beyond the orbit of Neptune and contains many small icy bodies. By studying Pluto and at least one object in the Kuiper Belt, it is hoped to learn more about the process of planetary formation.

Some of the spacecraft listed in the table have been used to observe comets during their missions. There have also been dedicated comet missions. ESA launched the Giotto spacecraft in 1985 on a mission to Halley's Comet, and it went on to visit Comet Grigg–Skjellerup. The agency is currently running the Rosetta mission. The original intention was to visit Comet Wirtanen but the launch window was missed. Scientists had to choose a new comet that could be reached with the spacecraft and it was decided to launch in 2004 and visit Comet 67P/Churyumov–Gerasimenko in 2014. If all goes well it will enter an orbit around the comet, which is about 4 km wide, and release a lander onto its surface. In such a long-term mission it is important to have personnel covering a range of age and experience to ensure that the key personnel will not all have taken their pensions when the data come in.

4.2 THE MOON

The characteristics of our satellite, the Moon, are given in Table 4.2. It orbits Earth in the same time that it takes for it to rotate on its axis, so that, apart from minor libration effects, it always presents the same hemisphere toward Earth. It has a heavily cratered surface and is covered by a layer of fine gray soil called regolith[3] that was produced by extensive bombardment by meteorites. The regolith layer varies in thickness from about 5 to 10 m. The Moon has no intrinsic magnetic field, but there are localized regions of magnetic activity in the crust. It has no surface water and practically no atmosphere, so the only weathering is by impact processes. Should extensive human activity start on the lunar surface its appearance from Earth would probably change.

[2] AU is the abbreviation for astronomical unit, which is approximately the average distance between the Sun and the Earth, with a value of 149.6×10^6 km.

[3] Most of the lunar regolith material is less than 1 mm in size.

Table 4.2. Characteristics of the Moon.

Characteristic	
Equatorial radius	1,738 km
Acceleration due to gravity at the surface	$1.6\ \mathrm{km/s^2}$
Escape speed	2.4 km/s
Diurnal temperature range	Approx. −250°F to +250°F
Surface pressure (night)	3×10^{-15} bar
Semi-major axis of its orbit around the Earth	384,400 km
Period of rotation around the Earth	27.3 days
Perigee	0.3633×10^6 km
Apogee	0.4055×10^6 km

The Moon is in an approximately circular orbit that is inclined only a few degrees to the plane in which Earth orbits the Sun. The orbit has a radius of about 60 Earth radii which means that the Moon travels in and out of the Earth's magnetosphere on each orbit, passing through the solar wind well upstream of the Earth's bow shock and through the magnetotail on the night side (Fig. 4.1). When the Sun, Earth, and Moon are appropriately aligned, a total or partial solar or lunar eclipse can be observed from Earth.

4.2.1 Unmanned exploration

Over 100 spacecraft have been launched towards the Moon, with varying degrees of success. Some exploded shortly after launch, some were either stranded in Earth orbit, or failed to set the correct trajectory to reach the Moon. In 1958 the U.S.A. and U.S.S.R. together made seven attempts to send probes to the Moon. The Soviet Union modified the rocket used to launch the first Sputnik into Earth orbit in 1957 in order to be able to send a spacecraft on a direct ascent trajectory to the Moon in 1958. Their lunar effort was initially plagued with failures which were not announced publicly and only came to light much later in declassified CIA documents, but there were also some spectacular successes. Table 4.3 presents a chronology of the most significant lunar missions. Luna 1 was the first spacecraft to successfully fly past the Moon and the first of this series to be publicly announced. It revealed that the Moon does not possess an intrinsic magnetic field and made important measurements of the solar wind. Luna 2 gathered further scientific data and hit the Moon. Fig. 4.2 shows the Luna 3 spacecraft, which returned the first images of the far side of the Moon. In the pre–digital era this

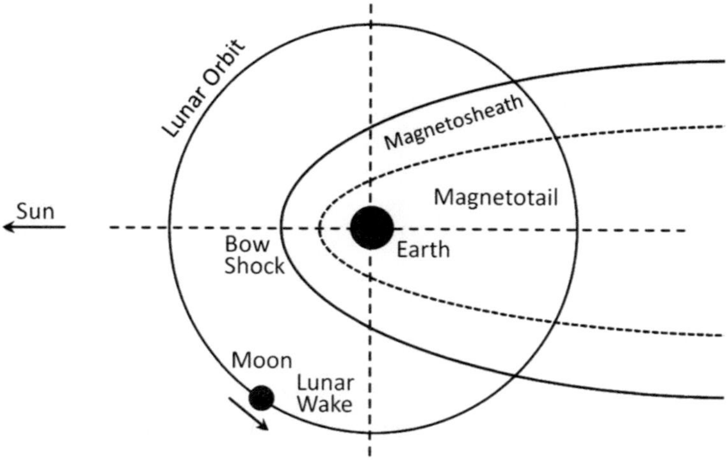

Fig. 4.1. Orbit of the Moon.

was quite a feat. A camera with 40 frames of film on board the spacecraft took pictures of the Moon. The film was then developed and, on command from Earth, passed through a scanner which encoded it for transmission to Earth. The first image that was taken is shown in Fig. 4.3. It revealed the far side to be mountainous, and hence different from the near side which is rather flat. The Luna series had many successes, including orbiters to study radiation in the lunar environment, landers that lowered drills to collect samples for return to Earth, and roving vehicles.

The U.S. attempted to send several Pioneer probes to the Moon in 1958 but they all failed. They had some success in 1959 with Pioneer 4. During the years 1961 to 1965 they launched a series of Ranger missions. The spacecraft were designed to crash on the lunar surface, taking pictures up until they impacted. The first six failed, but the final three successfully sent close-up pictures of different locations. The Ranger spacecraft were followed by a series of Surveyors designed to land on the Moon and gather data to help engineers decide whether the surface was suitable for a manned spacecraft to land on it. Five of the seven Surveyors were successful. Interleaved with these landing missions were five Lunar Orbiters. The primary goal of this program was to make a reconnaissance of candidate sites for Apollo landings. As a bonus they photographed 99% of the lunar surface at a resolution down to 1 m and provided better pictures of the far side than were obtained by the Luna missions. The orbiters were made to crash onto the Moon so that they would not pose problems for the Apollo missions in lunar orbit.

After a pause in lunar exploration missions of about 14 years, Japan became the third nation to reach the Moon, with the Hiten spacecraft in 1990. The U.S. renewed its interest with two orbiters equipped with a range of instruments including multispectral sensors: Clementine in 1994 and Lunar Prospector in 1998, both

Table 4.3. Chronology of unmanned spacecraft missions to the Moon valid until August 2012.

Name(s)	*Launch date*	*Mission*	*Country*
Luna 1	January 2, 1959	Flew past the Moon, was probably intended to impact the surface	U.S.S.R.
Pioneer 4	March 3, 1959	Lunar flyby, radiation measurements, partially successful	U.S.A.
Luna 2	September 12, 1959	First spacecraft to land on the Moon	U.S.S.R.
Luna 3	October 4, 1959	Lunar orbiter, first images of the far side of the Moon	U.S.S.R.
Ranger 1	August 23, 1961	Test flight in Earth orbit, partially successful	U.S.A.
Ranger 2	November 18, 1961	Test flight that became stranded in Earth orbit	U.S.A.
Ranger 3	January 26, 1962	Attempted lunar impact, the spacecraft missed the Moon by 36,800 km	U.S.A.
Ranger 4	April 23, 1962	Impacted the Moon, no images obtained	U.S.A.
Ranger 5	October 18, 1962	Attempted lunar impact, missed the Moon by 725 km	U.S.A.
Luna 4	April 2, 1963	Lunar flyby, technical problem made it miss the Moon	U.S.S.R.
Ranger 6	January 30, 1964	Impacted the Moon, no images obtained	U.S.A.
Ranger 7	July 28, 1964	Impacted the Moon, successfully obtained high-resolution images	U.S.A.
Ranger 8	February 17, 1965	Impacted the Moon, successfully obtained high-resolution images	U.S.A.
Ranger 9	March 21, 1965	Impacted the Moon, successfully obtained high-resolution images	U.S.A.
Luna 5	May 9, 1965	Impacted the Moon, missed its target	U.S.S.R.
Luna 6	June 8, 1965	Attempted soft landing on the Moon, missed the Moon by 161,000 km	U.S.S.R.
Zond 3	July 18, 1965	Lunar flyby and photography	U.S.S.R.

Name(s)	*Launch date*	*Mission*	*Country*
Luna 7	October 4, 1965	Attempted soft landing on the Moon, the spacecraft crashed	U.S.S.R.
Luna 8	December 3, 1965	Attempted soft landing on the Moon, the spacecraft crashed	U.S.S.R.
Luna 9	January 31, 1966	The first soft landing on the Moon, returned images from the lunar surface	U.S.S.R.
Luna 10	March 31, 1966	First lunar orbiter	U.S.S.R.
Surveyor 1	May 30, 1966	The first soft landing on the Moon for the U.S.A.	U.S.A.
Lunar Orbiter 1	August 10, 1966	Lunar orbiter, impacted on the Moon early	U.S.A.
Luna 11	August 24, 1966	Lunar orbiter	U.S.S.R.
Surveyor 2	September 20, 1966	Attempted soft landing on the Moon, the spacecraft crashed	U.S.A.
Luna 12	October 22, 1966	Lunar orbiter	U.S.S.R.
Lunar Orbiter 2	November 6, 1966	Lunar orbiter, photographed smooth surfaces for the evaluation of landing sites	U.S.A.
Luna 13	December 21, 1966	Lunar lander, achieved a soft landing for photography and science	U.S.S.R.
Lunar Orbiter 3	February 5, 1967	Lunar orbiter, photographed areas of the lunar surface for the confirmation of landing sites	U.S.A.
Surveyor 3	April 17, 1967	Lunar lander	U.S.A.
Lunar Orbiter 4	May 4, 1967	Lunar orbiter, photographic survey of the Moon	U.S.A.
Surveyor 4	July 14, 1967	Lunar lander, contact lost with the spacecraft two and a half minutes before touchdown	U.S.A.
Explorer 35 (IMP-E)	July 19, 1967	Lunar orbiter, from an elliptical orbit the spacecraft gathered scientific data about the Moon and its local space environment over 6 years	U.S.A.

(*continued*)

Table 4.3. Chronology of unmanned spacecraft missions to the Moon (*cont.*).

Name(s)	*Launch date*	*Mission*	*Country*
Lunar Orbiter 5	August 1, 1967	Lunar orbiter, further photography of landing sites and the far side of the Moon	U.S.A.
Surveyor 5	September 8, 1967	Lunar lander	U.S.A.
Surveyor 6	November 7, 1967	Lunar lander	U.S.A.
Surveyor 7	January 7, 1968	Lunar lander	U.S.A.
Luna 14	April 7, 1968	Lunar orbiter	U.S.S.R.
Zond 5	September 14, 1968	First successful Soviet circumlunar spacecraft, splashed down in the Indian Ocean. It had a biological payload and was a planned preparation for manned flight to the Moon	U.S.S.R.
Zond 6	November 10, 1968	Circumlunar flight, landed in U.S.S.R.	U.S.S.R.
Luna 15	July 13, 1969	Lunar orbiter, crashed attempting to land on the Moon	U.S.S.R.
Zond 7	October 7, 1969	Circumlunar flight, landed in U.S.S.R.	U.S.S.R.
Luna 16	September 12, 1970	First lunar sample return mission	U.S.S.R.
Zond 8	October 20, 1970	Circumlunar flight, landed in the Indian Ocean	U.S.S.R.
Luna 17	November 10, 1970	Landed on the Moon and deployed a lunar rover, Lunokhod 1, that was operational for 11 lunar days	U.S.S.R.
Luna 18	September 2, 1971	Lunar orbiter, crashed on the Moon	U.S.S.R.
Luna 19	September 28, 1971	Lunar orbiter	U.S.S.R.
Luna 20	February 14, 1972	Sample return mission	U.S.S.R.
Luna 21	January 8, 1973	Landed on the Moon and deployed a lunar rover, Lunokhod 2, that was operational for about 4 months	U.S.S.R.
Explorer 49 (Radio Astronomy Explorer-B)	June 10, 1973	Lunar orbiter, radio astronomical measurements of the planets, the Sun, and the galaxy	U.S.A.

Name(s)	*Launch date*	*Mission*	*Country*
Luna 22	May 29, 1974	Lunar orbiter	U.S.S.R.
Luna 23	October 28, 1974	Lunar lander and attempted sample return	U.S.S.R.
Luna 24	August 9, 1976	Sample return mission	U.S.S.R.
Hiten	January 24, 1990	Lunar flyby and impact on the Moon, subsatellite Hagoramo orbited the Moon	Japan
Clementine	January 25, 1994	Lunar orbiter that mapped the lunar surface, suggested the presence of ice	U.S.A.
AsiaSat 3/HGS-1	December 24, 1997	A commercial communications satellite, failed to enter the correct orbit, it was renamed HGS-1 and made two lunar flybys to get to a geosynchronous orbit	China
Lunar Prospector	January 7, 1998	From a low polar orbit the satellite gathered data about the Moon, measurements indicated water ice at both of the lunar poles, crashed near the south pole of the Moon	U.S.A.
Smart 1	September 27, 2003	ESA lunar orbiter. Test of solar electric propulsion, placed in a polar orbit around the Moon, looked for ice in the craters at the Moon's south pole, controlled crash on the Moon, September 2006	Europe
Kaguya	September 9, 2007	Lunar orbiter to make a global lunar survey, planned impact on the Moon at the end of its mission	Japan
Chang'e 1	October 24, 2007	Lunar orbiter, technology test, high-inclination orbit for scientific studies	China
Chandrayaan 1	October 22, 2008	The spacecraft was placed in a polar orbit around the Moon and released an impact probe that landed near the south pole. One of the goals was to search for water ice at the poles	India

(*continued*)

Table 4.3. Chronology of unmanned spacecraft missions to the Moon (*cont.*).

Name(s)	*Launch date*	*Mission*	*Country*
Lunar Reconnaissance Orbiter	June 18, 2009	Lunar orbiter to map the Moon and provide data for the assessment of possible landing sites for humans	U.S.A.
Lunar Crater Observation and Sensing Satellite (LCROSS)	June 18, 2009	Double impact on the Moon to search for evidence of water	U.S.A.
Chang'e 2	October 1, 2010	Lunar orbiter, testing technology in preparation for Chang'e 3, a lander and rover mission	China
Gravity Recovery And Interior Laboratory-A (GRAIL)	September 10, 2011	Twin spacecraft mission to perform gravity mapping and interior studies of the Moon from polar orbit	U.S.A.

of which found evidence for the presence of water ice at the lunar poles. In the new century Europe, Japan, China, and India all sent orbiters to study the Moon. In 2009 NASA launched the Lunar Reconnaissance Orbiter, which entered a polar orbit of the Moon at an altitude of about 50 km. A satellite cannot orbit Earth at that altitude owing to the presence of the atmosphere, but it is feasible in the case of the Moon because it has next to no atmosphere and hence there is no drag on a spacecraft. The Lunar Reconnaissance Orbiter is mapping the Moon in more detail than previously. Fig. 4.4 shows recent images of the Moon compiled from data supplied by the wide-angle camera at a resolution of about 100 m. The picture of the near side shows the smooth, dark maria, which are actually low-lying terrain flooded by lava, and the lighter toned highlands. The image of the far side shows the absence of maria and a pocked surface resulting from bombardment. It should be noted that impacting is an ongoing process, continually making new craters. As an adjunct to this mission, the LCROSS mission sent the final stage of the launch vehicle crashing into one of the permanently shadowed craters near the south pole. The impact was monitored by a second spacecraft which itself impacted the surface a few minutes later. Analysis of data from these spacecraft confirmed the presence of water ice.

4.2.2 Apollo

A chronology of the Apollo lunar landings was provided in Table 2.6. Here, Fig. 4.5 shows the Apollo landing sites in relation to those of the Luna and Surveyor missions. As can be seen, the Apollo landings were at equatorial and mid-latitudes—not in polar regions. Having landed on the Moon, the astronauts had

Fig. 4.2. Luna 3 spacecraft (source: U.S.S.R. Academy of Sciences).

only a short time in which to explore, gather samples for return to Earth, and deploy experiments. In the case of Apollo 11 the surface science, along with other tasks, had to fit into an allotted time of just 2 hours and 40 minutes out on the lunar surface. In order of priority the scientific tasks were

1. To collect at the earliest opportunity a so-called contingency sample of the lunar surface, so that at least this small amount of material would be returned to Earth if the astronauts had to make a hasty departure from the surface.
2. To quickly collect approximately 10 kg of lunar material to take back to Earth for analysis.
3. To deploy experiments on the surface of the Moon. These were (a) Passive Seismic Experiment and Lunar Dust Detector, (b) Laser Ranging Retro-reflector, and (c) Solar Wind Composition Experiment.
4. To collect a second set of carefully selected samples.

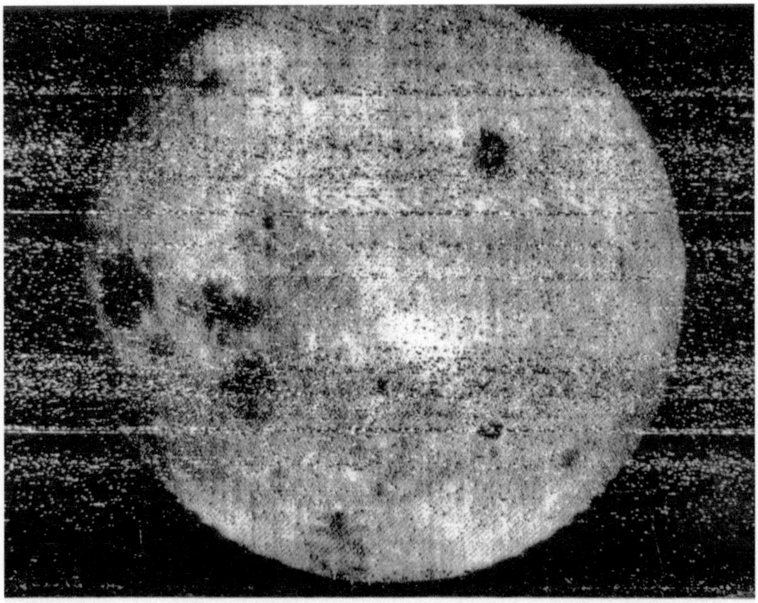

Фотография 1

Fig. 4.3. The first image of the far side of the Moon taken by Luna 3 on October 7, 1959 (source: U.S.S.R. Academy of Sciences).

The Passive Seismic Experiment was to detect meteor impacts and moonquakes to provide information about the interior of the Moon. The Lunar Dust Detector (which was attached to the Seismic Experiment) was to measure dust deposition on the experiments. The Laser Ranging Retroreflector consisted of an array of corner cubes to reflect a laser beam sent from Earth back again, so that the distance between the Earth and the Moon could be accurately measured. Since the Moon has no magnetic field and negligible atmosphere, when it is outside of the Earth's magnetosphere the solar wind can impinge directly on its surface. The Solar Wind Composition Experiment was an elegantly simple experiment consisting of a roll of thin extremely clean aluminum foil with an area of $0.4\,m^2$ that was unrolled and placed upright on the lunar surface facing the Sun to collect heavy ions in the solar wind, particularly those of the noble gases. Fig. 4.6 shows Buzz Aldrin on the lunar surface with the Lunar Module "Eagle" in the background. On his right is the deployed Solar Wind Composition Experiment. After 77 minutes, the foil was rolled up again to be returned to Earth for analysis, which revealed isotopes of helium, neon, and argon had been collected. The experiment was repeated on Apollo 12, 14, 15, and 16; in the latter case for 45 hours.

During each of the following lunar landings the astronauts deployed an Apollo Lunar Surface Experiments Package (ALSEP) containing a selection of experiments, some of which were sent on several missions while others were sent

Fig. 4.4a. The near side of the Moon imaged by the Lunar Reconnaissance Orbiter (credit: NASA/GSFC/Arizona State University).

only once. In addition to the experiments deployed by Apollo 11, the suite included

- Lunar Surface Magnetometer, to measure the magnetic field
- Cold Cathode Gauge, to measure the density of neutral particles at the surface of the Moon
- Suprathermal Ion Detector Experiment, to study the ionosphere
- Solar Wind Spectrometer, to measure the composition of the solar wind
- Active Seismic Experiment, to investigate the uppermost 100 m of the lunar regolith
- Charged Particle Lunar Environment Experiment, to measure low-energy electrons and ions
- Heat Flow Experiment, to measure the rate at which the Moon loses heat
- Lunar Seismic Profiling Experiment, to study the uppermost kilometer of the lunar crust
- Lunar Atmospheric Composition Experiment, to measure the composition and variability of the lunar atmosphere

Fig. 4.4b. The far side of the Moon, imaged by the Lunar Reconnaissance Orbiter (credit: NASA/GSFC/Arizona State University).

- Lunar Ejecta and Meteorites, to measure ejecta from meteorite impacts and micrometeoroids
- Lunar Surface Gravimeter, to measure gravitational acceleration at the surface of the Moon and its variations and to search for gravity waves.

These experiments could be monitored from Earth after the astronauts had left the Moon.

Apollo 12 landed close to the Surveyor 3 spacecraft. The astronauts gathered parts of the probe, which had been on the lunar surface for 31 months, and returned them to Earth for analysis. Other experiments carried out during the Apollo missions between Earth and the Moon, in lunar orbit, and on the lunar surface included a study of the flashes of light perceived by astronauts when cosmic rays passed through their eyes, apparatus to study cosmic rays, a variety of spectrometers and radar experiments, a laser altimeter, and an investigation of micrometeoroid impacts in space by analyzing the Apollo command module windows pre and post flight. A total of 382 kg of lunar material was returned to

Figure 4.5. Luna (red), Apollo (green), and Surveyor (yellow) landing sites on the Moon (source: NASA).

Earth by the six Apollo lunar missions.[4] A special building was constructed at the NASA Johnson Space Center to store the samples in pristine condition. Scientists and educators can apply for the use of a sample but it remains the property of the U.S. government.

4.3 MARS

Mars, the farthest terrestrial planet from the Sun, is often referred to as the Red Planet because of its reddish hue, which is the result of iron oxide on the surface and in the atmosphere. Table 4.4 presents some of its characteristics in comparison with Earth. Although the low temperature and pressure imply that water cannot currently exist as a liquid on the surface, a variety of observations indicate that it was present earlier in the planet's history and might still be present beneath the surface. A key quest in the modern studies of Mars is to understand the

[4] In addition Soviet spacecraft returned 300 g of lunar material.

Fig. 4.6. Buzz Aldrin next to the Solar Wind Composition Experiment on the Moon during the Apollo 11 mission in 1969 (source: NASA).

evolution of water to investigate whether life ever developed there and, if so, whether it still survives, hidden under the surface.

The first detailed map of Mars (Fig. 4.7) was published by Giovanni Schiaparelli at the Observatory in Milan, Italy, in the late 19th century. He claimed to have detected a large number of *canali* on the surface. Due to inaccurate translation from Italian, this led to the popular belief that the inhabitants of the planet had excavated deep canals to transport water from the poles to the less arid regions. It was not until the 1960s that imagery from spacecraft began to unveil more accurate details of the surface environment.

4.3.1 Travel time to Mars

Mars has a more elliptical orbit around the Sun than that of the Earth, which is close to circular. When Earth is in a position such that the Sun and Mars appear on opposite sides of the sky then Mars is said to be at opposition, and when Mars is behind the Sun as seen from Earth it is at conjunction. The minimum distance at opposition is 55 million kilometers and the maximum distance at conjunction is

Table 4.4. Characteristics of planet Mars in comparison with Earth.

Characteristic	*Mars*	*Earth*
Equatorial radius	3,396 km	6,378 km
Acceleration due to gravity at the surface	$3.7\ km/s^2$	$9.8\ km/s^2$
Escape speed	5.0 km/s	11.2 km/s
Average temperature	−63°C	+15°C
Natural satellites	Phobos and Deimos	The Moon
Atmospheric composition	95% carbon dioxide	78% nitrogen, 20% oxygen
Average surface pressure	6.4 mbar	1 bar
Semi-major axis of its orbit around the Sun	228×10^6 km (1.52 AU)	150×10^6 km (1 AU)
Period of rotation around the Sun	687 Earth days	365 days
Perihelion	207×10^6 km	147×10^6 km
Aphelion	249×10^6 km	152×10^6 km
Period of rotation about its own axis, length of a Martian day	24.7 hours	24 hours

400 million kilometers. The time it takes for a spacecraft to travel to Mars depends on the relative positions of the planets in their orbits at the time of launch, and the trajectory selected. The minimum travel time with present technology is about 9 months.

The minimum energy trajectory is tangential to both the orbit of Earth at the start of the trip and to the orbit of the destination at the time of arrival. This is a Hohmann transfer orbit. Launch windows are regulated by the synodic period of Mars, which is 780 days (2.1 years).[5] A direct trip to Mars using such a trajectory is illustrated in Fig. 4.8. At the beginning, Mars is near conjunction. Once in Earth orbit a spacecraft is given sufficient energy to enter an elliptical path with the Sun at one focus, its perihelion at the orbit of Earth and its aphelion at the orbit of Mars. On arrival, the spacecraft maneuvers to achieve an orbit around its target. For a crewed mission the round trip Earth–Mars–Earth using such a strategy would take about 900 days, including 500 days spent at Mars.

[5] The synodic period of a planet is the time taken for the planet to return to the same position in the sky relative to the Sun as viewed from Earth.

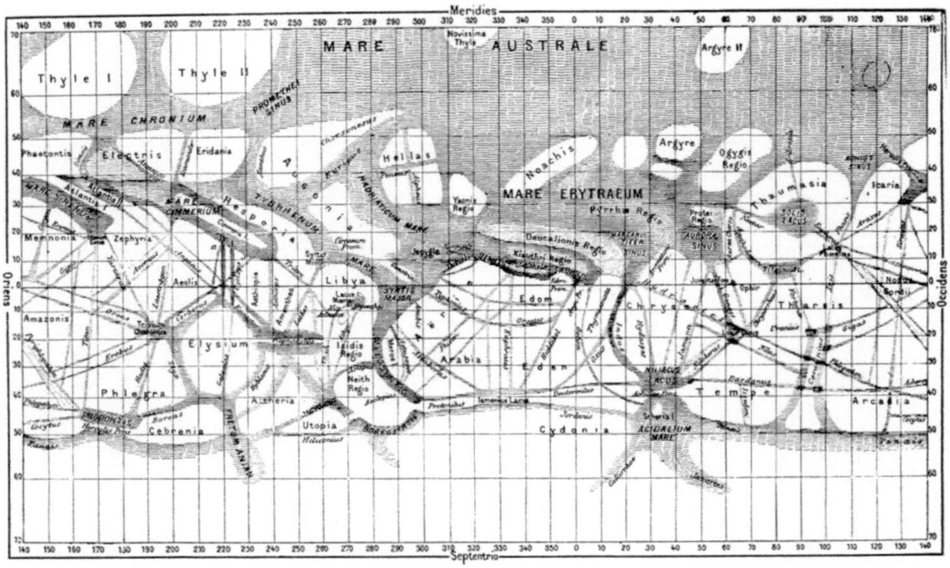

Fig. 4.7. Giovanni Schiaparelli's map of Mars, compiled over the period 1877–1886.

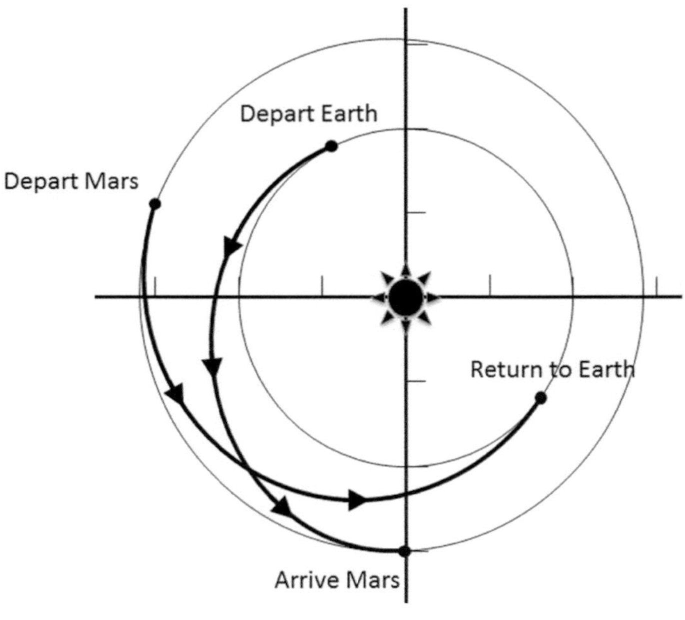

Fig. 4.8. Hohmann transfer orbits from Earth to Mars, and back again.

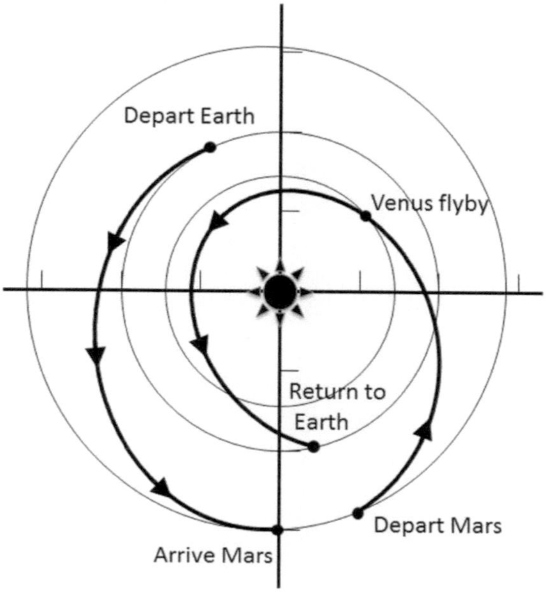

Fig. 4.9. An opposition-class mission to Mars.

Another possibility is an opposition-class mission using a Venus flyby (Fig. 4.9). The Venus flyby can occur on either the outward or return journey, depending on the relative positions of the planets at the start of the journey. Earth, Mars, and Venus have the same relative positions to one another every 2,338 days (6.4 years). In this case a crewed mission to Mars would take between 400 and 650 days, with 30–90 days spent at Mars. Although the total mission time is shorter, this strategy has the disadvantage that it requires more energy than the one in Fig. 4.8. A combination of the two schemes may be better, with cargo using the minimum energy strategy and humans using the faster strategy. A longer term solution is to develop a method of propulsion that will allow spacecraft to travel faster in order to shorten travel times.

4.3.2 Unmanned exploration

Missions with orbiters and landers are prerequisites for human exploration of Mars, because an understanding of the local environment is necessary to enable engineers to design and build spacecraft to take humans to Mars and return them safely to Earth. As can be seen from the chronology shown in Table 4.5, the exploration of the planet by unmanned spacecraft has been a hit-and-miss business. About half of the missions have failed, in many cases due to technical malfunction, in some cases due to human error, and in a few cases it was not possible to determine precisely what went wrong. One particularly tragic case is the Mars Climate Orbiter, which was lost because one engineering team was using

Table 4.5. Chronology of Mars exploration missions valid until August 2012.

Name(s)	*Launch date*	*Mission*	*Country*
Marsnik 1	October 10, 1960	Attempted flyby, launch failure	U.S.S.R.
Marsnik 2	October 14, 1960	Attempted flyby, launch failure	U.S.S.R.
Sputnik 22	October 24, 1962	Attempted flyby, the satellite was damaged before leaving Earth orbit	U.S.S.R.
Mars 1	November 1, 1962	Flyby, contact lost en route to Mars	U.S.S.R.
Sputnik 24	November 4, 1962	Attempted lander, the spacecraft was destroyed during the burn to transfer to a Martian trajectory	U.S.S.R.
Mariner 3	November 5, 1964	Attempted flyby, failed shroud separation	U.S.A.
Mariner 4	November 28, 1964	First successful Mars flyby, obtained the first images of the Martian surface	U.S.A.
Zond 2	November 30, 1964	Flyby, contact was lost prior to the flyby	U.S.S.R.
Zond 3	July 18, 1965	Photographed the Moon, Mars test vehicle	U.S.S.R.
Mariner 6	February 24, 1969	Flyby	U.S.A.
Mariner 7	March 27, 1969	Flyby	U.S.A.
Mars 1969A	March 27, 1969	Attempted orbiter, Proton booster failed	U.S.S.R.
Mars 1969B	April 2, 1969	Attempted orbiter, Proton booster failed	U.S.S.R.
Mariner 8	May 8, 1971	Attempted orbiter, Atlas-Centaur booster failed	U.S.A.
Cosmos 419	May 10, 1971	Orbiter, failed to leave Earth orbit	U.S.S.R.
Mars 2	May 19, 1971	Orbiter, lander crashed on Mars	U.S.S.R.
Mars 3	May 28, 1971	Orbiter, lander successfully descended to the surface of Mars, but ceased to function almost immediately	U.S.S.R.
Mariner 9	May 30, 1971	Orbiter, first spacecraft to orbit Mars, mapped 70% of the Martian surface	U.S.A.

Name(s)	*Launch date*	*Mission*	*Country*
Mars 4	July 21, 1973	Intended orbiter, made Mars flyby due to malfunction	U.S.S.R.
Mars 5	July 25, 1973	Orbiter, gathered data but functioned for just a few days	U.S.S.R.
Mars 6	August 5, 1973	Flyby, lander reached Mars surface, but little data were received	U.S.S.R.
Mars 7	August 9, 1973	Flyby and lander, a malfunction caused the lander to miss the planet	U.S.S.R.
Viking 1	August 20, 1975	Orbiter and lander, the orbiter was operational until August 1980 and the lander until November 1982	U.S.A.
Viking 2	September 9, 1975	Orbiter and lander, the orbiter was operational until July 1978 and the lander until April 1980	U.S.A.
Phobos 1	July 7, 1988	Mars orbiter and Phobos landers, failed en route to Mars due to an erroneous command which turned the spacecraft so that the solar panels did not face the Sun and the batteries were depleted	U.S.S.R.
Phobos 2	July 12, 1988	Mars orbiter and Phobos landers, the spacecraft successfully orbited Mars for 2 months but was lost shortly before it was due to release its two landers towards one of the Martian moons, Phobos	U.S.S.R.
Mars Observer	September 25, 1992	Attempted orbiter, contact lost a few days prior to orbit insertion	U.S.A.
Mars Global Surveyor	November 7, 1996	Orbiter, global mapping mission that successfully gathered data for 10 years	U.S.A.
Mars '96	November 16, 1996	Attempted orbiter and landers, failed to leave Earth orbit	U.S.S.R.

(*continued*)

Table 4.5. Chronology of Mars exploration missions valid until August 2012 (*cont.*).

Name(s)	*Launch date*	*Mission*	*Country*
Mars Pathfinder	December 4, 1996	Mars Pathfinder Lander & Sojourner rover, an airbag system was used to cushion the landing on Mars, the system was operational for over 2 months, in excess of expectation	U.S.A.
Nozomi (Planet B)	July 3, 1998	Orbiter, flew past Mars and failed to enter orbit around the planet	Japan
Mars Climate Orbiter	December 11, 1998	Attempted orbiter to study the Martian climate that went into too low an orbit due to a navigation mistake and was destroyed	U.S.A.
Mars Polar Lander and Deep Space 2	January 3, 1999	Attempted Mars lander to investigate the region close to the south pole of Mars and two lightweight probes sent to penetrate the surface of Mars, contact was lost as they approached Mars	U.S.A.
2001 Mars Odyssey	April 7, 2001	Orbiter, further Mars studies including the study of potential radiation hazards to human missions, communications relay for landers	U.S.A.
Mars Express	June 2, 2003	Mars Express Orbiter and lander, Beagle 2. The orbiter is continuing science operations in 2012 and the current mission is extended until December 2014. The lander was lost when it entered the Martian atmosphere	ESA
Spirit (Mars Exploration Rover A)	June 10, 2003	Rover, landed in Gusev Crater south of the equator in January 2004 and was operational until March 2010, well in excess of its planned lifetime	U.S.A.
Opportunity (Mars Exploration Rover B)	July 7, 2003	Rover, landed at Terra Meridiani, close to the equator, January 2004 and has exceeded expectations by still being operational in 2011	U.S.A.

Name(s)	*Launch date*	*Mission*	*Country*
Mars Reconnaissance Orbiter	August 12, 2005	Orbiter designed to be operational for a Martian year, arrived at Mars in March 2006 and still operational November 2011. Mission objectives include climate studies, the continuing search for water, and to identify landing sites for future missions	U.S.A.
Phoenix	August 4, 2007	Lander, spent 5 months investigating the environment of its landing site in the northern polar region of Mars, in particular searching for water and signs of habitability. Confirmed the existence of frozen water	U.S.A.
Phobos-Grunt/ Yinghuo 1	November 8, 2011	Tandem spacecraft: Russian Phobos-Grunt, a Martian moon Phobos lander, Phobos sample return mission, and the Chinese Yinghuo 1 Mars orbiter. The spacecraft did not perform the engine burn to transit to Mars and was stranded in Earth orbit	Russia/ China
Mars Science Laboratory	November 26, 2011	Curiosity rover landed inside Gale Crater August 2012, with a Primary mission of one Martian year	U.S.A.

the U.S. customary system of units while another was using the metric system. This caused the spacecraft to approach Mars at an altitude of about 60 km instead of the intended 150 km, with the result that it broke up in the Martian atmosphere. Almost all the spacecraft sent to Mars to date have been built by either the U.S. or U.S.S.R. These two nations competed in the race to Mars at the start of the space age, just as they competed to send their citizens into space. The Russian spacecraft have in general been very heavy (e.g., Sputnik 22 had a launch mass of 6,500 kg and an orbiter mass of 894 kg). U.S. spacecraft sent to Mars have typically had masses in the range 260–1,000 kg. This difference is partly due to the size of the available launch vehicles and partly due to the relative sophistication of spacecraft technology.

Japan's first mission, Nozomi, suffered a number of problems en route and failed to enter orbit around the planet. In contrast the European Mars Express orbiter was a tremendous success. It was a relatively low-cost mission that was built and launched quickly, hence its name. The Russian Phobos-Grunt became stranded in Earth orbit. It was carrying China's first mission to the planet, the

Fig. 4.10. "Twin Peaks" viewed by Mars Pathfinder (source: NASA/JPL/Dr.Timothy Parker).

Yinghuo 1 orbiter. Both spacecraft reentered the Earth's atmosphere 2 months after launch.

Mariner 9 mapped a large proportion of Mars and revealed a varied landscape, but the first mission to map the entire planet and to send back large numbers of images from the ground was Viking. These missions disclosed volcanoes, canyons, lava plains, and channels that suggested that water once flowed across the surface of Mars. It transpired that the northern hemisphere was a low plain and the southern hemisphere was cratered highlands. Fig. 4.10 is an image taken by Mars Pathfinder of hills in the Ares Vallis region, which is a former flood plain in the northern hemisphere.

The Mars Orbiter Laser Altimeter (MOLA) on the Mars Global Surveyor orbiter provided the data for a very high resolution shaded relief map of the planet.[6] The map (Fig. 4.11) is shown as a Mercator projection to latitude 70° north and south and uses colors to represent the range of elevation; low elevations in blue, high elevations in red, and peaks in white. It illustrates the smooth surface in the northern hemisphere and the heavily cratered surface in the southern hemisphere, which is about 5 km more elevated than the north. The leftmost white spot represents Olympus Mons, the largest volcano in the solar system rising to a height of 26 km with its base spanning 600 km. Instrumentation on the orbiter detected hematite (Fe_2O_3) in some locations, notably in an equatorial plain named Meridiani Planum. This was of interest because gray hematite on Earth is often an indication of the presence of water in the past. The Opportunity rover landed there and confirmed the existence of hematite in the form of spherical concretions several millimeters in size, nicknamed "blueberries" by scientists. The hematite

[6] Laser pulses are sent to the planet's surface. The time taken for a pulse to return to the instrument is noted and the distance the pulse traveled calculated in order to produce a detailed topographic map.

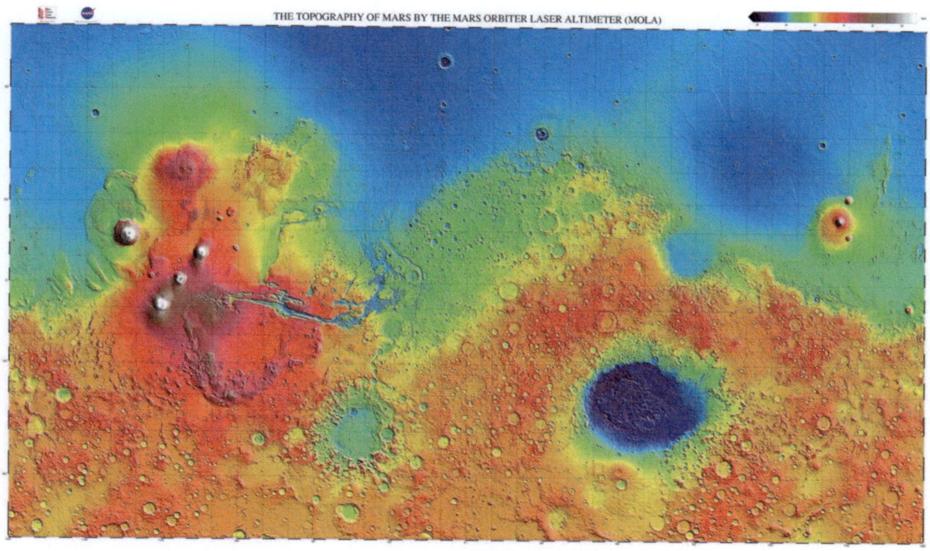

Fig. 4.11. A topographic map of Mars (credit: MOLA Science Team)

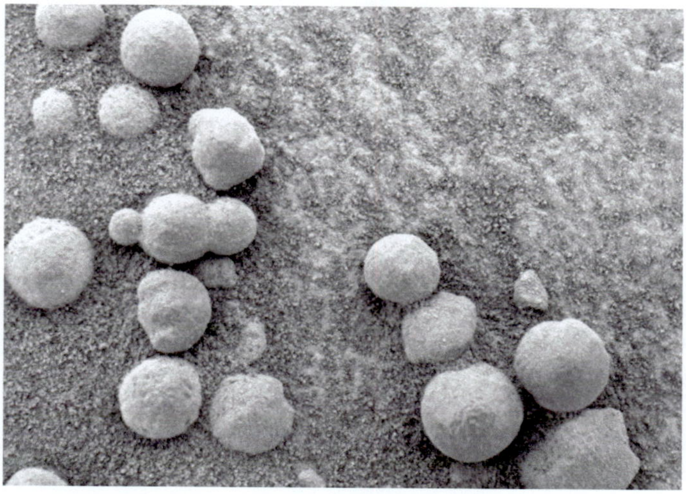

Fig. 4.12. A microscopic image showing rounded grains dubbed "blueberries" born in rock near to the Mars Exploration Rover Opportunity's landing site (source: NASA/JPL/Cornell/USGS).

was present in a rock formation that formed the plain (Fig. 4.12). The Spirit rover found other evidence for the past presence of water in Gusev Crater near the equator on the other side of the planet.

During the development of the solar system all of the planets were subjected to bombardment. In the case of Earth, smaller meteoroids burn up in the Earth's

Fig. 4.13. A meteor crater on the surface of Mars photographed by the Mars Orbiter Camera on Mars Global Surveyor in July 1998 (credit: NASA/JPL/Malin Space Science).

atmosphere. The surface is renewed by the process of plate tectonics so that the scars of most larger meteoroid impacts do not last long. In contrast Mars is pocked by craters dating right back through the planet's history. Fig. 4.13 shows a simple crater formed by the impact and subsequent explosion of a meteorite. A ring of ejected material that has been thrown outward by the force of the impact surrounds the rim of the crater.

Mars has polar ice caps. Measurements made with instrumentation on the NASA Mars Odyssey orbiter revealed the existence of vast amounts of water ice just below the surface in the polar regions. The Phoenix Mars lander had a robotic arm that could dig half a meter down into the surface. A scoop at the end of the arm was used to deliver samples to instruments for analysis, and the results verified that there is indeed water ice just below the surface. In the spring of 2011 it was announced that the NASA Mars Reconnaissance Orbiter had gathered evidence that salty water flowed on Mars in the warmest periods of the year. The picture shown in Fig. 4.14 was obtained by combining images taken by the Mars Reconnaissance Orbiter with three-dimensional computer modeling of a slope inside Newton Crater located at 42°S 202°E, depicting the slope from a vantage point inside the crater with a synthetic sky added. In spring and summer the dark

Fig. 4.14. Warm-season flows on a slope in Newton Crater, based on data gathered by the Mars Reconnaissance Orbiter in May 2011 (credit: NASA/JPL-Caltech/University of Arizona).

elongated features appear and extend down some slopes of 25–40° at several locations in the southern hemisphere. They fade in cold seasons, and return again the following spring. Salty liquid water under the surface may be the explanation. Salts can lower the freezing point of water and reduce its evaporation rate, so a salty solution would be more stable than pure water on the Martian surface.

The U.S. strategy for ongoing Mars exploration is to "follow the water" in order to determine whether the planet has been, is currently, or may one day be habitable. The NASA Mars Science Laboratory was launched in November 2011 with the Curiosity rover. After a guided entry into the Martian atmosphere, the spacecraft was slowed using a parachute. Retro-rockets controlled the descent, and on final approach a sky crane was used to gently lower the rover (Fig. 4.15) into Gale Crater. It is planned that the car-sized rover, which has a mass of 775 kg, will remain operational for at least one Martian year, during which it will

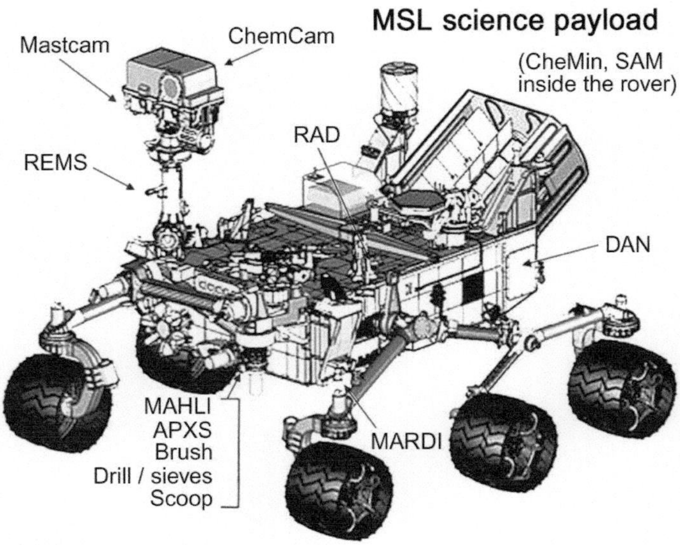

Fig. 4.15. NASA's Mars Rover Curiosity. Mastcam is the mast camera instrument, ChemCam is the Chemistry and Camera instrument, RAD is the Radiation Assessment Detector instrument, CheMin is the Chemistry and Mineralogy instrument, SAM is the Sample Analysis at Mars instrument, DAN is the Dynamic Albedo of Neutrons instrument, MARDI is the Mars Descent Imager instrument, MAHLI is the Mars Hand Lens Imager instrument, APXS is the Alpha Particle X-ray Spectrometer instrument, the brush, drill, sieves, and scoop are tools on the rover's robotic arm, REMS is the Rover Environmental Monitoring Station (source: NASA/JPL-Caltech).

travel up to 20 km from the landing site. The overall scientific goal of the mission is to investigate the habitability of a local region of the planet. It will also study the spectrum of ionizing radiation at the surface, which is a key issue in relation to human visits to the planet.

ESA plans to collaborate with Russia on the ExoMars program which will involve two missions, the first having an orbiter and an entry, descent, and landing demonstrator, and the second, which is expected to be launched two years later, featuring a rover. It is planned that the rover will carry a drill capable of penetrating to a depth of 2 m to investigate the subsurface environment.

4.3.3 Human exploration

In his speech on space exploration at the Kennedy Space Center on April 15, 2010, President Barack Obama made it clear that returning astronauts to the Moon was not currently of interest to the United States because it has been visited before. It is more important to prioritize new destinations. He stated

> Early in the next decade, a set of crewed flights will test and prove the systems required for exploration beyond LEO. And by 2025, we expect new spacecraft designed for long journeys to allow us to begin the first-ever crewed missions beyond the Moon into deep space. So we'll start by sending astronauts to an asteroid for the first time in history. By the mid-2030s, I believe we can send humans to orbit Mars and return them safely to Earth. And a landing on Mars will follow. And I expect to be around to see it.

Scientists from many nations are working on projects related to achieving a successful human mission to Mars. The risks to the health and safety of the crew of such a long-duration space mission can be broken down into the following categories

- danger due to possible malfunctions during launches from Earth and Mars, orbital maneuvers, and landings
- long-duration exposure to the microgravity environment and the radiation environment inside the spacecraft during transit (see Chapter 3)
- planetary surface environment: one third of the gravitational force found on Earth, greater radiation exposure than on the Earth's surface, dependence on space suits when on the surface
- psychological challenges due to confinement over long periods, separation from Earth, monotony of the trip, operational challenges, and dangers of the task.

There are limited data on the psychological challenges of long-duration spaceflight simply because so few people have experience of them. Most spacefarers have flown missions that lasted less than a fortnight, and only a handful have been continuously in space for more than a year. An operational spacecraft is an example of an isolated and confined extreme (ICE) environment. Other examples with similarities to a space mission are polar exploration expeditions and submarines at sea. By studying groups of people in terrestrial ICE environments we can postulate on the likely behavior of the crew of a mission to Mars. Despite the rigorous procedures used to select a crew for a space mission, there have been occasions when individuals have shown symptoms of stress while in space, when the relationships between crew members have been tense, and when there have been disagreements with ground control. Some astronauts have experienced problems after their return. The aim is to minimize the risk of similar situations occurring on a mission to Mars because that crew will experience challenges faced by no previous spacefarers.

To complement studies in extreme terrestrial environments, space missions can be simulated on Earth. A number of isolation studies have been carried out to investigate the psychological effects of being confined for an extended period with a small group of people. Many of these studies have been led by the Russian Institute for Biomedical Problems in Moscow, which began isolation experiments in the late 1960s. In 1994 three men were isolated for 135 days for the Human Behavior in Extended Spaceflight (HUBES) study that was done in a mock-up of the Mir space station, and was modeled on the then-imminent EUROMIR 95 mission. From 1999 to 2000 the Simulation of Flight of International Crew on Space Station (SFINCSS) was undertaken, a complex simulation involving three primary crews in interconnecting modules and four crews simulating Soyuz and Space Shuttle visits. Four Russian men spent 240 days isolated, mimicking a Mir space station schedule. After about a month a second all-male group of three Russians and one German entered the simulation to simulate ISS assembly. They were succeeded by another four-person crew consisting of a woman from Canada and three men, one Austrian, one Japanese, and one Russian, undertaking a similar mission to the second group. At one point in the study relations deteriorated to such an extent that one participant quit and the connecting door linking two of the crews was locked for a time until they could face one another again. The SFINCSS simulation highlighted some potential problems of heterogeneous crews. The crews of early space missions were homogeneous, usually military men of a single nationality who had much in common. The expansion and internationalization of space activities led to heterogeneous crews of mixed genders, nationalities, ages, cultures, mother tongues, and occupations. It is generally presumed that deep-space exploration crews will be heterogeneous.

Analyses of the data gathered during such simulations contribute to the design of psychological countermeasures for a human mission to Mars. There are four stages to be considered

- selection of a crew
- pre-flight training
- in-flight monitoring and support
- post-flight support.

During the selection process it is essential to exclude candidates who are at risk of developing a psychological illness. This can require careful examination of the candidate, because many people might not realize that they are at risk or may not be inclined to reveal their problems. At the same time there is a need to identify those candidates with desirable personality characteristics for the mission. Having made a selection of suitable candidates, a crew is formed by identifying the optimal combination of good team players with complementary skills and characteristics. The selection of the crew is probably the most significant factor in the psychological aspect of the mission. Pre-flight training is designed to meld the crew into a team with mutual understanding of one another's needs and capabilities in regard to all aspects of the mission. In-flight monitoring and support on space station missions includes ground-based monitoring, uplink of news, e-mail, private family conferences, psychological conferences, crew packages, on-board entertainment, and visiting crews. Astronauts on the ISS can even make telephone calls to the ground. A mission to Mars will impose limitations on in-flight monitoring because of the time delays associated with the tremendous distances. A contributing factor to the in-flight experience will be the interior environment of the vehicle. Typically space habitats are cramped, noisy, the sealed environment can have unpleasant odors, and there is a lack of privacy. These stresses can be difficult for individuals to accept. Post-flight support will be needed to help the crew readapt to life on Earth.

4.3.4 Mars analogues on Earth

Analogues on Earth can simulate aspects of living and working on Mars. Those run by the Mars Society and the Mars 500 program, both of which are focused on developing tools for exploration missions, are described below.

Mars Society

The Mars Society is an international non-profit organization that promotes the human exploration and settlement of Mars. Robert Zubrin, one of its directors, is an advocate of Mars Direct, a proposal for a program with a low-cost, live-off-the-land approach. It calls for using existing launch technology to send an Earth Return Vehicle (ERV) to Mars with equipment that will be used to produce methane fuel by drawing carbon dioxide from the atmosphere. The crew follows later in another spacecraft that serves as a base for operations on the surface. The fuel that is produced on Mars is used for the return journey of the crew to Earth. At the same time as the first crew launches, a second ERV is dispatched to Mars to be used by the next crew for their return journey to Earth, and so the cycle continues. A significant advantage of this strategy is that the fuel for the return journey does not have to be launched from Earth, carried on the outward trip, and landed on Mars, which is a significant mass saving.

The Mars Society has established two Mars analogue research stations. They are the Flashline Mars Arctic Research Station (FMARS) in the Canadian Arctic (Fig. 4.16) and the Mars Desert Research Station (MDRS) in southern Utah. In 1996 a group of researchers mounted an expedition to Devon Island in the Canadian Arctic in search of a place that had an environment similar to that of Mars. Devon Island is the largest uninhabited island on Earth, with a rocky desert and the Haughton Impact Crater. The local environment there has been nick-named "Mars on Earth" due to its similarities to Mars. In the summer, large parts of the island are snow free and accessible by plane. The region around the impact crater proved interesting as a Mars analogue site, and the idea of establishing a station to study the region while at the same time simulating the daily routine of a crew on the surface of Mars was born. FMARS was constructed in 2000. MDRS was built in 2002, and it has the advantage of being accessible all year round. The stations were designed with the Mars Direct concept in mind. Each station consists of a prototype of a habitat that could be used both by a crew en route to Mars and then on the surface. The constructions are cylindrical units on landing struts. Each has a diameter approximating that of the Saturn V rocket, which is the most powerful rocket thus far employed to reliably send humans into space. The idea is that a space-approved habitat of the same configuration could be put inside the shroud of a rocket similar to a Saturn V for

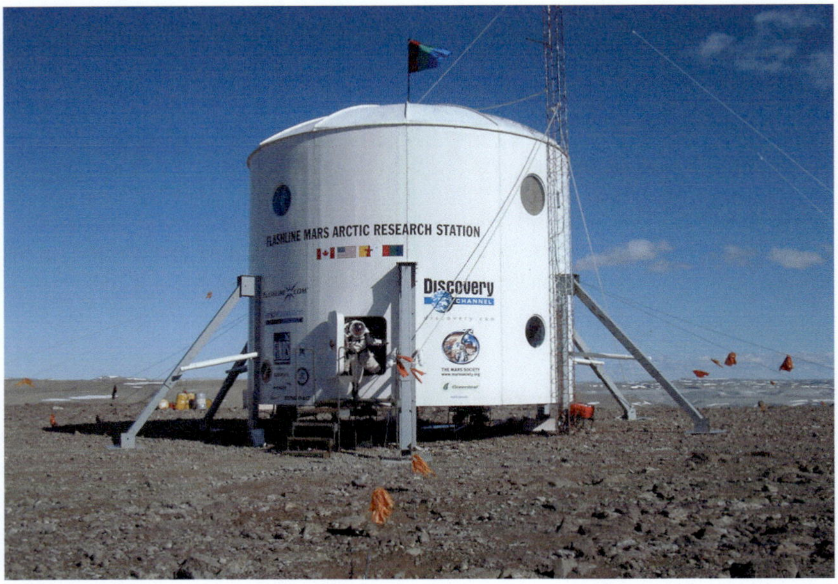

Fig. 4.16. The Flashline Mars Arctic Research Station in the Canadian Arctic (source: Ella Carlsson, FMARS Crew 8),

launch to Mars. Each station has two decks. The lower deck has two entrances, each of which can operate as an airlock. This level has a work area with laboratories for researchers, a workshop, toilet, and shower. The upper level contains six sleeping berths, one for each crew member, workstations, kitchenette, and dining area. Each station can support a crew for weeks to months living and working in a Mars analogue environment. During a field season at MDRS, crews typically change every two weeks. The participants are drawn from a wide variety of occupations, including engineers, biologists, geologists, and journalists, many of whom come with their own experiments. When the "Marsonauts" work outside they don prototype space suits to improve the fidelity of the simulation.

Mars 500

The most extensive space isolation study so far was Mars 500, a collaborative project led by the Russian Institute for Biomedical Problems. The goal of the project was to determine the key psychological and physiological effects on an international crew as a result of being in a confined environment for a period of time equivalent to a mission to Mars. The facility consisted of four hermetically sealed interconnected modules, plus a module representing the Martian surface (Fig. 4.17). The interconnected modules were a medical module, a habitat module including individual crew cabins, each of roughly $3\,m^2$ area, a utility module, and a simulator of a Mars landing module operated by three of the six crew members. The total volume of the facility was $550\,m^3$. The program began with a 14-day

Fig. 4.17. An exterior view of the Mars 500 isolation facility at the Russian Institute for Biomedical Problems in Moscow (source: ESA).

isolation of a crew of five men and one woman in November 2007 to test the functionality of two modules. The full simulation was preceded by an initial 105-day isolation period from March to July 2009, in which six male crew members participated, four Russians and two Europeans. A crew of six men, three Russians, two Europeans, and one Chinese (Fig. 4.18), were the subjects for the full simulation that began June 2010 and ended November 2011. They simulated for the first time a complete mission to Mars, including a flight to Mars lasting 250 days, a 30-day stay with surface operations, and the return flight of 240 days (i.e., a total of 520 days). The full timeline of the isolation project is presented in Table 4.6. During the experiment the crew only had direct voice contact with the simulated ground control, all contact with their families was via e-mail. The issues of communication on a deep-space mission were simulated by introducing a delay that varied depending on how far from home the mission was supposed to be, with a maximum one-way delay of 20 minutes at the farthest point from Earth.

Fig. 4.18. Mars 500 crew inside the isolation facility, December 2010. Back row from left: Yue Wang (China), Alexandr Smoleevskiy (Russia), Diego Urbina (Italy). Front row from left: Sukhrob Kamolov (Russia), Alexey Sitev (Russia), Romain Charles (France) (source: ESA).

Crew selection varied for the different nationalities in the group. The procedure for the Europeans was similar to that for astronaut selection, including extensive medical and psychological tests. Prior to the start of the isolation, the crew participated in training activities for about four months. During isolation the crew passed their time with daily exercise, maintenance of the facility, and scientific experiments in which they were often the test subjects. One of the experiments was to study the effect of blue-enhanced light on alertness and sleep–wake behavior. Recent research indicates that humans have an increased sensitivity to blue light. It is believed to have a positive effect on circadian rhythms. Members of the crew were stimulated with blue-enriched light for several 2-week periods. The actual light exposure of each individual was measured and the effects evaluated by use of questionnaires, computerized tests, and detectors to measure sleep–wake activity. During the sorties on the simulated Martian surface the crew wore space suits to carry out a variety of representative tasks, which included surface experiments and driving a rover.

At the conclusion of isolation it appeared that the crew had been on good terms throughout the mission. Towards the end they grew bored with the repetitive work of the experiments. The menu was based on a weekly cycle with some variation, but the fact that they consumed their favorite foods early in the mission led to a monotonous menu towards the end. The new data set gathered during the project will be invaluable in planning a real mission. A follow-up study in microgravity is under discussion. It could involve astronauts staying for an extended

Table 4.6. Timeline of the 520-day simulation of a human mission to Mars.

Date	*Activity*
March 23, 2010	European candidates for the crew announced
May 18, 2010	The complete crew was announced
June 3, 2010	The start of the isolation study, hatch closed, lift-off. Travel began towards low Earth orbit to dock with the ISS
June 15, 2010	The simulated spaceship undocked from the Space Station
July 25, 2011	The end of direct contact with Earth due to the distance between the spaceship and Earth
February 1, 2011	The spaceship entered a circular orbit around Mars. The hatch to the Mars lander was opened to prepare for landing
February 8, 2011	The preparation of the lander was completed. Three members of the crew transferred to the lander module and the hatch was closed
February 12, 2011	Three members of the crew (from China, Europe, and Russia) transferred to the lander module, which was undocked and landed on Mars. The simulated landing site was at Gusev Crater close to the equator on Mars
February 14, 18, and 22, 2011	Three egresses on the Martian surface on three different days by two crew members each time wearing specially adapted Orlan space suits
February 23, 2011	Takeoff from Mars, return to Mars orbit
February 24, 2011	The lander docked with the main spacecraft
February 27, 2011	After a short period of quarantine, the hatch to the lander was opened
March 1, 2011	Having loaded the lander with waste the hatch to the lander was once again closed
March 2, 2011	The journey back to Earth began
November 4, 2011	Landing on Earth, the end of the simulation

period on board the ISS to simulate aspects of a Mars mission under microgravity conditions.

4.4 ASTEROIDS

Asteroids are small irregularly shaped rocky bodies that orbit the Sun. Most of them occupy a belt that lies between the orbits of Mars and Jupiter, at distances between 2 and 4 AU from the Sun. Ceres with a diameter of about 1,000 km is the largest known asteroid in the asteroid belt. It was recently classified as a dwarf planet. The smallest asteroids are less than 1 km in size. The asteroid belt is fairly stable, but occasionally its members can be perturbed into orbits so eccentric that they cross the orbits of the inner planets, posing the risk of collision. There is routine tracking of objects whose trajectories approach Earth; these are known variously as near-Earth objects or potentially hazardous objects. We know very little about asteroids, but they may contain useful mineral resources that could one day be mined and used either in space or on Earth.

The spacecraft that have investigated asteroids are listed in Table 4.7. Galileo was deployed by the Space Shuttle Atlantis and flew by asteroids Gaspra and Ida en route to Jupiter. It found that Ida had a moon. NEAR-Shoemaker orbited asteroid Eros and landed on it. Cassini flew by asteroid Masursky en route to Saturn. Deep Space 1 was a technology test flight that investigated asteroid Braille

Table 4.7. Unmanned missions to asteroids.

Name	*Launch date*	*Asteroids* (year of encounter)	*Country*
Galilieo	October 18, 1989	Gaspra (1991) Ida (1993)	U.S.A. and Europe
NEAR-Shoemaker	February 17, 1996	Mathilde (1997) Eros (2000)	U.S.A.
Cassini	October 15, 1997	Masursky (2000)	U.S.A.
Deep Space 1	October 24, 1998	Braille (1999)	U.S.A.
Stardust-NExT	February 7, 1999	Annefrank (2002)	U.S.A.
Hayabusa (Muses-C)	May 9, 2003	Itokawa (2005)	Japan
Rosetta	March 2, 2004	Steins (2008) Lutetia (2010)	Europe
Dawn	September 27, 2007	Vesta (2011) and Ceres (planned for 2015)	U.S.A.

Fig. 4.19. Asteroid Lutetia viewed at closest approach by the Rosetta spacecraft (credit: ESA 2010 MPS for OSIRIS Team MPS/UPD/LAM/IAA/RSSD/INTA/UPM/DASP/IDA).

as its primary science goal. Stardust-NExT flew by asteroid Annefrank en route to collect samples of a comet. The Japanese probe Hayabusa rendezvoused with asteroid Itokawa and released a small lander, Minerva, but that missed its target. The main spacecraft then touched down twice on the asteroid and managed to collect a sample that was returned to Earth for analysis. The Rosetta comet mission obtained the close-up image of asteroid Lutetia shown in Fig. 4.19 in July 2010 from a distance of about 3,600 km, having already flown past asteroid Steins. Lutetia has a maximum diameter of about 130 km. In July 2011 the Dawn spacecraft entered orbit around asteroid Vesta to conduct a thorough study. In 2012 it will depart from Vesta and travel to asteroid Ceres.

In 2010 President Obama expressed an interest in sending a human mission to an asteroid as a precursor to a mission to Mars.

4.5 REFERENCES/RECOMMENDED LITERATURE

Carlsson, E.: *Framåt Mars*, Fahrenheit (2006) [in Swedish].

Harland, D.M. and Lorenz, R.D.: *Space System Failures, Disasters and Rescues of Satellites, Rockets and Space Probes*, Springer/Praxis (2005).

Kanas, N. and Manzey D.: *Space Psychology and Psychiatry*, Springer-Verlag (2008).

McEwen, A.S., Ojha, L., Dundas, C.M., Mattson, S.S., Byrne, S., Wray, J.J., Cull, S.C., Murchie, S.L., Thomas, N., and Gulick, V.C.: Seasonal flows on warm Martian slopes, *Science*, **333**, 740–743 (2011).

Mohanty, S., Fairburn, S., Imhof, B., Ransom, S., and Vogler, A., *Survey of Past, Present and Planned Human Space Mission Simulators*, SAE Technical Paper 2008-01-2020, doi: 10.4271/2008-01-2020 (2008).

Pyle, R.: *Destination Moon: The Apollo Missions in the Astronaut's Own Words*, Smithsonian, HarperCollins (2007).

Roach, M.: *Packing for Mars*, Norton (2010).

Sullivan, T.A.: *Catalog of Apollo Experiment Operations*, NASA RP-1317, NASA (1994).

Space Task Group, *The Post-Apollo Space Program: Directions for the Future*, report to the President, September 15, 1969.

Zubrin, R.: *The Case for Mars*, Free Press (2011).

United Nations treaties and principles on outer space and related General Assembly resolutions Addendum *Status of international agreements relating to activities in outer space as at 1 January 2010*, United Nations, Ref.: Sales No. E.08.I.10 ST/SPACE/11/Rev.2/Add.3 (2010).

Apollo 11 Preliminary Science Report, NASA SP-214, NASA (1969)

Websites

ESA: *www.esa.int*

International Astronomical Union: *www.iau.org*

Mars 500: *http://mars500.imbp.ru/*

Mars Institute: *www.marsinstitute.info*

NASA Astromaterials Acquisition & Curation Office: *http://curator.jsc.nasa.gov/*

NASA, Mars Exploration program: *http://mars.jpl.nasa.gov/*

NASA, National Space Science Data Center: *www.nssdc.gsfc.nasa.gov*

The Mars Society: *www.marssociety.org*

United Nations Office for Outer Space Affairs: *www.oosa.unvienna.org*

5

Life support systems

Åke Ingemar Skoog

ABSTRACT

Basic human needs—a breathable atmosphere, water, food, and waste removal—are all natural life support functions and in normal life they are basically provided by our planet's ecological system. These basic needs are the same for humans in space as on Earth, but nature must be copied by mechanical and physicochemical equipment. In particular, the limited volume and the very high cost of transportation require the application of individual weight, volume, and power-optimized systems for any given mission duration. The conditions are very similar to those in a submarine, apart from the fact that in an emergency the submarine can reach a safe condition in a short time. The similarities and systems used in submarines were amongst the first inputs for the design of spacecraft life support systems.

Concepts optimized to mission duration are mandatory because human needs and space missions are both very much time dependent. In addition, the stability of these systems, just as for the Earth's ecological system, cannot be realized by large buffer capacities but must be achieved by sophisticated design solutions and high reliability. Modern life support systems can handle the recovery of oxygen and water to a very high degree, which reduces the resupply mass quite considerably. For future systems the food loop must be at least partially closed by growing plants from waste products in order to further reduce the difficult resupply or initial high transportation costs.

This chapter provides an overview of possible regenerative functions and system concepts, life support systems employed in the first 50 years of human spaceflight, and options for future biological life support systems for long-term space exploration.

5.1 INTRODUCTION

The system solution is individual for each and every spacecraft, station, or planetary base and is defined by the mission duration, crew size, power sources, and potential resupply opportunities. The first life support systems for Vostok and Mercury were very simple, and they did provide a breathable atmosphere and an adequate thermal environment. A small supply of drinking water was an extra. With larger crew sizes and increasing mission durations, food and waste management became standard requirements with water for drinking, food preparation, and hygiene.

Today's ISS is equipped with a life support system which recovers basically all oxygen and water, and in-orbit experiments to grow food are starting on a larger scale. But even for a six-person crew such a system has a large volume and mass. It must be kept in a stable balance without large buffer capacities, despite all inadvertent human "negative" efforts to change it. In addition, safety and medical care must be taken into consideration.

5.2 HUMAN NEEDS? REQUIREMENTS AND DESIGN PHILOSOPHIES

The basic requirements for the well-being of a human are defined by metabolic rates (Fig. 5.1). These very soon sum up to considerable mass and volume penalties for a space mission, for example, about 750 kg for a six-person crew for one week and 15,000 kg for a six-person crew for a flight of 90 days. This is the average duration of the crew turnaround on the ISS, and this amount of life support consumables must be covered by resupply missions or in-space regeneration. Thus, needs must be reviewed for time dependence and life support system concept options must include an increasing amount of regenerative functions the longer the mission and the larger the crew.

A careful review of the primary requirements reveals clear time dependence for each of these needs. Obviously a breathable atmosphere with sufficient oxygen is essential for any mission, no matter how short or long it is; even for an emergency situation. Air revitalization by CO_2 removal must be provided at the latest after a few hours. Water is indispensable for a flight longer than about six hours, and at the same time the need for waste management is evident. Food becomes a basic need for flights longer than a day. Thus the life support functions required and the resulting system complexity for a particular mission are very much time dependent (Fig. 5.2).

To handle the system size and to reduce the potential resupply requirements for any given mission, the life support concept design must take into consideration the interaction of individual life support functions, just as in the case of the Earth's ecological system, and regenerative alternatives must be optimized (Fig. 5.3). For each life support function various technical design solutions are available, and system analysis must achieve the right overall concept for each and

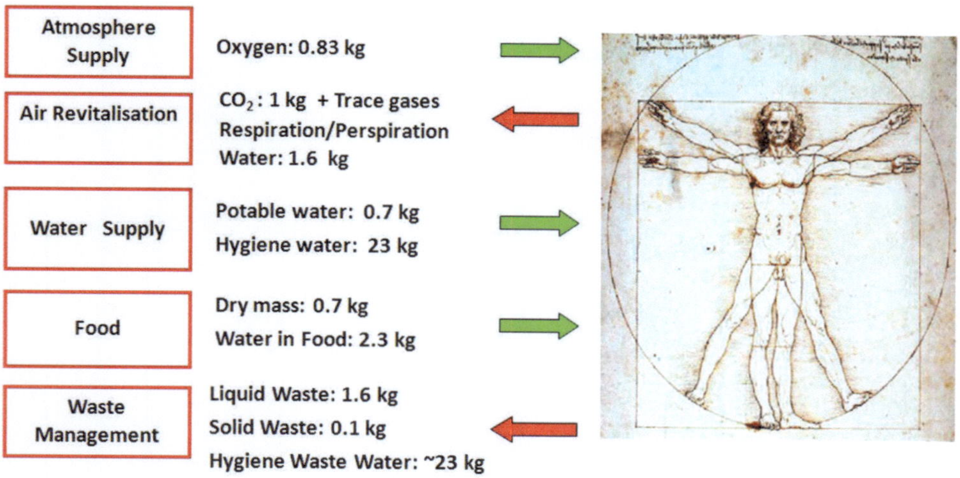

Fig. 5.1. Human metabolic rates per person and day (source: Åke Ingemar Skoog).

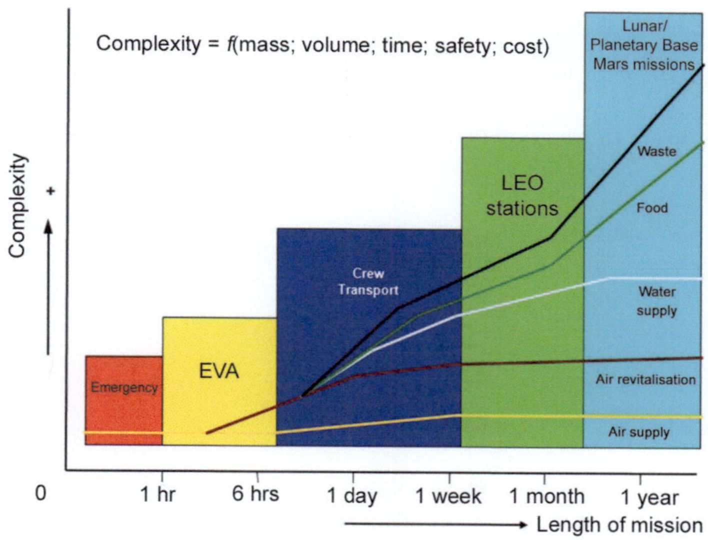

Fig. 5.2. Complexity of life support functions required for basic types of space missions as a function of mission length (source: Åke Ingemar Skoog).

every individual mission depending on the mission drivers and the technical requirements detailed.

These basic needs and functional interrelations result in a top set of mission drivers that must be defined and considered before any life support system can be optimized for a given mission. Only then is it possible to consider options for the individual life support functions to begin to define alternative overall system concepts. The mission drivers are

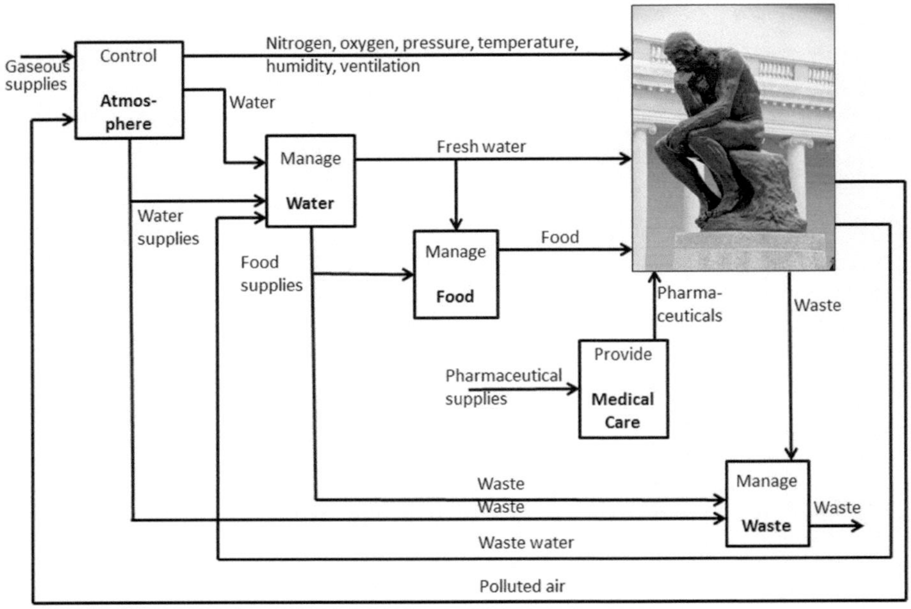

Fig. 5.3. Interaction of basic life support system functions (source: Åke Ingemar Skoog).

- mission target or location (Earth orbit, Moon, Mars, ...)
- crew size
- size of habitat(s)
- mission duration
- resupply opportunities and frequency
- type of power supply
- external radiation
- safety/safe haven/reliability
- gravity
- possible local resources.

The mission drivers provide the first inputs for system concept trade-offs. In the final selection of the system design, more detailed but stringent requirements must be considered. Air quality, water quality, and crew comfort requirements are today well defined in the life support standards to be used for any given space mission; for example, the NASA Man-System Integration Standards (NASA-STD-3000), the Russian GOST Standards, and the ESA Standards for Environmental Control and Life Support (PSS-03-40).[1] Over the past five decades these requirements have evolved to become more and more Earth-like in order to improve conditions for crews of long-term missions, in particular on space stations. Table 5.1 shows the gradual change of requirements for the atmosphere

[1] With several standards and continuous updates over the years, the exact values might vary by a small amount depending on the standard and issue used.

Table 5.1. Evolution of basic atmosphere requirements for space station life support (source: NASA).

Parameter	*Earth*	*Skylab (1973–1974) 84-day max. stay*	*Mir (1986–2001) 365-day max. stay*	*SSF (1992) 90+-day design stay*	*STD-3000 (1995)*	*ISS (1996) 90+-day design stay*
Total pressure (kPa)	101.3	34.5	88.0–114.7	100.0–102.7	100.0–102.7	97.9–102.7
O_2 partial pressure (kPa)	21.0	22.8–26.9	20.0–26.7	19.5–23.1	19.5–23.1	19.5–23.1
CO_2 partial pressure (Pa)	31–100	670 (nominal)	800	400	400	Average = 707, peak = 1,013
Diluent gas	N_2/Ar	N_2	N_2	N_2	N_2	N_2
Temperature (K)	292.5–300.5	285.9–305.4	291.1–301.1	291.5–299.8	291.5–299.8	291.5–302.6
Dew point (K)	274.9–289.9	280.9–288.7		277.6–288.7	277.6–288.7	277.6–288.7
Relative humidity (%)	20–85	25–85	30–70 (g)	25–70	25–70	25–70
Ventilation (m/s)	<0.1		0.2–0.5	0.076–0.203	0.076–0.203	0.051–0.203
Particulates >0.5 μm (particles/m^3)				<3.5E6	= 3.5E6	Average < 3.5E6, peak < 7.1E7
Particulates >0.5 μm (μg/m^3)	8–9 (nominal)					Average < 50, peak < 1,000
Microorganisms, CFU/m^3 (h)	200–600			1,000	500	Average < 1,000
Trace contaminants	Variable			SMAC (i)	SMAC	SMAC

Standard Atmosphere Parametric Limits - Dry Bulb, Dew Point, Relative Humidity

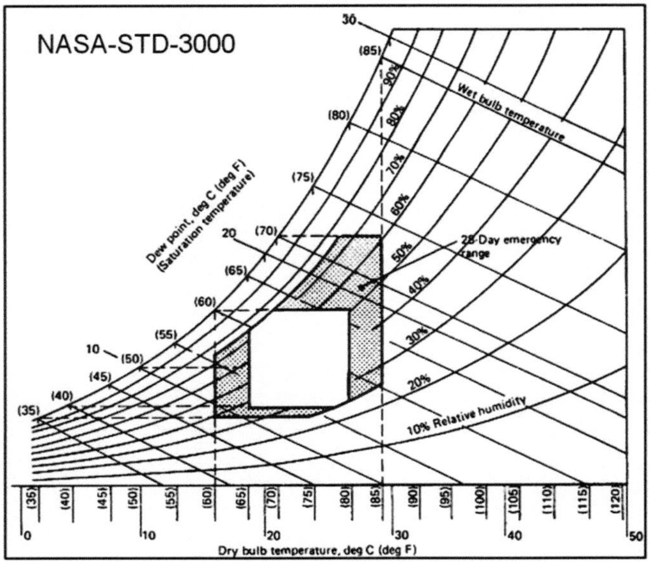

Fig. 5.4. The "crew comfort box" for the atmosphere requirements in NASA-STD-3000 (source: NASA).

of the cabin ranging from the low-pressure Skylab concept to the present standard atmosphere on the ISS.

A typical example of the Earth-like standard atmosphere is the "crew comfort box" which defines the interrelation of the atmosphere parameters temperature, dew point, and relative humidity (Fig. 5.4).

As the detailed design progresses, other requirements such as spacecraft maximum allowable concentration (SMAC) of trace gases and odors (some 60 in the standard lists) and water pollutants (some 30 in the official listing) must be taken into account. Other detailed design requirements in the standards are noise limitations, light, touch temperatures, materials selection, monitoring and control, maintenance and repair concepts, gravity/vibration influence, fire detection and suppression, radiation, and safety conditions. They all influence the final design.

5.3 TYPES OF LIFE SUPPORT SYSTEMS

Various means exist to realize any combination of life support functions for short or long-duration missions with small or large crews. The simplest way is to use some combination of physicochemical processes. This is practicable for short to medium-term missions with small or medium-size crews, but for long-term missions with large crews (e.g., a base on the Moon or Mars) the optimal solution might be a controlled ecological life support system that copies nature on Earth.

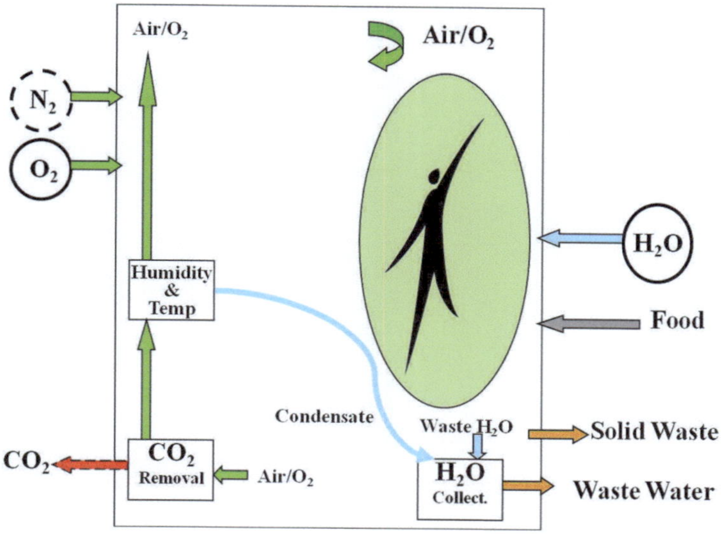

Fig. 5.5. Open-loop LSS (source: Åke Ingemar Skoog).

The simplest system is the open loop in which all consumables (oxygen, nitrogen, water, and food) are supplied from on-board storage or produced on board as part of another process, such as water from the fuel cell process that generates power. Waste products (carbon dioxide, water, and solid waste) are dumped overboard or stored for return to Earth. Also, chemicals for the removal of CO_2 must be stored for the total flight time (Fig. 5.5).

These open-loop systems are only feasible for short missions and small crews due to their mass and volume penalties. Open-loop systems are used in the Soyuz spacecraft, the decommissioned Space Shuttle (with water produced by the fuel cell process) and in space suits. To reduce mass penalties, regenerable functions for CO_2 removal and subsequent reduction and oxygen generation can close the air loop to a high degree, but at the cost of increased power consumption. Some products like methane or solid carbon will still be produced as residual waste, but process water can either be used as such or as input to electrolysis to recover the oxygen (Fig. 5.6).

Closed air loop systems are feasible for any crew size on longer missions (e.g., permanent space stations such as the ISS or a lunar base with regular scheduled crew exchange flights). Some 70% of the total resupply mass is water, so the greatest saving effect can be achieved by partial or full closure of the water loop (Fig. 5.7). Humidity and water from CO_2 reduction, as well as water used for washing, can be reclaimed by simple filtering methods to potable and hygiene water, as today on the ISS or in the past on Mir. However, recovering water from urine is a larger problem, both in terms of processing and the psychological aspect of using it as potable water. If possible, the potable water loop of reclaimed condensate and water from CO_2 reduction ought to be kept separate from the

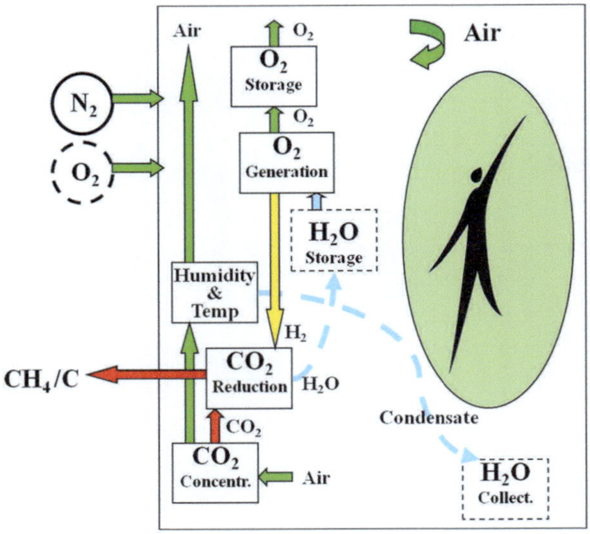

Fig. 5.6. Closed air loop LSS (source: Åke Ingemar Skoog).

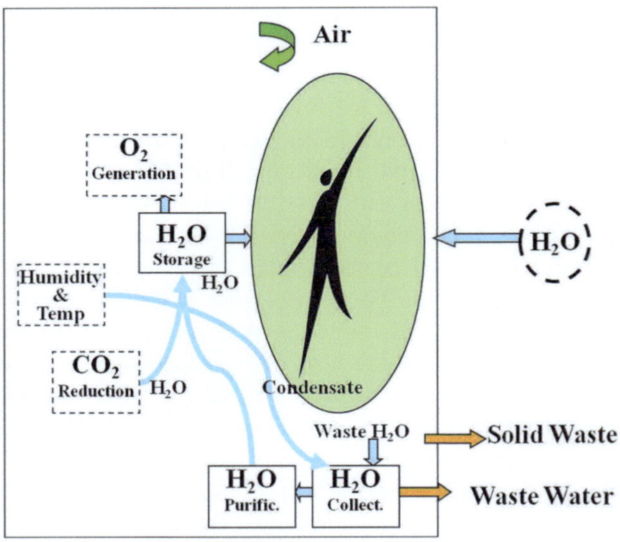

Fig. 5.7. Closed water loop LSS (source: Åke Ingemar Skoog).

hygiene loop. The water recovered from urine might be used for oxygen generation.

With water as the common element between the air and water loops, it is logical to combine them into a partially closed life support system (Fig. 5.8), with the only basic resupply being food and nitrogen (if using a normal atmosphere). In

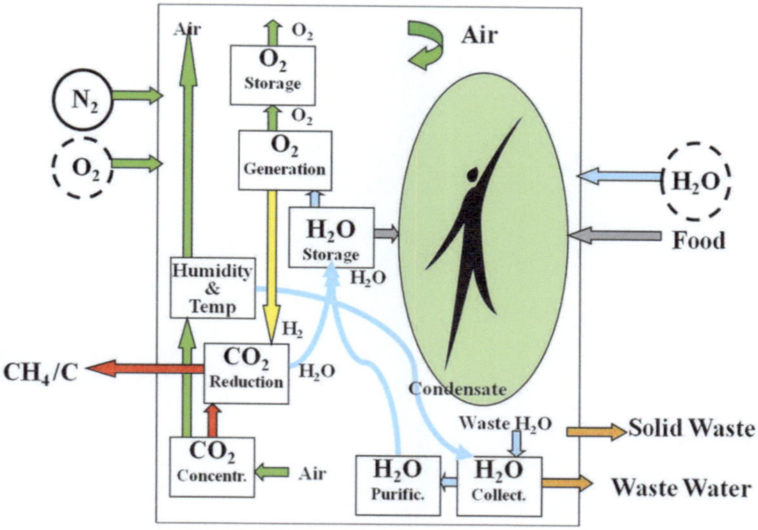

Fig. 5.8. Partially closed LSS (source: Åke Ingemar Skoog).

this case, a reduction of 90% of the resupply mass is possible. If a full mass balance cannot be achieved due to, for example, leakage, docking with other vehicles which draw upon consumables, or a payload/experiment that draws upon consumables, then a resupply of oxygen and/or water might be necessary. A final waste product in these systems will be human solid waste, which might be processed for simpler treatment and final removal from the spacecraft or habitat. A basic presumption for a partially closed life support system is the availability of a considerable amount of power for all steps of regenerative physicochemical life support functions.

Complete closure of all functions of a life support system will require closure of the food loop as well, and the basic copying of all natural environmental functions on Earth into a Controlled Ecological Life Support System (CELSS)[2] or Biological Life Support System (BLSS) (Fig. 5.9). Such a life support system would then handle all loops (air reclamation, water reclamation, and food production) in the life support system by means of plants and possibly animals, and use physicochemical processes and equipment only as required for safety backup. Even if 100% closure of the life support system is not practical, a partial closure of the food loop would, for missions with larger crews and lasting years, mean a lot to the well-being of crew members by providing fresh plant products. However, the production of, for example, meat or fish is a very large and costly final step for total closure of all life support functions. The implementation of a CELSS

[2] This is preferable to the earlier term *Closed* Ecological Life Support System, because for the foreseeable future it is unlikely we will be able to completely close the food loop and recycle all waste products on a spacecraft or a base on another celestial body.

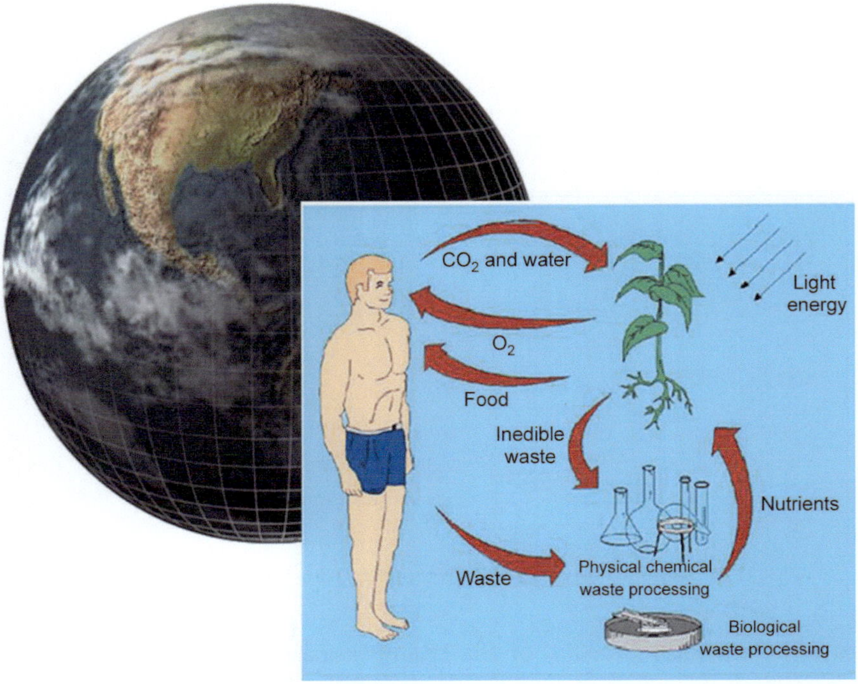

Fig. 5.9. The basics of a Controlled Ecological Life Support System (source: Åke Ingemar Skoog).

requires obeying stringent environmental requirements (i.e., atmosphere properties, temperature, humidity, water supply, light sources, and plant nutrients) as well as proper function of biological species in a zero-gravity or low-gravity environment.

5.4 TYPES OF LIFE SUPPORT FUNCTIONS

Over the past 50 years of developing and operating life support systems for manned spacecraft, system design has grown from open-loop systems (Vostok, Voskhod, Mercury, and Gemini) to increasingly more partially closed physicochemical systems (Mir and ISS; Fig. 5.10). For each and every one of the functional elements of the life support system several technical options have been developed, and all functions of a partially closed physicochemical life support system are now in an operational state. Sometimes the customer has investigated and developed alternative technologies, and sometimes competing manufacturers have decided to pursue alternative technologies.

The implementation of a partially closed life support system is now a question of optimization at the system and subsystem level of all parameters for the spacecraft (i.e., mass, power, crew size, flight duration, resupply frequency, cost, and maintenance).

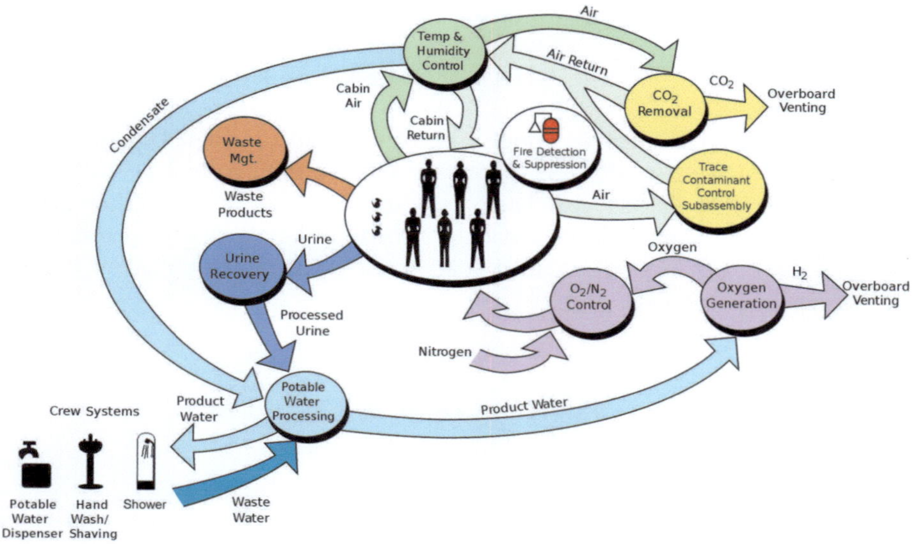

Fig. 5.10. ISS partially closed regenerative LSS (source: NASA).

5.4.1 Air supply and revitalization

Air circulation and pressure control

Air circulation in a spacecraft is performed by fans placed in each of the modules. These suck the air out of the cabin and then blow it into CO_2 removal devices and/or temperature and humidity control hardware. In larger space complexes, such as the ISS, fans may also transport the air from peripheral modules into ones having central processing for CO_2 removal, temperature and humidity control, plus the removal of trace gases. Fans are used in single compartment spacecraft and the safety critical modules of stations in parallel redundant configurations. Cabin pressure control is performed by hardware that monitors total cabin pressure and the partial pressures of oxygen, nitrogen, and CO_2. This hardware has pressure regulators and valves to automatically control and maintain not only nominal conditions but also specified emergency levels to reduce cabin pressure leakage or selected reduced cabin pressures to support a spacewalk.

CO_2 concentration

Carbon dioxide is a by-product of the breathing process and thus it is found at low concentrations in nature. However, in increased concentrations it is toxic and even lethal ($\sim$8%). In a confined environment such as a space habitat, CO_2 would continuously increase unless methods were implemented to achieve its removal. Non-regenerative methods have been used since the very first manned spaceflights, and are still in use in short-duration missions, space suits, and as backup for regenerative functions. By tradition the lithium hydroxide concept has been used in the West and the potassium superoxide concept in Russia.

The lithium hydroxide system uses the chemical process

$$2LiOH + CO_2 \rightarrow Li_2CO_3 + H_2O$$

and 1 kg of LiOH can remove about 1 kg of CO_2 and yield 0.4 kg water. This concept has been used in all U.S. spacecraft and space suits, and is still a safety backup for the ISS. It has also been used in the European Spacelab and the Russian Orlan space suit, and as a backup on Mir. The alternative process is the potassium superoxide one, also called solid fuel oxygen generation, and it uses a three-step chemical process

$$4KO_2 + 2H_2O \rightarrow 4KOH + 3O_2$$

$$4KOH + 2CO_2 \rightarrow 2K_2CO_3 + 2H_2O$$

$$2K_2CO_3 + 2H_2O + 2CO_2 \rightarrow 4KHCO_3$$

which can remove about 0.3 kg of CO_2 per kg of KO_2 and yield about 0.34 kg of oxygen. The water needed for the process is taken from air humidity. It has been used in all Soviet/Russian spacecraft, and still serves as the backup for the Russian section of the ISS.

Non-regenerative processes impose a high mass (and hence volume) penalty on longer missions, so regenerative processes are necessary for at least CO_2 removal even if the liberated oxygen is not recovered. A number of regenerative processes are available, many of which have already been utilized in space. The first system to be implemented (Skylab), and also the most widely used one (Mir, ISS), is the molecular sieve (Fig. 5.11). This system consists of two parallel loops, each with one canister for humidity control and a second for CO_2 removal. These canisters contain zeolites composed of silicon and aluminum, and can be either vacuum or heat desorbed. The system is cycled between the two loops, one adsorbing humidity and CO_2, while the other is being desorbed. A canister with 100 grams of zeolite cleans about 10 grams of CO_2 (at 1 kPa $ppCO_2$). The mass characteristic of a molecular sieve system makes it favorable for longer missions in combination with the reduction of CO_2 in order to recover oxygen.

The regenerable solid amine system with water desorption (SAWD) had been used on board the Shuttle. Solid amine is an ion exchange resin matrix produced by a polymerization of resins which contain amino groups. Space suit systems for the ISS (Orlan-M and EMU/ISS) use regenerative CO_2 removal by means of metal oxides, a process that is reversed by applying heat. Silver oxide is presently used

$$Ag_2O + CO_2 \rightarrow Ag_2CO_3$$

and the system can remove 0.12 kg of CO_2 per kg of Ag_2O. Also tested, although not yet implemented, is the electrochemical depolarized concentrator (EDC) system, an alkaline fuel cell with an aqueous carbonate electrolyte which consumes oxygen and hydrogen to produce water and power.

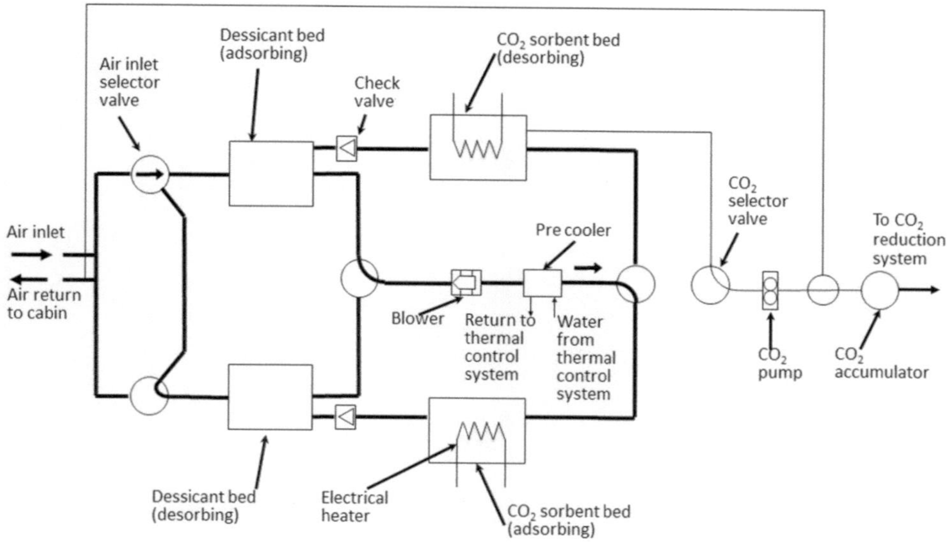

Fig. 5.11. Four-bed molecular sieve system concept (source: NASA).

CO_2 reduction

For longer missions the recovery of oxygen is essential, and two competing methods for CO_2 reduction in space vehicles are the Sabatier and Bosch reactors.

The Sabatier process (Fig. 5.12) is the most advanced system and is favored for the ISS. It uses an exothermic reaction to convert CO_2 by means of heat and hydrogen to water and methane

$$CO_2 + 4H_2 \rightarrow 2H_2O + CH_4$$

using a catalyst of ruthenium on an alumina support. If the hydrogen in the methane is not to be recovered, the methane can be used for vehicle thrusters. If the hydrogen is to be fully recovered, the Sabatier process can be combined with a high-temperature pyrolysis process (Carbon Formation Reactor, CFR) with water and solid carbon as the final products.

The Bosch reactor produces water and solid carbon in a catalytic high-temperature reaction

$$CO_2 + 2H_2 \rightarrow 2H_2O + C$$

with a catalyst of iron, nickel, cobalt, or alloy. The Bosch system has a higher mass and power consumption than a combination of Sabatier and CFR, and needs periodic cartridge replacement to remove the carbon powder on the steel wool.

Trace gases and odor control

The basic method for short missions is a combination of particulate filters and active charcoal, and for longer missions the addition of a catalytic burner

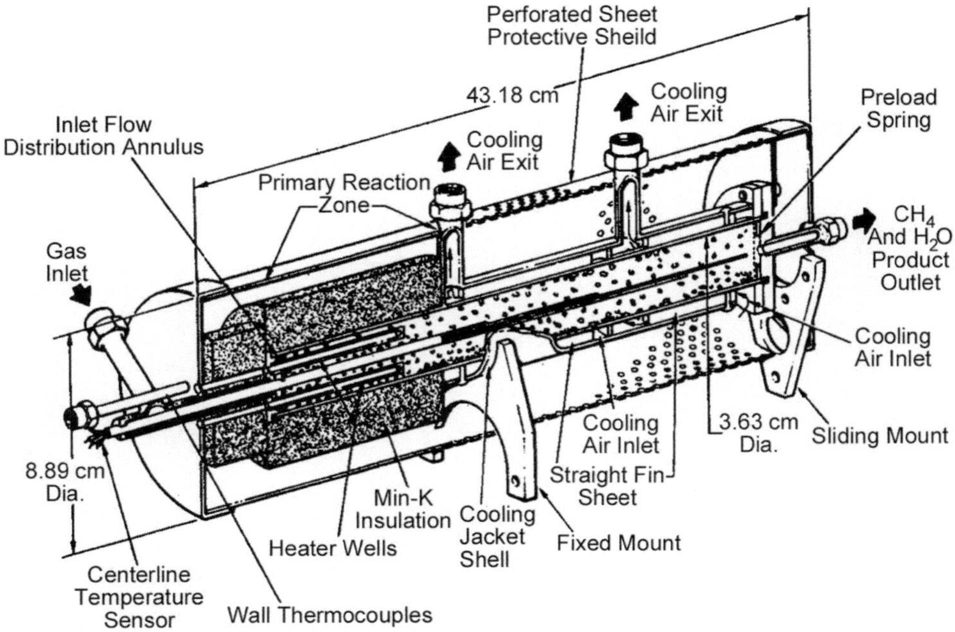

Fig. 5.12. Sabatier reactor for CO_2 reduction (source: NASA).

(Shuttle, Mir, and ISS). The active charcoal is either in a separate canister or integrated into the CO_2 removal unit (e.g., the LiOH cartridge). For longer missions, using a separate canister permits regenerative operation by exposing the charcoal to vacuum. The control of trace gases and odors at the source is a central element of spacecraft design. By appropriate selection of materials and the design of ventilation and of water and waste management systems, trace gases can be reduced significantly. Also, operational procedures such as spillage control can reduce the amount of trace gases and odors.

Humidity and temperature control

The standard method, which is used in all spacecraft, is a condensing heat exchanger in combination with a water separator. To separate condensate from the main airstream in zero gravity, the heat exchanger is provided with a hydrophilic coating. Final condensate collection is done either by a rotating centrifugal water separator or by a porous wick system.

Air supply

The primary atmospheric constituents, oxygen and nitrogen, are either supplied from storage or, in the case of oxygen, possibly generated on board. The oxygen supply must provide metabolic oxygen and make up for any leakage. Oxygen for short missions is mainly from high pressure gas (200–500 bar) storage or from cryogenic storage (e.g., Apollo and Shuttle). In Soviet/Russian spacecraft (Vostok, Voskhod, Soyuz, and Salyut, and as a backup on Mir and the Russian section of the ISS) chemical storage for oxygen has been favored, but with high-pressure oxygen as an emergency backup.

The implementation of CO_2 reduction systems provides the opportunity for oxygen production by electrolysis. Electrolysis of either waste water or water from the CO_2 process produces the required oxygen. At the same time, hydrogen is recovered for the CO_2 removal process. Favored U.S. concepts for the ISS are static feed water electrolysis and solid polymer electrolysis, whereas Mir and the Russian section of the ISS are equipped with water vapor electrolysis equipment.

Nitrogen is required for spacecraft using a normal or near-normal atmosphere (all Soviet/Russian ones, Skylab, Shuttle, Spacelab, and the ISS) and it is usually supplied as a high-pressure gas, but in some cases the nitrogen supply for leakage compensation has been performed by a normal air mixture.

Atmosphere monitoring and control system

The spacecraft atmosphere must be monitored for a large number of parameters from pressure and temperature to trace gases and odors, humidity, particles, bacteria, and fungi. Where a rapid change might threaten the lives of the crew, the main parameters are monitored by active sensors for pressure (transducers), temperature (thermistors), carbon dioxide (electrochemical or infrared sensors), oxygen content (electrochemical sensors or fiber optics), and trace gases (mass spectrometer and gas chromatograph). Other control processes are indirect (humidity by heat exchanger function or trace gases by a charcoal filter) or by sampling and analysis.

5.4.2 Water management

Next to a breathable atmosphere, potable water is the most important life support element for a crew. Basic survival requires 2 to 3 liters of water per person per day, of which about two thirds is contained in the food of modern space menus. This supply could be handled by on-board storage for crews of three or four persons on missions of several weeks, but partial or full reclamation of waste water is essential for larger crews and longer flights. Water reclamation processes were not introduced until the space stations in the 1980s, from Salyut 6 on. The water balance for a mission must therefore take into account on-board storage, reclamation processes, and water that is obtained as a by-product of, for example, a fuel cell system.

Water storage and resupply

All early U.S. and Soviet spacecraft had a full water supply on board, filled prior to launch or by additional supply from the fuel cell process (Apollo, Shuttle). Potable water was stored mainly in metal tanks. Also important is the handling of waste and excess water in case of storage only or additional supply from fuel cells. The venting of waste water overboard has been practiced successfully in many former spaceflight systems, but in many cases this had to be limited due to external sensors and scientific instruments, and in particular due to ice formation near the spacecraft.

Even in a partially closed life support system, the full closure of the water loop is impractical because there is some water in final waste products, and because the evaporative cooling systems of space suits consume water. Replenishment of potable water on Mir was by transportable water bags delivered by Progress vehicles. This is also the case for the ISS, although here the water supply was additionally supplemented by the Shuttle transferring any excess of water from its fuel cells before departure.

Water reclamation

The first water recovery system was implemented on Salyut 6 and 7 by producing potable water from condensate recovered from the cabin atmosphere. The condensate was filtered through ion exchange resins and active charcoal and other filters. A final treatment was performed in an ambient temperature catalyst process, as an integral part of the multifiltration system. Minerals were then added to the reclaimed water. This concept was implemented in the Mir station by two additional water reclamation loops. Hygiene water was collected in a second loop and treated in the same way as condensate water, but the loops were separate and the recovered hygiene water was only used for hygiene purposes. Another separate loop collected urine, and a vapor diffusion distillation process was used to supply the water for the electrolysis process that generated oxygen. This overall concept has also been carried over to the Russian section of the ISS.

Water recovery was first introduced by the U.S. in the ISS and, unlike the Russian segment, all water and urine is processed to potable water in a single overall system. Urine is passed through a distillation process and then handled together with condensate and hygiene water in the multifiltration process. The U.S. section of the ISS uses a water vapor compression distillation (VCD) process for urine. Waste water is distilled by evaporating filtered water from a thin film of waste water at low pressure (3.4 to 5.5 kPa) and ambient temperature (32 to 43°C) on the interior of a centrifugal drum. The latent heat of condensation is recovered by first compressing the water vapor in order to raise its saturation temperature and pressure, and then condensing the vapor on a surface that is in thermal contact with the evaporator. An alternative for urine and waste water recovery is the thermoelectric integrated evaporation membrane system (TIMES), which can efficiently recover potable water from waste liquids. It has advantages such as low power consumption, compactness, and gravity-insensitive operation. The system comprises a vacuum distillation system combining a hollow-fiber polysulfone membrane evaporator with a thermoelectric heat pump and condenser.

In the multifiltration process, waste water is treated in a number of filter beds. First particulate material is removed by filtration, then beds with ion exchange resins, active carbon, and other sorbents are used to adsorb salts and organics. Low-molecular-weight species which are not effectively removed by absorption are post-treated in a low-temperature phase catalytic oxidation process.

Water quality control

Water quality is maintained by the addition of iodine (U.S.) or silver ions (Russia) in combination with heat sterilization. Several parameters are measured to monitor the quality of the water (e.g., pH, conductivity, free gases, and microbial counting).

5.4.3 Food and waste management

Nourishment is of great importance not only from the physiological point of view, but also for the psychological well-being of the crew during longer missions. A balanced menu with a variety of dishes including fresh fruits and vegetables is essential, and this is reflected in the evolution of the "space kitchens" of both the Soviet/Russian and the U.S. space programs (Fig. 5.13). On the very early short missions, the food supply consisted of a paste in a tube that was heated by the injection of hot water. Later, a wider variety of food and beverages was introduced in the form of dehydrated and canned products. In today's food management, each member of a crew has a personal menu based on a combination of canned, frozen, and fresh products selected from a large variety of food elements, many commercially available off the shelf. The food is stored in freezers and refrigerators, and prepared in an oven. To reduce both handling and waste, the

Fig. 5.13. "Space kitchen" evolution. Note the "three-dimensional" table setting in the ISS illustrations (source: NASA).

food elements of the daily menu are packed in single-serve sizes. Fresh fruits and vegetables are brought to the ISS with each Progress resupply flight.

With longer missions and a continuously improving food system, the requirement for an adequate human waste disposal system is obvious. Due to weightlessness and the increasing need to close the water loop, urine and fecal collection is undertaken separately. In the early flights, urine was collected by receptors and a transfer system to plastic bags or tanks. A number of alternatives for ultimate disposal of urine were utilized prior to the implementation of water recovery on Mir and the ISS. In some cases urine was stored on board (Mercury and Apollo Lunar Module), used in a boiler system for heat rejection (Gemini), or vented overboard (Apollo Command Module, Skylab, Shuttle, and early Soviet/Russian spacecraft). In the case of Mir and the ISS, urine is processed to fresh water and only "a brine of concentrated waste water products" is collected in tanks and discarded by the Progress resupply vehicles. On all early flights, feces were collected in plastic bags, sealed, and stored; except for the Salyut stations where feces bags were regularly ejected into space. With the Shuttle and Mir, a commode system was introduced. To overcome the absence of gravity, a forced air stream was used and the fecal products were collected in plastic bags, or in the Shuttle by centrifugal force in a rotating collection drum for return to Earth (Fig. 5.14). On Mir, feces were discarded along with other solid kitchen waste in the Progress resupply vehicles. This is also the case for the ISS.

5.4.4 System safety

Certain safety-critical functions such as fire detection, fire suppression, and control of trace gases, bacteria, and fungi are handled as part of the life support system because they are closely dependent on, and interface with, other functions of the life support system.

A number of sensors are available for fire/smoke detection. In early spacecraft the main "detector" was the human sense of smell, but active sensors were introduced with Skylab. Thermistors, ultraviolet detectors (Skylab), ionization detectors (Shuttle and Spacelab), photoelectric sensors (the U.S. section of the ISS), and optical sensors (Mir and the Russian section of the ISS) are the main alternatives used today. These sensors are mounted in air ducts, electronics racks, and in the cabin. In places where a detector cannot be used, such as payloads, temperature parameters are monitored as part of the data-handling system.

Because the crews of early spacecraft wore space suits, fire suppression relied on either cabin depressurization or, in some cases, water such as that available for food preparation. Skylab was also provided with an extinguisher that used aqueous gel. In addition to the option of depressurization, the Shuttle and Spacelab had a built-in fire suppression system and hand-held extinguishers that used Halon 1301. The U.S. section of the ISS employs CO_2 for fire suppression, both for a fixed system and for hand-held extinguishers. In addition to the option of sealing off and depressurizing a module, the Russian section uses a system with a water-based foaming agent.

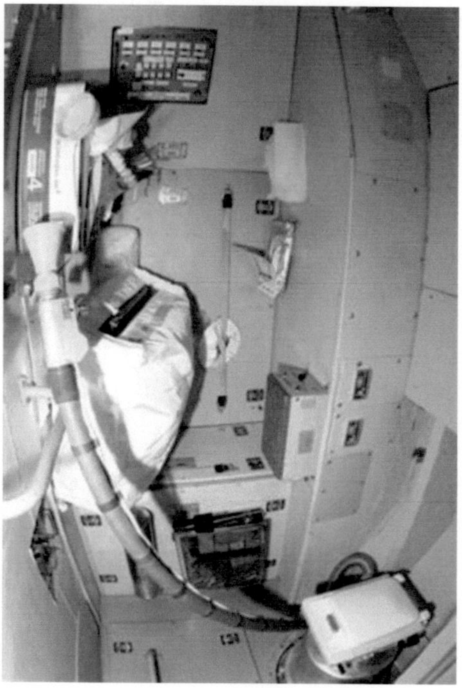

Fig. 5.14. ISS toilet (lower right) with urine receptor left (source: NASA).

The control of bacteria and fungi with slow rates of buildup is performed by bio-sampling and analysis in orbit and on Earth. The main alternatives for removing these elements are placing charcoal filters in the airstreams and wiping surfaces clean.

5.5 PHYSICOCHEMICAL LIFE SUPPORT SYSTEMS

The design of a life support system is an iterative process to find an optimal concept for a given mission scenario. Based on the major design parameters (mission target or location, size of crew, size of habitat(s), mission duration, resupply opportunities and frequency, type of power supply, feasibility of overboard dumping, alternative safe heavens, gravity, and possible use of in situ resources) system trade-offs are made to identify, for example, mass, power consumption, crew time, cost and resupply needs for various concepts with different degrees of loop closure. Clearly a small crew for a short mission will find an optimal concept with very few or no regenerative functions, while large crews and a longer duration, perhaps without resupply possibility (such as a Mars mission), will only be feasible with a very high degree of loop closure.

Centralized versus distributed life support systems

A large space complex like an orbital space station, or a base on the Moon or Mars, may include a significant number of modules for habitation, research, command and control, and repair and maintenance equipment, each one with a different degree of crew presence and dedicated activities. In a centralized system, most functions like air revitalization, oxygen generation, and water processing are done in one place (e.g., in a dedicated life support module) and food and hygiene functions are located in a habitat module. Generally, one large unit for processing of all carbon dioxide or producing all oxygen would be lighter and more energy efficient than several smaller units. However, from a safety point of view, redundancy and backup equipment are necessary and several smaller units distributed across a number of modules might make the overall system more efficient. The breakdown of one single module in case of fire or leakage could have life-threatening consequences if it held vital life support equipment. Thus in the trade-offs various degrees of centralized versus distributed life support functions must be reviewed in order to find an optimal solution that balances safety against mass and power penalties.

Mass balance

The mass balance for life support system consumables in a given mission enables the determination of requirements for resupply of non-regenerative functions and the preliminary sizing (mass, volume, and power) of equipment for various regenerative alternatives. Fig. 5.15 shows an example of mass balance in a life support system concept. Analysis of the degrees of loop closure, and technical alternatives for the required life support functions, leads to individual life support system concepts with unique overall figures for mass, volume, power, and cost (equipment versus resupply transportation cost).

Technology maturity

An important trade-off parameter is the maturity of different technical options for the various functional needs. If any particular option is not available off the shelf, then a degree of technology development will be required and this might become a cost and schedule issue. The final selection of a configuration for the life support system must then be made at the project system level by taking into account mass, power, schedule, cost, and safety constraints.

5.5.1 The first spacecraft life support systems

The very first manned spacecraft (Vostok and Mercury) used open-loop life support systems (Fig. 5.16). Flight duration was from fractions of an hour up to five days maximum for a single person. The available technology at that time (the early 1960s) was based in part on high-performance aircraft to support the exploration of what was an unknown domain, the void of space, and also on

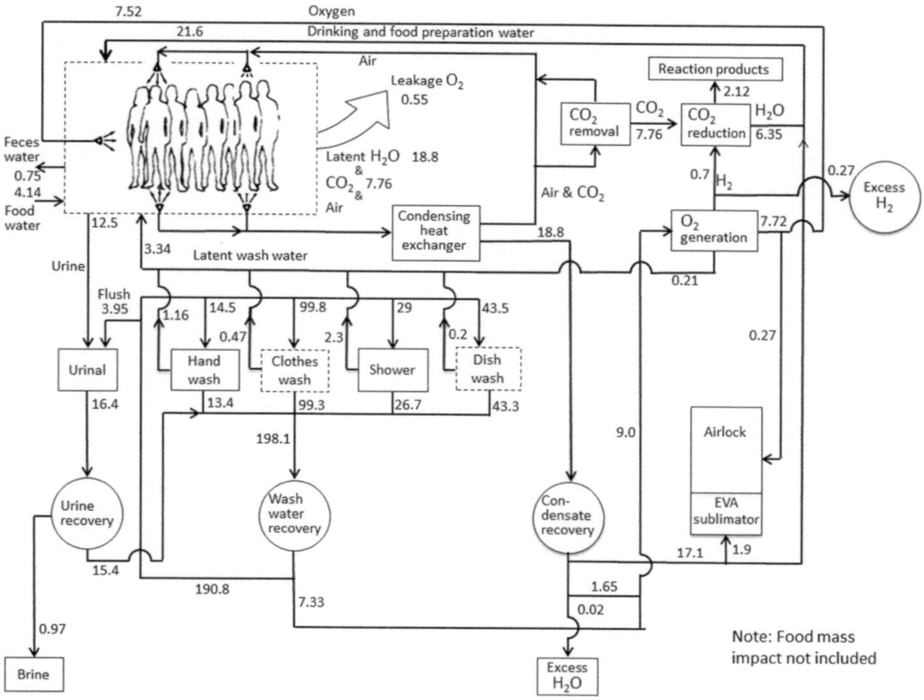

Fig. 5.15. Mass balance of human consumables in a life support system for a space station concept. The numbers are in units kg/day and apply to an eight-person crew (source: NASA MSFC).

submarine life support systems supporting larger crews and longer submerged operation time. To ensure the safety of the crew, the system comprised not only a pressurized capsule but also a suit that was partially integrated into the overall atmosphere supply system. The Vostok spacecraft used an ambient air atmosphere, with CO_2 removal by KOH generating the oxygen to compensate for crew respiration and cabin leakage; plus a separate oxygen supply for the suit during an emergency. The Mercury spacecraft used an oxygen atmosphere at 345 hPa (5 psia) with a LiOH CO_2 removal system. In both systems the atmosphere was also used for in-suit cooling.

The basic open atmosphere loop life support system concept was used for the next generations of spacecraft with the addition of an open water loop for drinking and the preparation of food. Even the Apollo Lunar Module used the same basic concept with a combined cabin and suit loop system for a pure oxygen atmosphere (Fig. 5.17). In the early 1970s the Skylab space station was the first spacecraft with a complete life support system for atmosphere (345 hPa) and water supply, hygiene facilities, and a new food supply including canned, frozen, and fresh products. Both the constituents of the atmosphere (3,520 kg of oxygen and nitrogen) and the water supply (2,700 kg) for six months were loaded on board for launch. Only CO_2 removal was regenerative, using a twin-canister molecular sieve

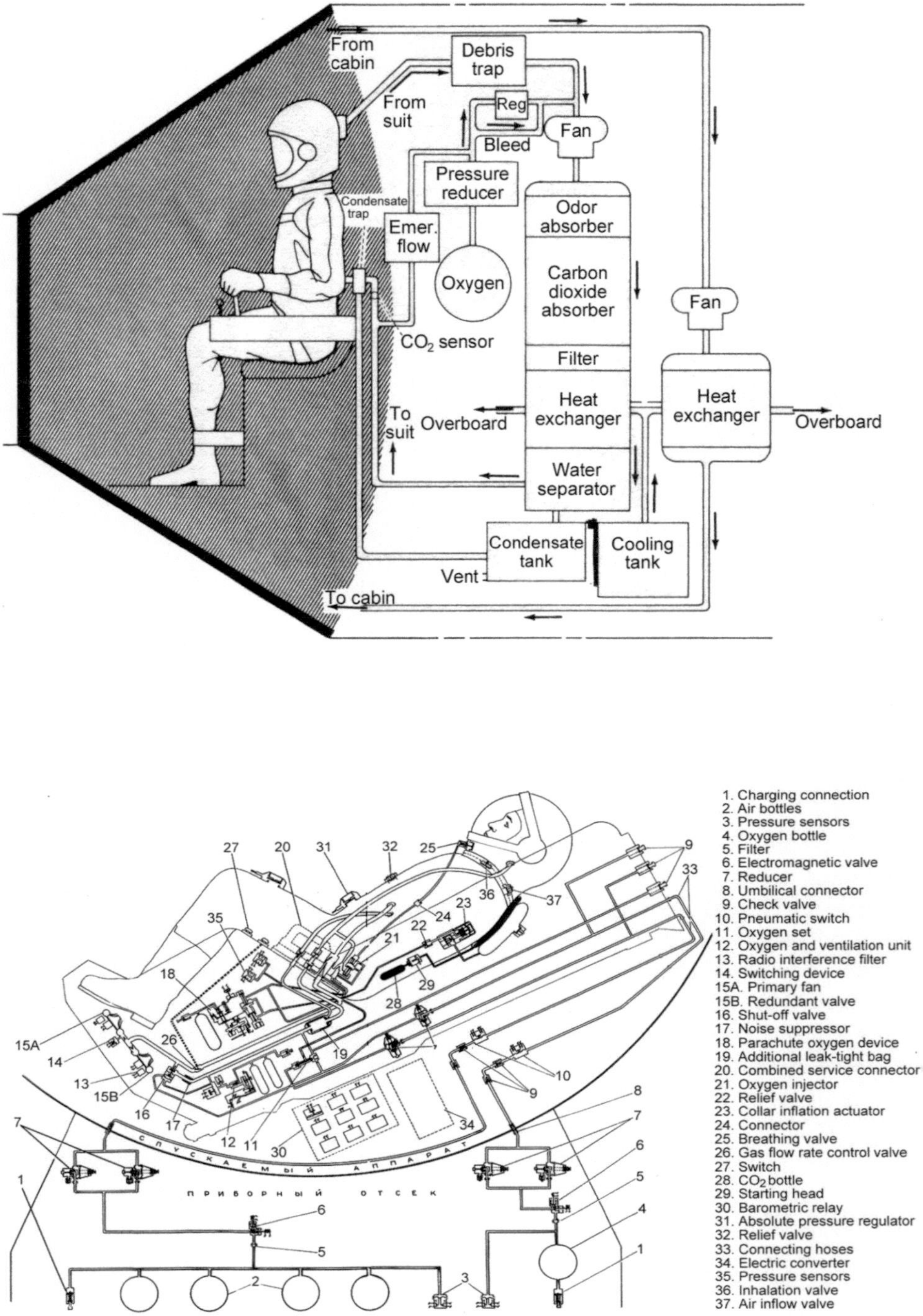

Fig. 5.16. The U.S. Project Mercury LSS (top) and U.S.S.R. Vostok LSS (bottom), both from 1961 (source: NASA and Zvezda).

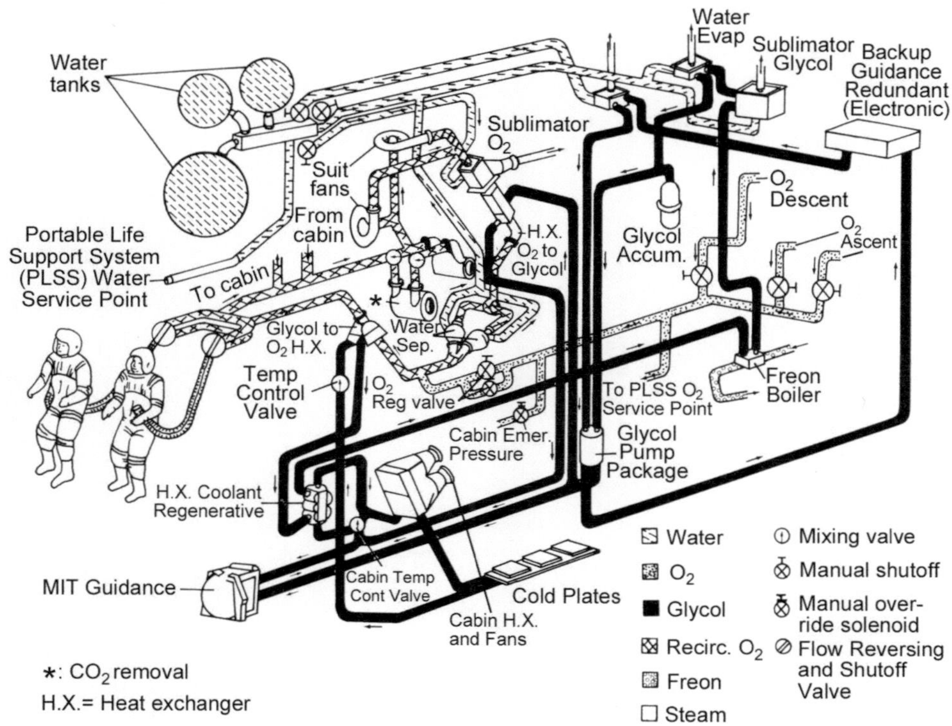

Fig. 5.17. Apollo LEM life support system and thermal cooling system (source: NASA).

system. All of these spacecraft had autonomous open-loop life support systems even when operated in a docked mode like Skylab and the Apollo spacecraft.

In the early 1980s the first space vehicle to have a partial life support system was introduced with the European Spacelab, designed to fly in the Shuttle cargo bay as an attached microgravity laboratory for life sciences and material sciences research. The Shuttle was designed for all habitable functions for the entire crew, and so Spacelab had only to provide a breathable atmosphere for the crew whilst they were in the module, plus cooling for the laboratory equipment (Fig. 5.18). The module was closed during launch and reentry, and in case of a failure or fire. Thus the atmosphere supply had to be partly autonomous with a separate nitrogen supply to make up for any leakage or a repressurization of the module, but the oxygen was provided from the Shuttle because (with the exception of module repressurization) the same amount of oxygen would be needed by the crew irrespective of where they were on board. Atmosphere revitalization was of the same design as on board the Shuttle, using an identical CO_2 removal system with LiOH cartridges for common use. Any excess of condensate water would be dumped overboard. A separate loop for the cooling of avionics and experiment racks was used in order to minimize the risk of potential contamination in the event of an experiment equipment failure. Spacelab also had its own fire detection and Halon suppression system.

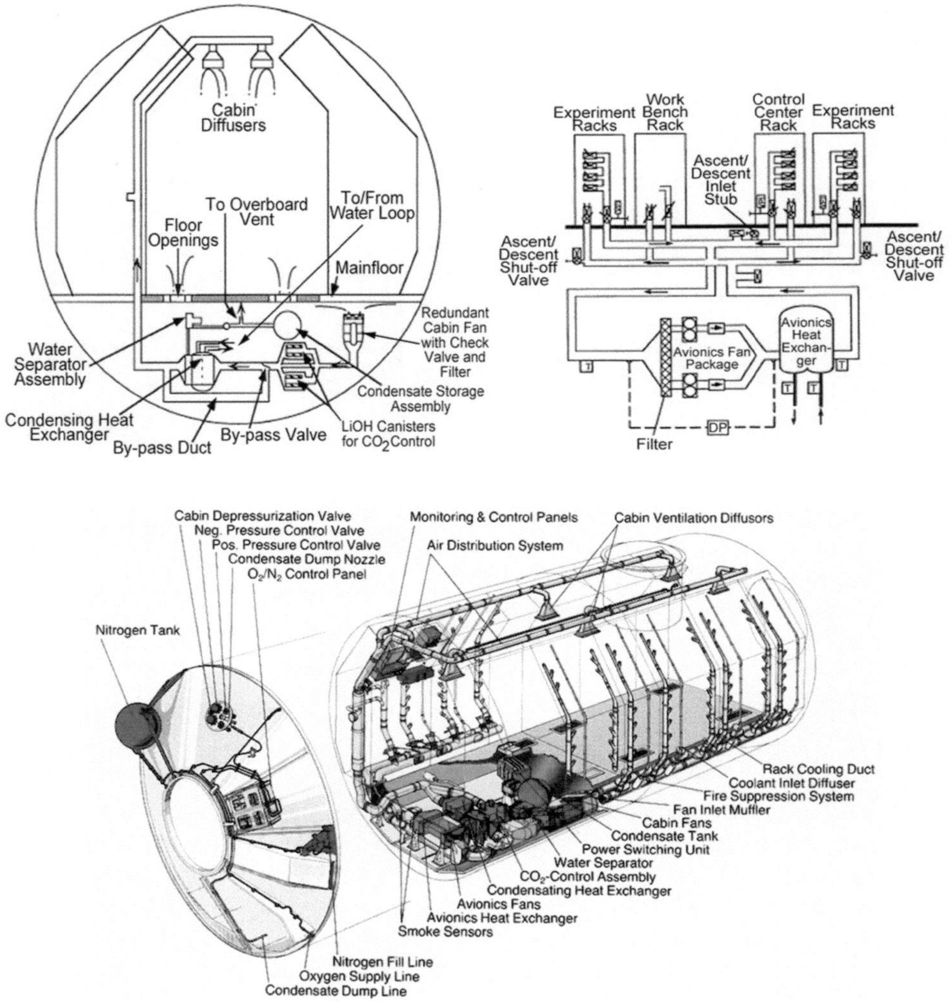

Fig. 5.18. Spacelab life support system, cabin air loop (upper left), and avionics loop (upper right) (source: Åke Ingemar Skoog).

These concepts for non-regenerative life support systems were basically used in all spacecraft until the mid-1980s, except for the Salyut 6 and 7 space stations in which condensate was recovered as potable water. The first real long-term spacecraft to come into operation was the Soviet–Russian Mir space station, whose assembly in orbit began in 1986. Mir was designed for an average crew of three persons over an operation life of 5 years, but was operated for fully 15 years with many repairs and improvements. Mir had a fully regenerative closed-loop physico-chemical life support system (Fig. 5.19), so only food and a resupply of water and nitrogen to make up for losses were provided by the Progress spacecraft every six months.

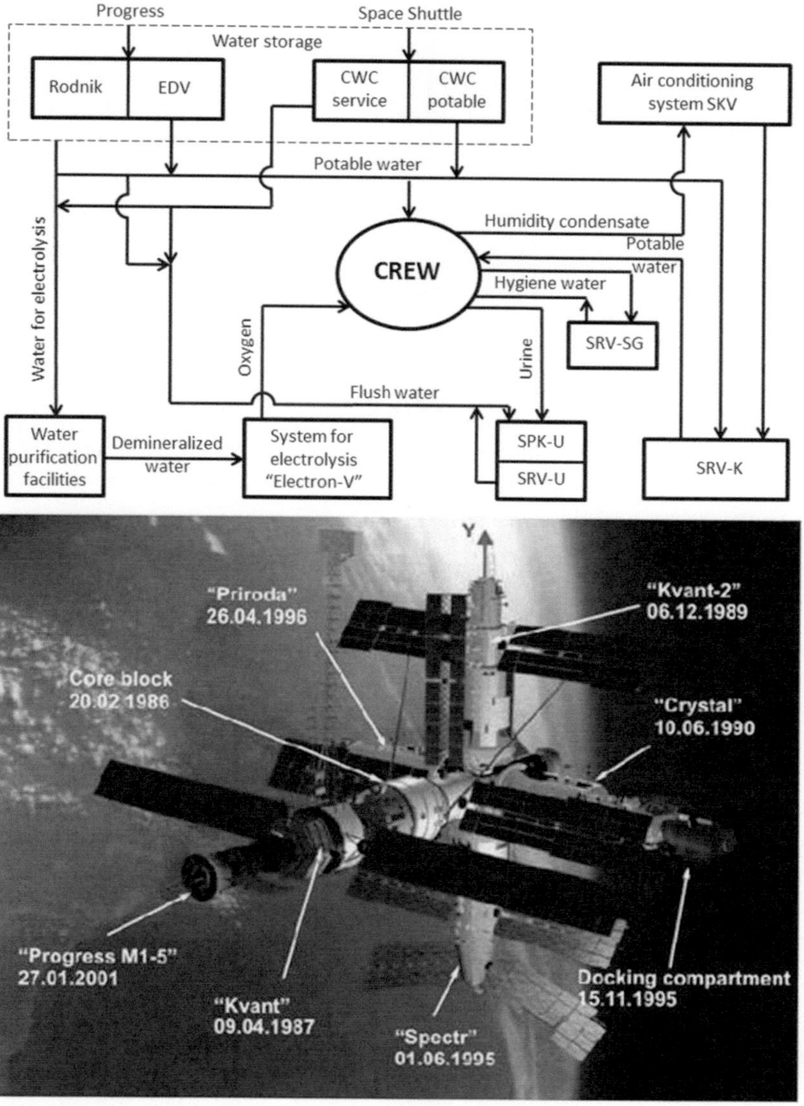

Fig. 5.19. Mir life support system water loops and oxygen generation system (source: NIICHIMMASH and TSuP).

Mir had three separate water loops. Condensate was filtered and used as potable water (SRV-K), hygiene water was recovered and after multifiltration (SRV-SG) was reused for that purpose, and urine was subjected to vapor diffusion distillation and filtration (SPK-U and SRV-U) and supplied to the "Elektron V" to generate oxygen by electrolysis. The system was not fully closed, as oxygen and nitrogen were lost due to leakage and airlock operations, oxygen was lost by venting CO_2 overboard, and there were water losses due to the water content in human solid waste and urine brine. Thus the water balance had to be maintained by resupply of potable water in plastic bags delivered by the Progress to the "Rodnik" storage system, and later also by the Shuttle to CWC storage. The CO_2 removed by the "Vozdukh" molecular sieve system was vacuum-desorbed overboard.

5.5.2 Life support systems of recent space vehicles

Rescue suits

The smallest life support system is that for the rescue suits (Fig. 5.20) where the suit, when attached to the spacecraft life support system by an umbilical, is supplied with oxygen and cooling for periods of up to several days. In the autonomous emergency mode the suit is only functional for some 30 minutes by means of an oxygen supply.

Space suits

The smallest fully autonomous life support systems are those of the U.S. EMU (Extravehicular Mobility Unit) and Russian Orlan space suits, which can be considered as equivalent to small spacecraft. These are closed-loop physicochemical life support systems with basically only CO_2 removal in a regenerative functional mode. It is essential that suit life support system functions be taken into account in designing a closed-loop regenerative life support system for a spacecraft because, depending on the selected suit concept, the influence on mass balance can be considerable over a long period of operation; for example, due to loss of water in the suit's cooling sublimation process or oxygen loss if CO_2 is not recovered by the suit system but instead vented to space. (For more details on the space suit life support system see Chapter 6 on space suits.)

Russian Soyuz TMA/TMA-M

The oldest operational spacecraft is the Soyuz, which was developed as the manned transportation system in the Soviet/Russian space program. It was introduced in the mid-1960s and was designed for a crew of up to three persons and missions lasting up to 18 days. It is now in its fifth generation with over 100 flights and serves as the crew taxi for the ISS. The life support design is non-regenerative (Fig. 5.21), and basically unchanged since the early flights. In many functions it uses the same concepts as the Vostok and Voskhod spacecraft. The design is based on a hermetically sealed cabin with a minor storage of compressed air to make up for leakage. Metabolic oxygen is produced by water in reaction with potassium superoxide (KO_2) and the reaction product (KOH) is used for CO_2 removal. Additional LiOH cartridges remove some 20% of the CO_2. Potable

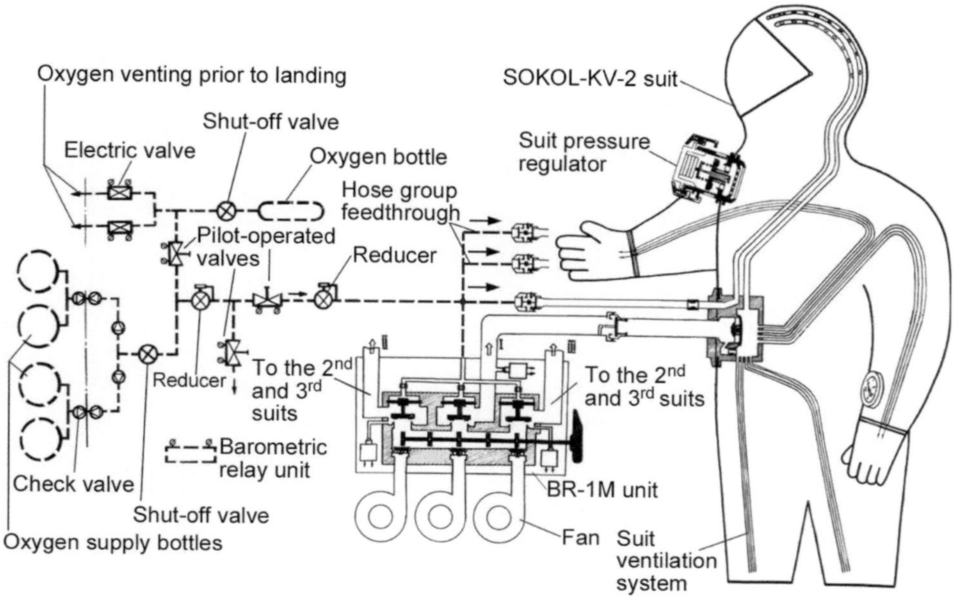

Fig. 5.20. The Sokol rescue suit (source: Zvezda).

water is stored for crew use, and there is a separate collection system for human waste products.

U.S. Space Shuttle

The Shuttle was the second oldest space transportation system in use until July 2011, when it was retired after 135 flights. The basic design of the life support system was from the 1970s. The atmosphere was a normal sea level one with revitalization by a non-regenerative process using LiOH cartridges for CO_2 removal, and a condensing heat exchanger for temperature and humidity control (Fig. 5.22). A separate loop was used to cool avionics equipment. With a nominal crew of seven persons and a mission duration of 7 to 10 days, water supply, food management, and hygiene facilities were required. All water needed for drinking and food preparation as well as some personal hygiene was of potable quality and a by-product of a fuel cells system similar to that of the Apollo spacecraft. On missions to Mir and the ISS, excess water from the fuel cell system was transferred to the station prior to undocking. Waste water and human waste were stored on board and returned to Earth.

International Space Station

The ISS comprises U.S., Russian, European, and Japanese habitation and laboratory modules and trusses assembled in orbit over a period of a decade. It is designed for a permanent crew of up to six persons, serving tours of roughly 90

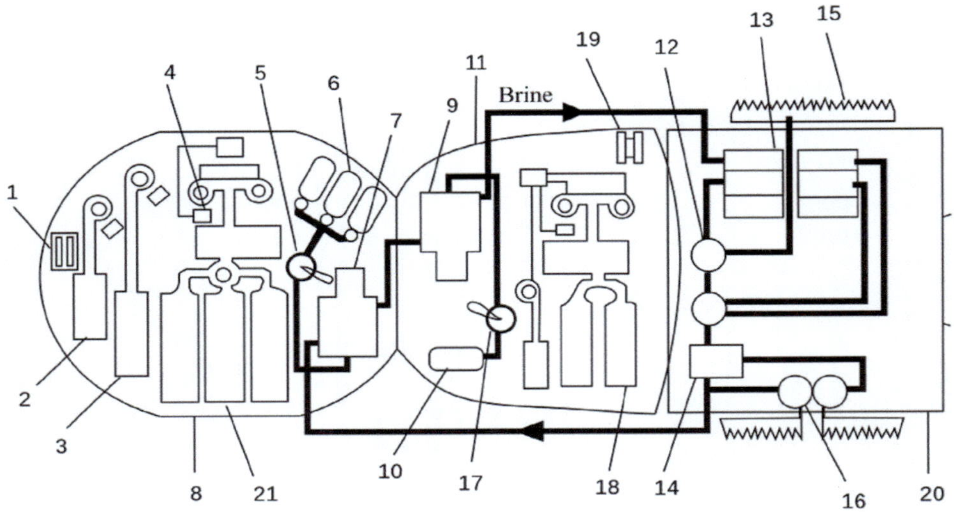

(1) Compressed air for leakage makeup
(2) LiOH for topping CO_2 removal
(3) KO_2 oxygen supply and primary CO_2 removal beds
(4) Flow meter and fans
(5) Manual pump
(6) H_2O storage tanks
(7) Condensing heat exchanger with wick-type H_2O separator
(8) Flight Module
(9) Condensing heat exchanger with wick-type H_2O separator
(10) H_2O storage tank
(11) Landing Module
(12) Temperature control valves
(13) Equipment cooler (primary and topping)
(14) Primary heat exchanger
(15) Primary space radiator
(16) Sequencing space radiators
(17) Manual pump
(18) Trace contaminant control bed
(19) Pressure relief valve
(20) Equipment module
(21) KO_2 beds for oxygen supply and trace contaminant removal with activated charcoal and bacteria filter

Fig. 5.21. Russian Soyuz TMA spacecraft (source: RSC Energia).

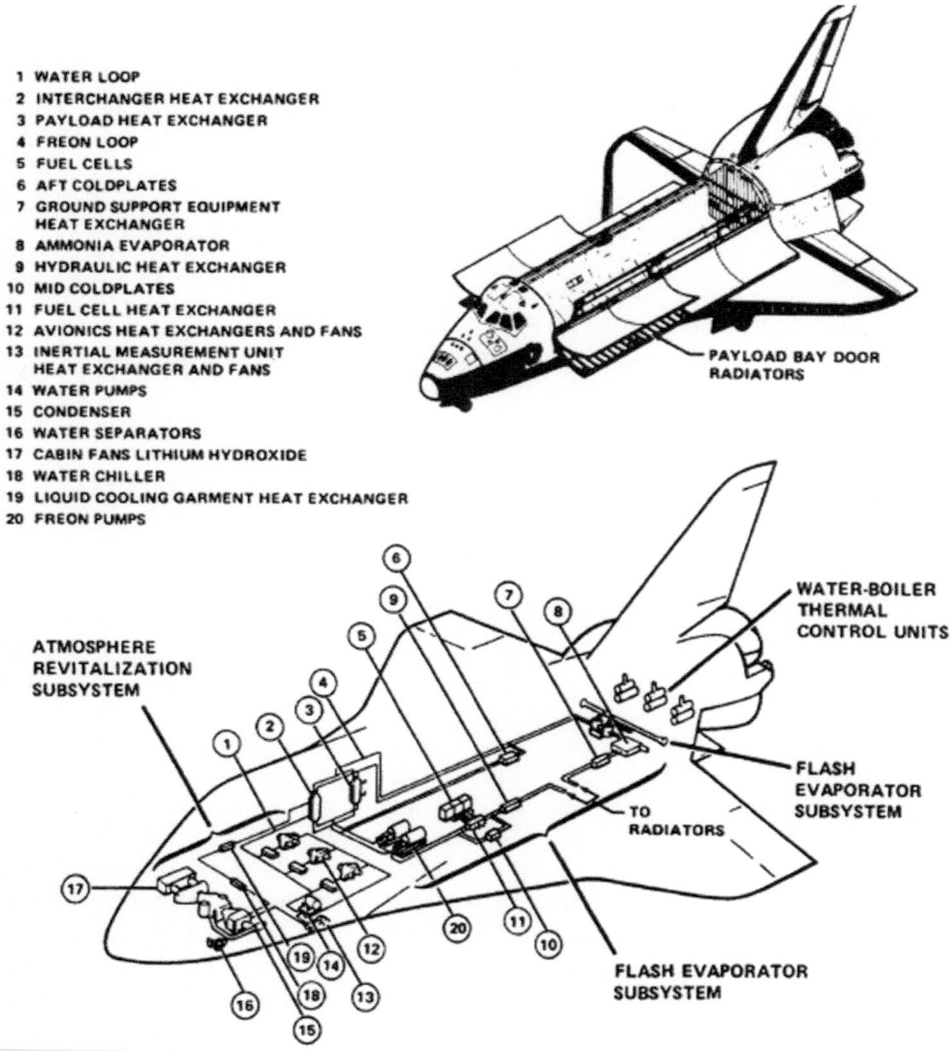

Fig. 5.22. Space Shuttle life support and thermal control systems (source: NASA Space Shuttle News Reference).

days. The final layout of the partially closed regenerative physicochemical life support system is the logical result of several space station concept changes and political redirection by including the Russian Federation as a partner in an ongoing program. From a partly distributed system in the early original U.S.–European–Japanese configurations, the design evolved to a more centralized life support system in the U.S. section, with only temperature and humidity control, air circulation, and fire detection and suppression equipment left in most of the individual laboratory and working space modules.

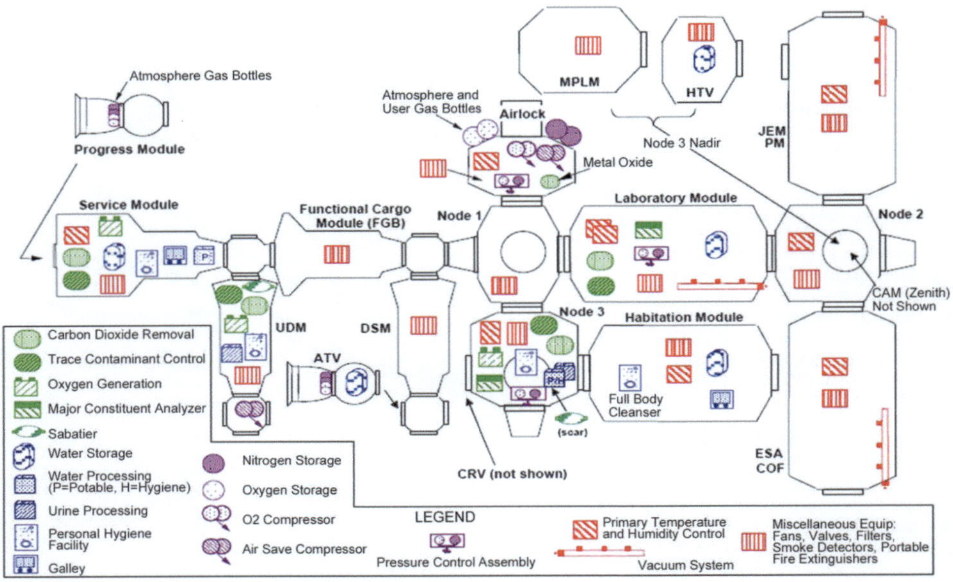

Fig. 5.23. The ISS life support system distribution (source: NASA).

The political decision to include Russia as a partner and to start the assembly with Russian modules led to an overall life support configuration having two complete life support systems, each one in itself redundant, the Russian ones in the Zvezda Service Module and the Universal Docking Module for the initial manning of the station with a crew of three (Fig. 5.23). These modules were originally planned for Mir 2, so their life support functions are the same as in the Mir space station and were built into the modules from the very beginning (Fig. 5.24).

The major life support functions for a crew of six in the U.S. section of the ISS are in the Destiny Laboratory and Node 3 Tranquility modules. This capability was built up step by step, with the Oxygen Generation System (680 kg) being launched by the Shuttle in 2006 and the Water Recovery System (1,500 kg) in 2008, facilitating a full crew of six by the end of 2008 (Fig. 5.25). The completion of the system to a partially closed-loop physicochemical life support system will come with the change from vacuum-desorbed molecular sieves in which the removed CO_2 is vented overboard, to a Sabatier reactor that recovers metabolic oxygen. To generate 5.4 kg of oxygen per day, a solid polymer electrolysis concept was selected. The Water Recovery System produces 130 kg of potable water per day by the low-pressure vacuum distillation of urine and the multifiltration and high-temperature catalytic oxidation of hygiene and condensate water. So, for the one-time transportation mass penalty of 2,200 kg these processes reduce the annual resupply to ISS by some 10,000 kg.

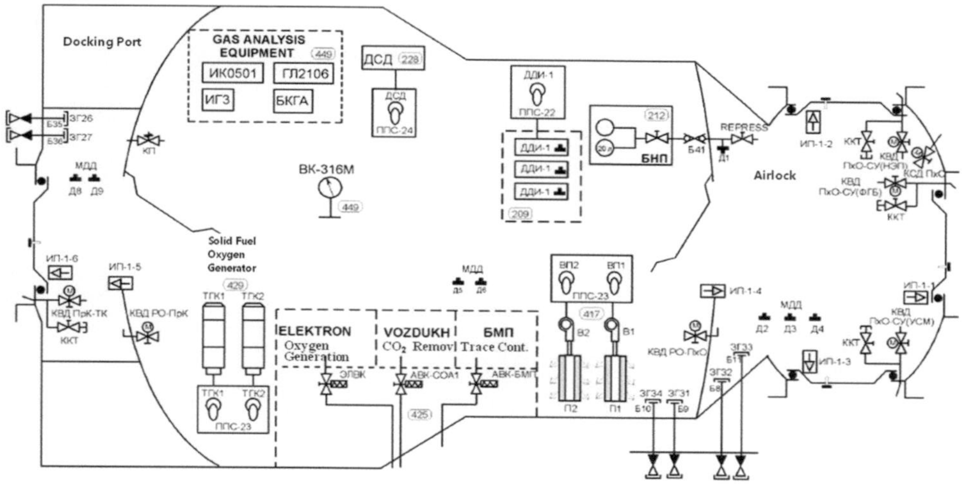

Fig. 5.24. Atmosphere control and revitalization system for Russian Zvezda Service Module on ISS (source: RSC Energia).

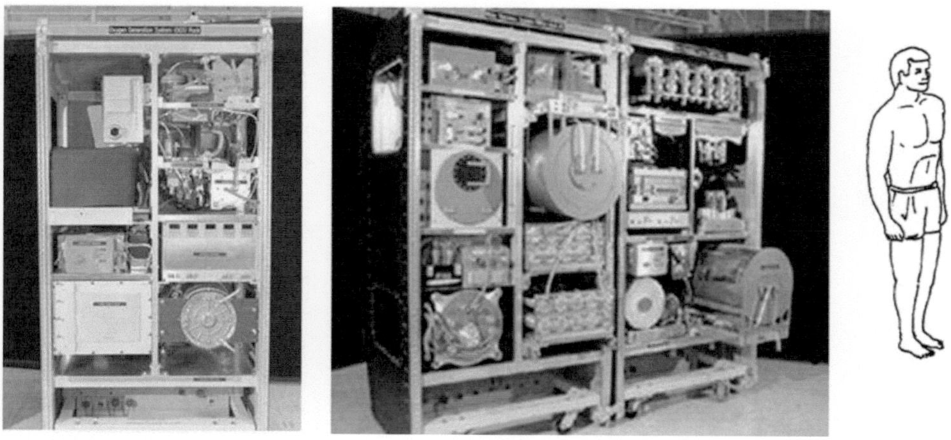

Fig. 5.25. ISS U.S. Oxygen Generation System (left) and Water Recovery System (right) (source: NASA).

European attached module Columbus

The European Columbus laboratory is a typical example of a module in a centralized life support system configuration. Only air circulation and temperature and humidity control are provided locally in the module (Fig. 5.26). Condensate and air for CO_2 removal are transported to the core modules of the station for treatment. Fresh air is supplied by the core modules.

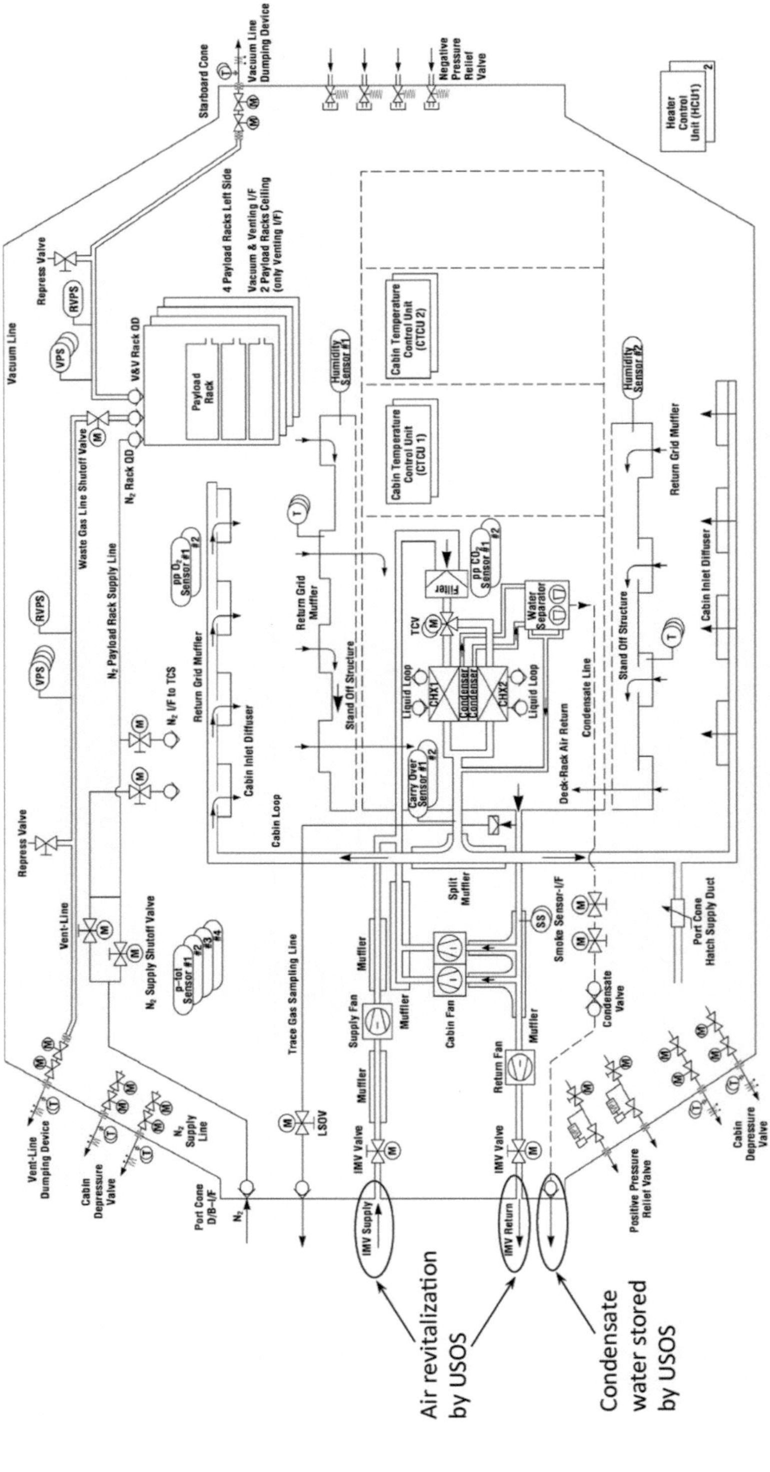

Fig. 5.26. European Columbus life support system (source: EADS Astrium).

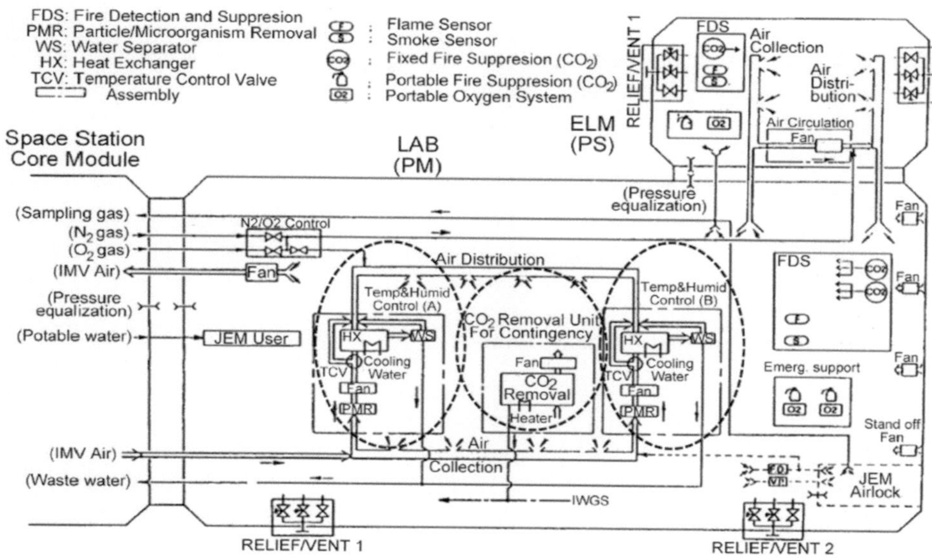

Fig. 5.27. JEM Kibo life support system (source: JAXA).

Japanese Experiment Module Kibo

The life support system of the Japanese Experiment Module (JEM) Kibo is almost the same as for Columbus, but it contains an additional contingency CO_2 removal unit and a separate pressure control panel for operations involving the attached logistics module and airlock (Fig. 5.27).

Resupply vehicles

Three autonomous unmanned resupply vehicles service the ISS: the Russian Progress spacecraft, the European ATV, and the Japanese HTV. All three provide a standard atmosphere for the cargo, so the life support functions are limited to pressure control using compressed air, air circulation by fans, air temperature monitoring, and fire detection and suppression. When docked, air is circulated from the station. In addition to spares and experiments, all three types of vehicle deliver water, nitrogen, and oxygen to the station life support system, and are loaded with station waste prior to undocking for destruction by burning up in the Earth's atmosphere.

5.6 CONTROLLED ECOLOGICAL LIFE SUPPORT SYSTEMS

Thus far, manned space systems have been designed for short (<10 days) to medium-term operations (exchanging crews every 3 to 6 months) for a small crew of three to seven persons in Earth orbit and short visits to the lunar surface.

Future missions will require systems to support long-term operations (1 to 2+ years) with limited (Moon) or hardly any (Mars) resupply possibilities. These new mission scenarios imply long-term isolation, changing gravity environments ($1\,g \rightarrow 0\,g \rightarrow 1/6\,g$ or $1/3\,g \rightarrow 0\,g \rightarrow 1\,g$), a very hostile environment, and a difficult or impossible rescue scenario.

To master future requirements for life support systems on long-duration missions, the food loop must be partially closed, taking mass and energy constraints and human well-being into consideration.

5.6.1 CELSS background and constraints

The relative resupply mass for long-term missions can be reduced by some 80–90%, in comparison with a completely open-loop system, with optimized regenerative physicochemical life support functions for air revitalization, oxygen production, and water recovery. The remaining 10–20% is mainly the food supply and inevitable losses due to leakage, human waste, and residuals in the regenerative waste and air processing. The major part, some 3 kg per person per day for food, represents a major resupply weight for large crews on long missions, so the potential for producing food in space must be considered. It is of psychological importance for the well-being of the crew on long missions that they be able to produce some fresh food in order to supplement frozen and canned foods.

Producing a portion of the food supply on board a spacecraft or a base on another celestial body will require an ecological life support system capable of hosting live plants, animals, and other species, and covering all relevant functions of a life support system—air revitalization, water recovery, food production, and waste management. Fortunately, there is sufficient sunlight available in space to drive photosynthesis in such a system. This concept has been under investigation since the late 1960s in a number of high-level studies (Fig. 5.28).

The options for CELSS include cultivation of higher plants, algae, aquacultures, microbiological cultures, and animals with an increasing degree of complexity in the implementation. The advantages are clear: a single system will generate oxygen and food from carbon dioxide, whilst simultaneously creating clean water. The drawbacks are high mass and power consumption, high degree of control complexity, and a very limited buffer capacity for small-scale systems. The design of a CELSS must take into account the fact that the average human intake of food is ~0.64 kg/day of dry mass, consisting of 12–15% protein, 50–55% carbohydrates, 30–35% lipids, and additional minerals, vitamins, and trace elements (NASA JSC 38571).

Higher plants

Higher plants are of major interest, both from a nutritional and psychological point of view. Plants such as wheat, rice, and potatoes have a high caloric density and are rich in carbohydrates and proteins. A high harvest index (i.e., the edible portion of the biomass) is also of interest, especially in combination with a

$$6CO_2 + 6H_2O \longrightarrow C_6H_{12}O_6 + 6O_2$$

Fig. 5.28. Photosynthesis and the NASA Space Settlement Design Study 1977 (source: NASA SP-413).

possible consumption in raw form. Plants such as strawberries, lettuce, and broccoli contain several important vitamins and have a very high psychological value in daily menus.

The cultivation of higher plants could close the carbon loop and at the same time provide the necessary air revitalization and water regeneration, but each one imposes a certain size requirement on the system

- required area for food production: 15–20 m^2/person/day
- required area for oxygen production: 6–10 m^2/person/day
- required area for water regeneration: ~5 m^2/person/day

whereby a 50% closure of the food loop would be sufficient to close the oxygen and water regeneration loops. Taking crop growing, yield, and safety aspects (buffer capacity) into consideration, a higher plant system would most likely require an estimated cultivation area of ~100 m^2 per person per day, with the corresponding power demand. The light needed for photosynthesis would require some 125 m^2 of solar arrays for the overall food production for one person per day at Earth's distance from the Sun; they would increase substantially for a mission to Mars.

The microgravity environment poses a number of questions for the cultivation of higher plants. Cultivation parameters like light characteristics (wavelength, source of artificial light, duration/cycle), temperature optimization and means of fertilization in microgravity, in addition to methods for cultivation (matrix, soil, or hydroponics) and water supply must be developed in parallel with basic research on plant growth under microgravity conditions (Fig. 5.29).

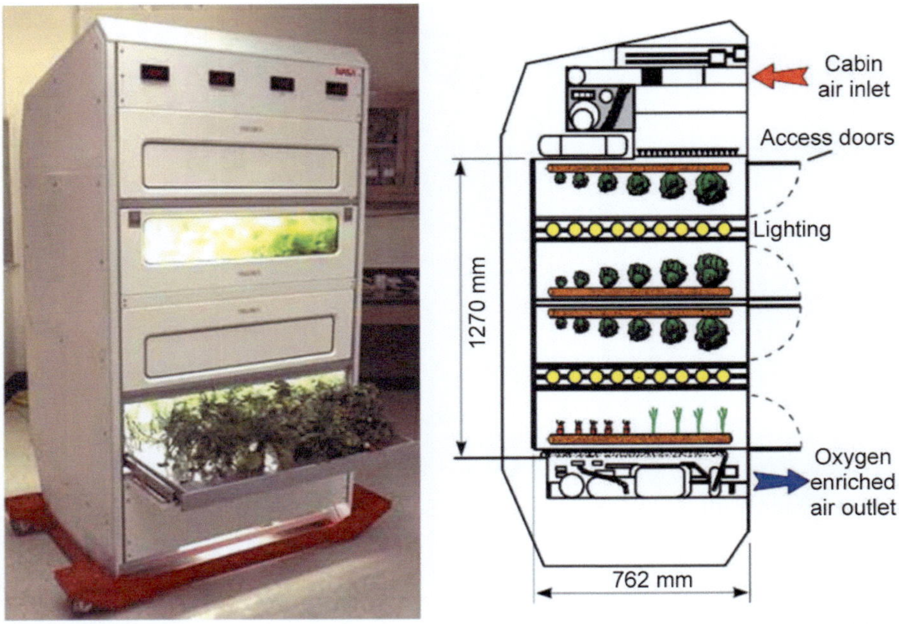

Fig. 5.29. The NASA Ames Salad Machine for testing of higher plant growth (source: NASA).

Algae

Experiments to use algae for food production in space were started in the 1960s in the U.S.S.R., U.S.A., and Japan. Algae are fast growing (a few days) with a controllable metabolism process. Owing to their rapid dynamic response in case of system failure or shutdown algae are an excellent biological life support alternative but they have a low carbohydrate and fat content, and require a well-controlled nutrient medium and controlled light intensity. Another advantage is the low volume for oxygen recovery from CO_2: 0.1 to $0.5\,m^3$ ($=0.5\,kg$ of biomass) per kg of CO_2 conversion. A drawback is their lower acceptability as a food element in many societies. Potential species are green algae such as *Scenedesmus*, *Chlamidomonas*, and *Chlorella*, and blue-green algae such as *Anacystis* and *Spirulina*. Fig. 5.30 shows a sample of *Spirulina* gnocchi developed by Ducasse Education (Training & Consulting division of Alain Ducasse Enterprise), whose chefs worked together with ESA to create recipes adapted for missions to Mars.

Aqua cultures

Systems with a combination of algae and water animals (e.g., fish and/or snails) have a higher nutrient value than a pure plant or algae system, and are controllable with a quick to medium dynamic reaction time. Facilities for terrestrial and in-orbit testing include the German Closed Equilibrated Biological Aquatic System (CEBAS) and the Japanese Aquatic Habitat.

Fig. 5.30. *Spirulina* gnocchi (source: Pierre Desgrieux for Ducasse Education).

Microbiological systems

Microbial processing in a bioreactor is a simple controllable process, but the potential risk of microbial contamination has to be considered. A further problem is that the biomass is not directly consumable. Nevertheless, a microbiological system provides a relatively simple alternative for human and plant waste processing.

Animal systems

Animals are the natural source of proteins as standard food, but breeding them is a very slow process and difficult to control. Since the breeding of animals also implies cultivation of food in space for these species and associated waste treatment, such systems become extremely large and require an elaborate food preparation process as well, unless aquatic animals are used. The tendency today is to select animals having a small mass and short life cycle, mainly aquacultures. Systems with larger animals are presently not under consideration for space applications because of their extreme complexity.

Waste management

The collection of human and plant waste in an advanced CELSS provides 1 to 1.5 kg per person per day of biodegradable waste that has to be processed in space. Possible waste treatment and stabilization processes are

- incineration (simple and efficient—safety concerns)
- biological composting/bioreactors (recovery of nutrients—simple but slow and high volume—safety concerns)
- supercritical wet oxidation (high temperature and high pressure with oxygen—safety concerns—oxidation products are N_2, H_2, and CO_2 which have potential use for life support resupply in general).

System complexity

Clearly a CELSS on a space station or a base on the Moon or Mars is a very complex and control-sensitive system; the higher the degree of food loop closure, the greater the overall complexity. Complexity is also directly reflected in the time until the break-even point in terms of crew size, habitat location, and degree of closure for the food loop. Several studies over the last 30 years have all reached similar conclusions on the feasibility of CELSS, or a combination of controlled ecological and physicochemical life support systems. For a crew of four in Earth orbit with a 50% food loop closure, the inclusion of CELSS functions is only feasible for durations greater than 5 to 7 years, and even with a 97% food loop closure the system becomes advantageous compared with the 50% closure only after some 11 to 12 years (Fig. 5.31). The break-even point can be achieved in a shorter time only with crews greater than about 1,000 persons. For a lunar base, the break-even point is 6 to 7 years for a crew of the order of 20 persons or more.

In situ resource utilization

For space bases on the Moon and Mars it is likely that some in situ resources can be used for advanced life support systems, thus reducing any residual resupply mass. On the Moon, oxygen is present in iron-rich surface minerals and glasses. This could be extracted simply by heating the material to temperatures above 900°C and exposing it to hydrogen gas

$$FeO + H_2 \rightarrow Fe + H_2O$$

and this process was made much more practical by the discovery in the 1990s by the Clementine mission that there are significant amounts of hydrogen-enriched regolith in the vicinity of the lunar poles. Recent missions have increased the likelihood of there being frozen water mixed in with the regolith (see Chapter 4 on space exploration).

The atmosphere of Mars is 95% CO_2, and methods are available to produce water and/or oxygen, and carbon, using solar energy. It remains to be seen whether there is water available at or just below the surface.

Safe haven functions

The small size of ecological life support systems in comparison with Planet Earth makes them more difficult to control and more sensitive to contaminations, and they are also susceptible to total breakdown. In this case, a safety concept must

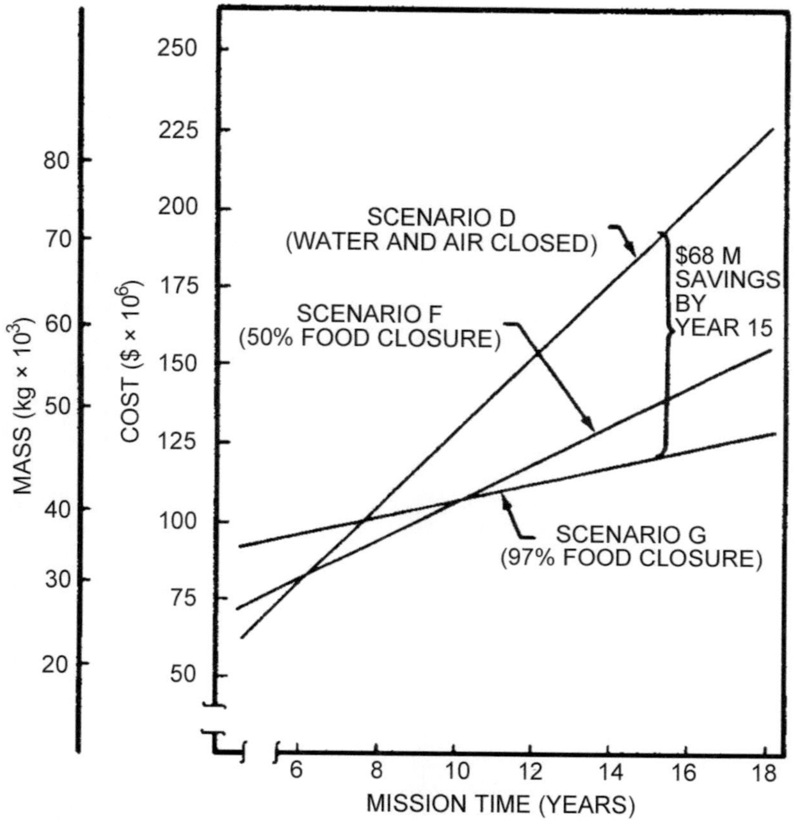

Fig. 5.31. Breakeven for different degrees of food loop closure compared with a physicochemical life support system (scenario D) for a crew of four in LEO (source: Boeing).

be included in the basic system design. A number of options and combinations of physicochemical and controlled ecological life support systems are possible

- Several separate modules of a CELSS can give buffer capacity for restarts.
- If an ecological life support system is providing a certain degree of oxygen and water recovery, this part of the system must be replicated by a physicochemical backup system. An alternative is to scale the physicochemical system (e.g., by modularity) for 100% of system requirements and then operate it at a fraction of its capability.
- With 50% food closure, the oxygen and water recovery requirements will be fulfilled, and the physicochemical system would be a true backup. Food storage (e.g., dry food) must be included.
- For defined emergency situations (e.g., 28 days in the case of the ISS) oxygen, water, and food corresponding to a reduced system performance level must be included, either by backup systems or by modularity.

5.6.2 Large-scale CELSS testing

Although a controlled ecological life support system has not yet been implemented as the basic life support system in a space project, and it will probably be 10 to 20 years before a partial CELSS is operational, several full-scale tests have been conducted in the U.S.S.R. and U.S.A. since the mid-1960s.

At the Institute of Biophysics in Krasnoyarsk, Siberia, a number of sealed manned experiments with ecosystems have been made since the mid-1960s. The first sealed manned experiment, *Bios 1*, was performed in 1965 when *Chlorella* algae were used to recycle air exhaled by humans in a closed facility of $5\,m^3$. The algae absorbed the exhaled carbon dioxide, replenished the air with oxygen, and provided about 10% of the food supply. In the *Bios 2* series of tests in 1968–1969, a single person occupied a $15\,m^2$ chamber for a few months with single-cell algae and higher plants. Up to 85% of the food was produced, but physicochemical components required balancing the total system. For the *Bios 3* experiments (1974–1985) the test facility was increased to $315\,m^3$ and accommodated two and three persons for a maximum of 180 days. This facility was divided into four equal parts. One had housing with individual cabins, a kitchen, a toilet, and a control room, equipment for food processing and repairs, and systems for additional purification of air and water (when necessary). The other three parts were for food plant growth (wheat, vegetables), other food plants, and cultures of *Chlorella*. Xenon lamps provided the lighting (400 kW). An overall system closure of 80–90% was achieved.

Biosphere 2 was built in 1986–1989 in Oracle, Arizona, as an airlock-sealed large-scale habitat. The project was defined as an ecosystem and miniature of the Earth's biotope with rain forest, savannah, ocean, desert with 3,000 species on $13{,}000\,m^2$. It had living quarters for eight persons and an internal control center. The energy source was sunlight. In Mission 1, the facility was closed and sealed, and the crew lived inside for two years from 1991 to 1993. A second mission in 1994 lasted 6 months. The diet was nutritious and diverse, utilizing over 80 crops together with goat's milk, eggs, and some animal meat. The average daily intake of about 2,200 calories included 70 grams of protein and 32 grams of fat. The project provided experience and data, but failed to find a dynamic steady state.

The NASA Ames Research Center Crop Growth Chamber is about $12\,m^2$. Its energy requirement is about 15 kW, of which 13 kW provides the light needed for plant growth. It takes about two weeks to grow lettuce from seed to harvesting. At present all kinds of food are grown—including strawberries, chives, passion fruit, potatoes, some herbs, spinach, and tea—with varying degrees of success. The present system provides complete nutritional needs for one person (a vegetarian). The goal is to grow plants which produce at least 50% edible biomass with the remainder of the biomass being recycled to provide plant nutrients.

In 1986 NASA KSC introduced the Biomass Production Chamber (BPC), which is a sealable large steel chamber approximately 3.5 m in diameter and 7.5 m tall, with two floors each having multiple racks and banks of lamps, ducting for

Fig. 5.32. NASA KSC Biomass Production Chamber (source: NASA).

airflow, and apparatus to control temperature and humidity. The total chamber plant-growing area is $20\,m^2$ (Fig. 5.32).

The NASA Johnson Space Center Lunar–Mars Life Support Test Program (LMLSTP) series of four-person, 90-day manned tests were conducted in 1995–

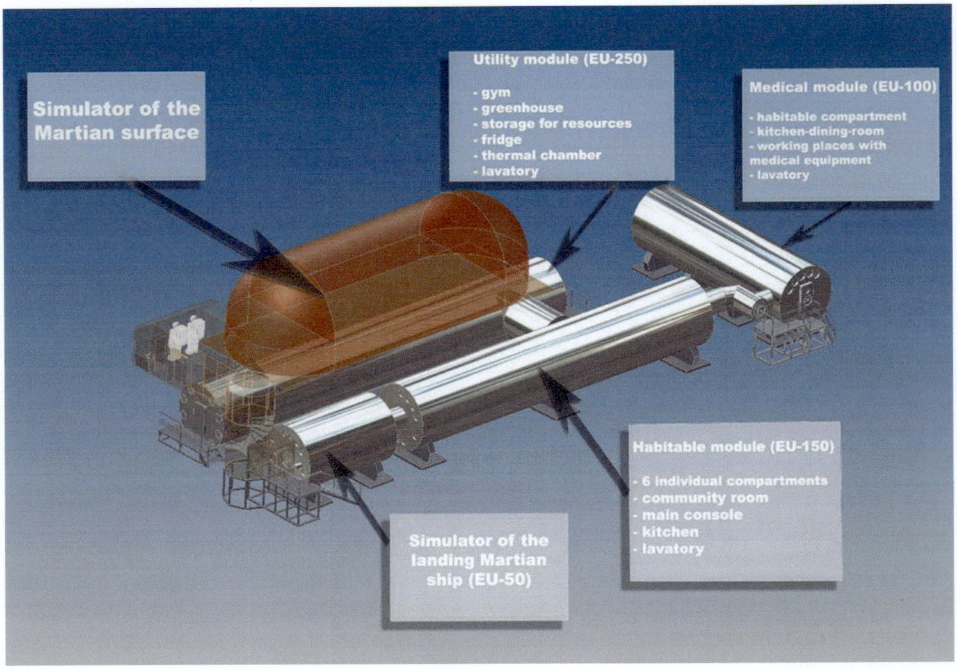

Fig. 5.33. The Mars 500 test facility at the Institute for Biomedical Problems: EU-50 Simulated Mars Lander, EU-100 Medical Module, EU-150 Habitat, EU-250 Utility Module with greenhouse, and on top of EU-250 the simulated Mars surface test area (source: IBMP).

1998, with a physicochemical regenerative air revitalization system and a biological water recovery system using the Integrated Life Support System Test Facility with a test chamber 6 m in diameter as the habitat and the Variable Pressure Growth Chamber for hydroponics cultures. Potable water was recovered from urine, hygiene water, and human condensate.

In June 2010 the Mars 500 international isolation experiment was started at the Institute for Biomedical Problems (IBMP) in Moscow to study the interaction of a crew of six during a simulated 520-day flight to Mars and back (see also Section 4.3.4). This provided data about the state of health and work capability of a crew spending a long time in total isolation inside an hermetically closed and confined volume during simulation of the main parameters of a Martian flight (duration, autonomy, the varying signal delay of communication with the Earth, and limited consumables). The facility consisted of three "flight" modules, a lander, and a simulated Martian surface for extravehicular activity (Fig. 5.33).

With a volume of 500 m^3 distributed over the three living, medical, and greenhouse modules, the Mars 500 facility (Fig. 5.34) contained

- living quarters, kitchen and dining room, lavatory, and gym
- working places with installed medical equipment

Fig. 5.34. Mars 500 living quarters, gym, private quarters, and greenhouse (images: IBMP/Oleg Voloshin).

- transfer tunnels with hatches connecting the modules
- life support systems, freezer for storage of food products
- storage with shelves for foods that do not require special conditions, and disposable plates and dishes, and clothes
- room for experimental greenhouse
- lock chamber for discharging waste
- life support systems.

5.6.3 CELSS flight experiments

In the last 40 years many smaller plant growth chambers with higher plants, algae, and sometimes fish have been tested both on Earth and in space, in particular in short trials on board the Shuttle and longer term trials on the Mir space station.

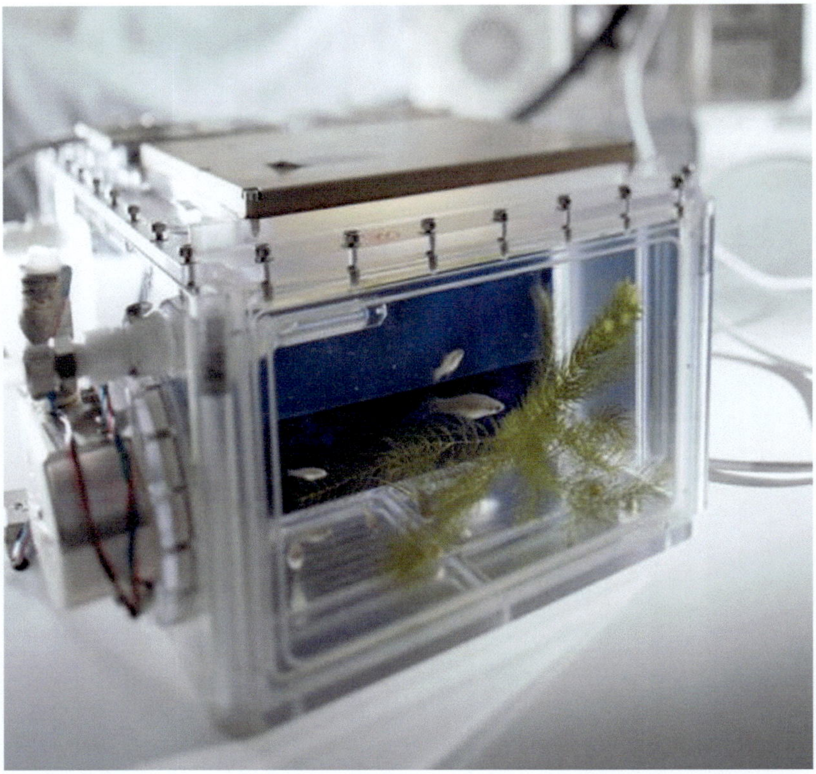

Fig. 5.35. CEBAS test chamber (image: OHB System AG).

The SVET plant-growing experiment is a joint undertaking in the framework of the Interkosmos program by the Space Research Institute in Bulgaria and the Institute for Biomedical Problems in Russia. It was used for the first time on Mir in 1990 for a period of 54 days, and used artificial lighting by fluorescent lamps to grow lettuce and radishes. Both vegetables grew much more slowly in space than on Earth, and did not reach their regular size as a food complement. The aim of the project is to develop biotechnology to grow higher plants in microgravity in support of future long-term manned missions.

CEBAS (Closed Equilibrated Biological Aquatic System) developed by the DLR, Ruhr University, and OHB Systems in Germany in cooperation with NASA (Fig. 5.35) consists of four chambers in a closed ecosystem. One chamber contains fish (*Xiphophorus helleri*) and water snails (*Biomphalaria glabrata*), the next chamber contains rootless water plants (*Ceratophyllum demersum*), another chamber has microorganisms, and finally there is a bacterial filter chamber for ammonia reduction. The fish and plants serve as food supply. A mini-module (8.6 liters) was flown three times on the Space Shuttle in 1998 and 2002, with the longest flight being six days. A larger module (150 liters) has been tested on Earth for periods of up to a year.

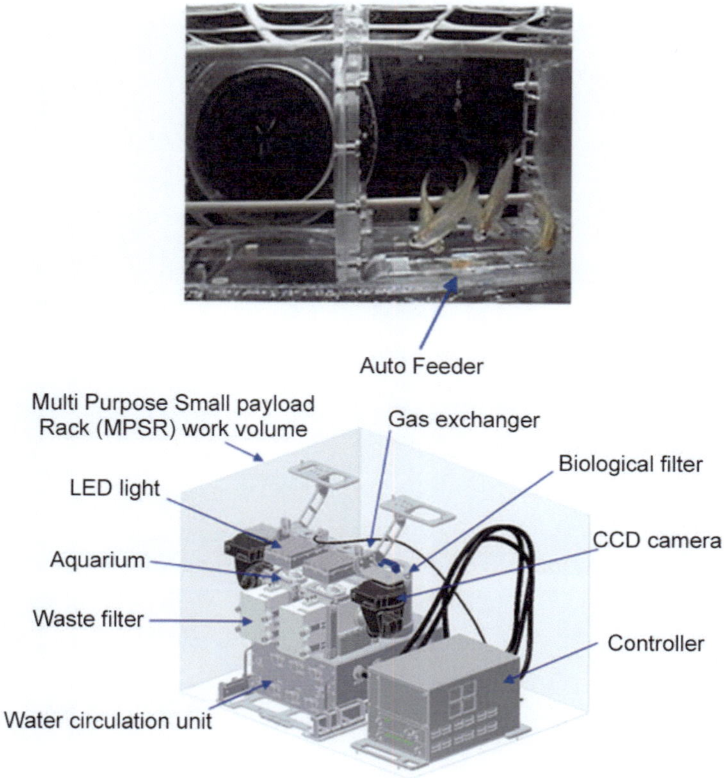

Fig. 5.36. Japanese Aquatic Habitat for ISS Kibo (source: JAXA).

The Japanese Aquatic Habitat (Fig. 5.36) is able to accommodate small freshwater fish, which have many advantages as vertebrate species. It is located in the Kibo module of the ISS and is for experiments that investigate how microgravity and space radiation affect living species, including humans, over the long term. The habitat comprises one closed water circulation system with two aquariums. An aquarium has an automatic feeding system, LED lighting for the day/night cycle, and CCD camera for observation, and its environment will be maintained by controlling the water flow rate, water temperature, dissolved gas exchange with the air, and biological/physical filtrations in order to run experiments lasting up to 90 days.

The Micro-Ecological Life Support System Alternative (MELiSSA) is a program started at ESA to study an artificial ecosystem that employs microorganisms to process waste into plant nutrients. Its recycling system has five separate but interconnected compartments. In three of them, waste will be progressively broken down by different fermentation processes. In the fourth compartment, either algae or plants will grow to produce food, oxygen, and water. The fifth "compartment" is where the consumers (i.e., the astronauts) will live on a real mission to Mars (Fig. 5.37). The different components were studied and built all

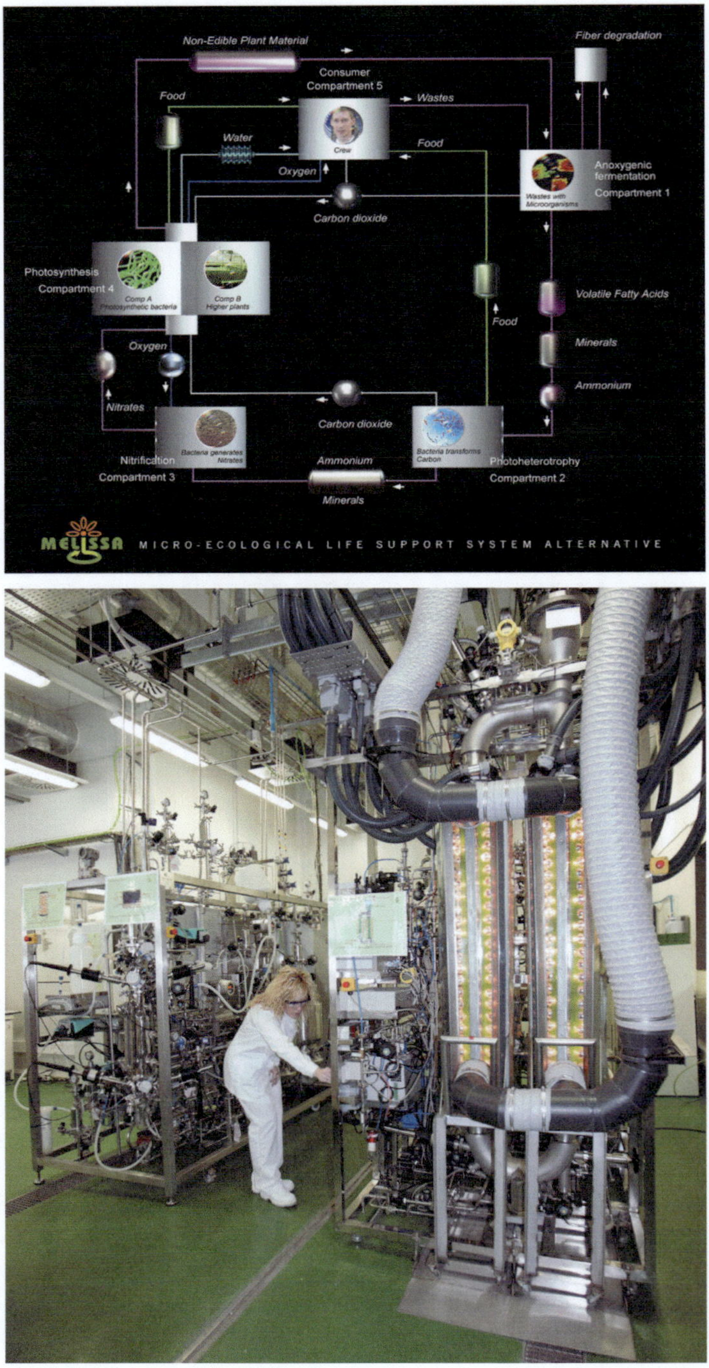

Fig. 5.37. The MELiSSA CELSS loop (top) and a photosynthesis compartment, 4A (bottom) (source: ESA/MELiSSA).

over Europe and Canada, and then in 2009 were assembled into a pilot facility in Barcelona designed to demonstrate the concept using animals as its "crew". One of the novelties of the MELiSSA approach is the compartmentalized structure of its closed-loop life support system. The aim is to be able to support a real human crew by 2020 to 2025.

5.7 FUTURE DEVELOPMENT

The technologies envisaged in the early 1980s for regenerative physicochemical life support systems have been developed, but their transformation into operating flight systems has been considerably delayed beyond their expected introduction. However, with the final completion of the ISS a fully regenerative, partially closed physicochemical life support system will be in operation for the first time. The pace and direction for renewed manned space exploration is currently vague, but the possible goals are defined and have in common an increased demand for crew comfort during still longer missions, which in many respects can only be implemented by means of improved life support systems. Given presently envisaged programs—Earth-orbiting stations, missions to the Moon and Mars—the life support systems in coming decades will be of the partially closed regenerative physicochemical type, but with a wider dynamically controlled performance range to satisfy the new requirements for crew comfort and human factors. In terms of crew comfort and well-being, next-generation systems will have to be complemented with CELSS for partial fresh food production. Controlled ecological life support systems with a larger degree of closure can only be envisaged for very much larger permanent bases, with expected corresponding long operational life.

5.8 REFERENCES/RECOMMENDED LITERATURE

Eckart, P.: *Spaceflight Life Support and Biospherics*; Springer Science and Business Media (1996).

Johnson, R.D. and Holbrow, C. (Eds.): *Space Settlements: A Design Study*, NASA SP-413, NASA (1977).

Sulzman, F.M. and Genin, A.M.: Life support and habitability. In: *Space Biology and Medicine, Volume 2*, American Institute of Aeronautics & Astronautics (1994).

Website

NASA: Space Settlements; spreading life throughout the solar system: *http://settlement.arc.nasa.gov/*

6

Space suits

Åke Ingemar Skoog

ABSTRACT

A spacefarer, a fragile human being, working outside a spacecraft in space or outside a habitat on the surface of another celestial body such as the Moon or Mars, needs a breathable atmosphere, protection against vacuum or very low pressures, adequate mechanical, thermal, and radiation protection, communications, a means to move from worksite to worksite, and a way to use diversified tools. To justify complex systems satisfying all these human-specific needs, it is desirable to exploit the advantages of a human being at work (i.e., his dexterity, tactility, vision, experience, and intelligence). This can only be accomplished by providing the spacefarer with a suit that is tailored to the mission concerned.

The first space suits 50 years ago fulfilled the single simple objective to provide a breathable atmosphere for an hour or two. In less than a decade, suits for use on the lunar surface had been developed and today spacewalkers annually spend hundreds of hours undertaking the construction and necessary maintenance of a space station. For a future mission to Mars, the challenging task is to design a suit with a mass less than half that of the suits used by the Apollo astronauts on the Moon.

This chapter gives a historical overview of past and present space suits, and the important accomplishments achieved, and analyzes the problems in designing future suits and the technologies required.

6.1 INTRODUCTION

The first cosmonauts and astronauts of the Vostok and Mercury programs used a suit designed to protect them against the decompression of their capsule—what we would today refer to as a "rescue suit". Soon, in both the U.S.S.R. and the

U.S.A., the initial developments of suits for activities outside a spacecraft were started. The first steps in 1965 were simply to prove that a human being could exist and move around outside a vehicle in the weightless vacuum of space. However, the real goal at that time was to provide suits appropriate to the lunar programs of the two nations, a goal which both had achieved by 1968–1969. The Apollo suits were the first, and indeed the only ones to be used operationally in space and on the Moon.

In the decades from the late 1970s until today the focus has been on suits for use on board the Space Shuttle and at the space stations Salyut, Mir, and the ISS in support of construction, maintenance, and safety-related tasks. During this period, over 3,700 hours of spacewalking (more accurately referred to as extra-vehicular activity, EVA) were accomplished. With the completion of the ISS and the decommissioning of the Space Shuttle in 2011, the design and development of a new generation of space suits for general use in space and on celestial bodies has been initiated.

6.2 WHAT IS A SPACE SUIT? BACKGROUND AND REQUIREMENTS

The world's first spacewalk was performed by Soviet cosmonaut Alexei Leonov from Voskhod 2 on March 18, 1965 and it lasted 12 minutes (Fig. 6.1). Its purpose was to test how a man could exit his spacecraft and move in open space. A similar test was performed by American astronaut Edward White from Gemini 4 on June 3, 1965, in this case lasting 36 minutes (Fig. 6.2). The first step taken on

Fig. 6.1. First EVA ever, Alexei A. Leonov, March 18, 1965 (source: Zvezda).

Fig. 6.2. First U.S. EVA, Edward White, June 3, 1965 (source: NASA).

another celestial body was by Neil Armstrong on July 20, 1969, during the Apollo 11 mission to the Moon (Fig. 6.3).

A number of spectacular EVAs have been performed over the past four decades, in particular the capture and retrieval of malfunctioning satellites (Fig. 6.4), the four repair missions to the Hubble Space Telescope (Fig. 6.5), the ongoing maintenance of Mir to extend its operational life by five years (Fig. 6.6), and the construction of the ISS (Fig. 6.7).

The standard operational use of space suits in the years since the Apollo missions has focused on the maintenance and repair of the Skylab, Salyut 6, Salyut 7, and Mir space stations and the construction and servicing of the ISS (Fig. 6.8). To date, more than 3,700 hours of EVA time has been accumulated over 670 individual spacewalks. The pace picked up with the advent of the ISS, whose assembly involved 1,500 hours of EVA by its completion in mid-2011, with another 400 hours devoted to continuous maintenance. Thus, today's human presence in space cannot be accomplished without EVA and the use of space suits.

Modern space suits provide an environment to enable their users to operate in an autonomous mode for 6 to 8 hours. A space suit must contain all the subsystems and carry out all the functions of a general spacecraft. Like any manned spacecraft, a suit must have the following subsystems: structural protection (a pressure-tight structure or enclosure providing an adequate pressure, passive thermal protection, mechanical and radiation protection), a life support and thermal control system, a power supply, communications, data management incorporating system monitoring and control, and a propulsion/maneuvering system. However, in a space suit system all these functions must have a very low mass

Fig. 6.3. First moon landing, Neil Armstrong, July 20, 1969 (source: NASA).

Fig. 6.4. STS-51A retrieval of the Palapa communication satellite by Joe Allen, November, 1984 (source: NASA).

Fig. 6.5. Repair of Hubble Telescope by Claude Nicollier during the STS-103 mission in December 1999 (source: NASA).

Fig. 6.6. Repairing Mir in 1998 (source: Zvezda).

Fig. 6.7. Christer Fuglesang participates in the construction of the ISS, December, 2006 (source: NASA).

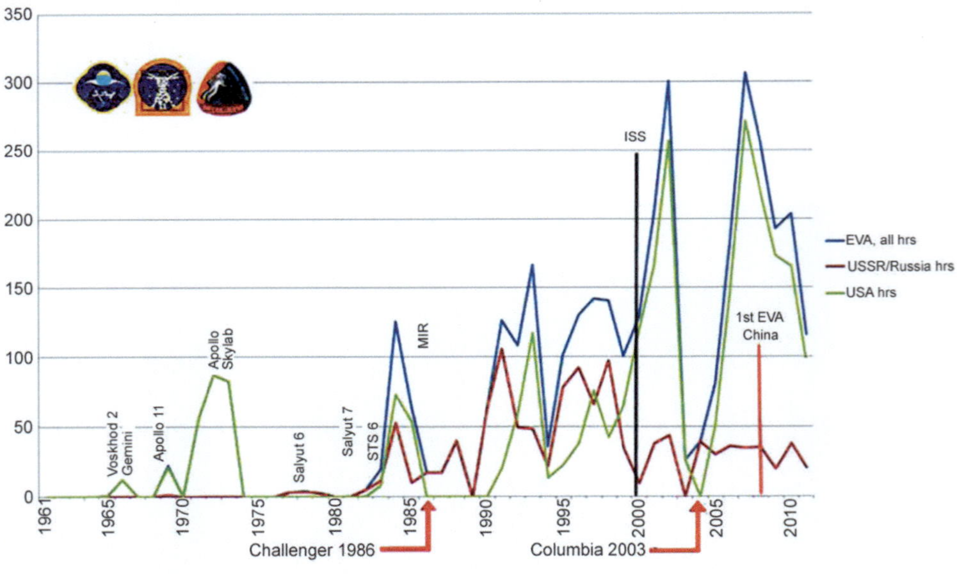

Fig. 6.8. Annual EVA hours until completion of ISS mid-2011 (source: Åke Ingemar Skoog).

and be contained in a very limited volume. Furthermore, the space suit system must provide these functions whilst allowing the user sufficient mobility, dexterity, and tactility to perform the required tasks in open space. All these general requirements make the design, development, and testing of any new space suit system a challenging task.

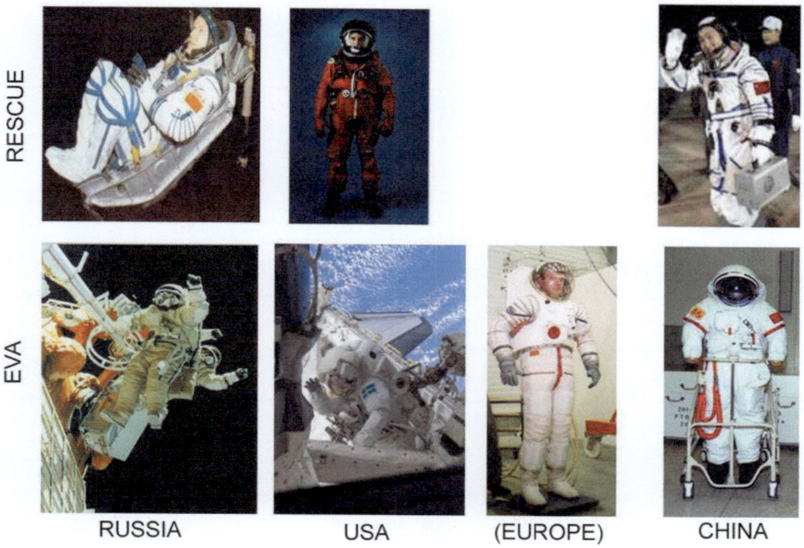

Fig. 6.9. Present operational space suit systems (sources: Zvezda, NASA, ESA, and CAST).

General definitions

Space suit is the overall term for equipment to provide a human being with personal environmental protection and life support during *any* part of a space mission. Space suits can be subdivided (Fig. 6.9) into

- extravehicular suits for use in Earth orbit and the vacuum of space in a state of weightlessness
- intravehicular suits for use inside a depressurized spacecraft (so far not implemented)
- lunar/planetary suits for use on other celestial bodies with partial gravity
- rescue suits for use during launch and reentry.

The *space suit system* consists of a *suit enclosure*, a *life support system*, and a *communication and data management system*.

Basic requirements

The design and development of a space suit is guided by a wide variety of basic requirements (i.e., mission objectives, user anthropometry, mobility and tactility, suit pressure, metabolic rates, and safety). Each project has its own specific requirements based on the intended mission and planned tasks, but many of these requirements are of a standard type. Most of the standard requirements are documented in handbooks such as the *NASA Man-System Integration Standard* (NASA-STD-3000) and the Russian GOST standards.

Table 6.1. Suit sizing requirements.

Parameter	*U.S.A.*		*Russian Federation*	
	Max.	*Min.*	*Max.*	*Min.*
Body height (cm)	188.7	163.5	190	164
Chest circumference (cm)	114.9	85.0	112	92
Suit height (cm)	191.9	171.5	194	171
Suit breadth, arms (cm)	84.8	—	80	—
Airlock diameter (cm)	100	—	80	—

The general EVA mission objectives are defined as *"EVA can provide an effective means for service, maintenance, repair or replacement of space equipment without the need to remove it to a pressurized environment, return it to Earth or abandon it."*

There are two types of EVA: planned EVAs which require specific training, and unplanned EVAs that are not in the standard mission planning and so require general basic training. Depending on the importance of the success of the mission, the tasks of an EVA are divided into three criticality classes: mission enhancement, mission success, and safety critical.

Early space suits were manufactured individually tailored for each astronaut and cosmonaut, but modern suit systems have modular or size-adjustable elements whose fit can be set to accommodate people spanning a significant range of body shape and size (Table 6.1).

Mobility is clearly defined as the number of body movements and allowable forces applied for each movement (Fig. 6.10). Tactility is of major importance for the design of pressurized gloves, and the smallest object that should be able to be grasped is 8 to 12 mm.

Metabolic rates and EVA durations, along with the predefined suit pressure, are the principal design parameters for the life support system and for thermal control. Modern spacewalks have a duration of 6–8 hours each, a mean metabolic rate of 300–350 watts, and peak values of some 600 watts for periods of 15 minutes. During these varying metabolic rates, the in-suit temperature may vary within the range about 10 to 27°C, with the possibility of individual temperature adjustment. The suit pressure and the atmospheric composition in the suit are major design drivers. A high suit pressure matching that at sea level on the Earth's surface would allow simple use of the suit, but such a high pressure renders all soft parts very stiff and thus dramatically reduces mobility. The lower the suit pressure, the higher the mobility and tactility, but it requires an in-suit atmosphere that has a very high, or possibly pure, oxygen content. Using pure oxygen reduces the complexity and mass of the life support system, but it raises safety issues. With

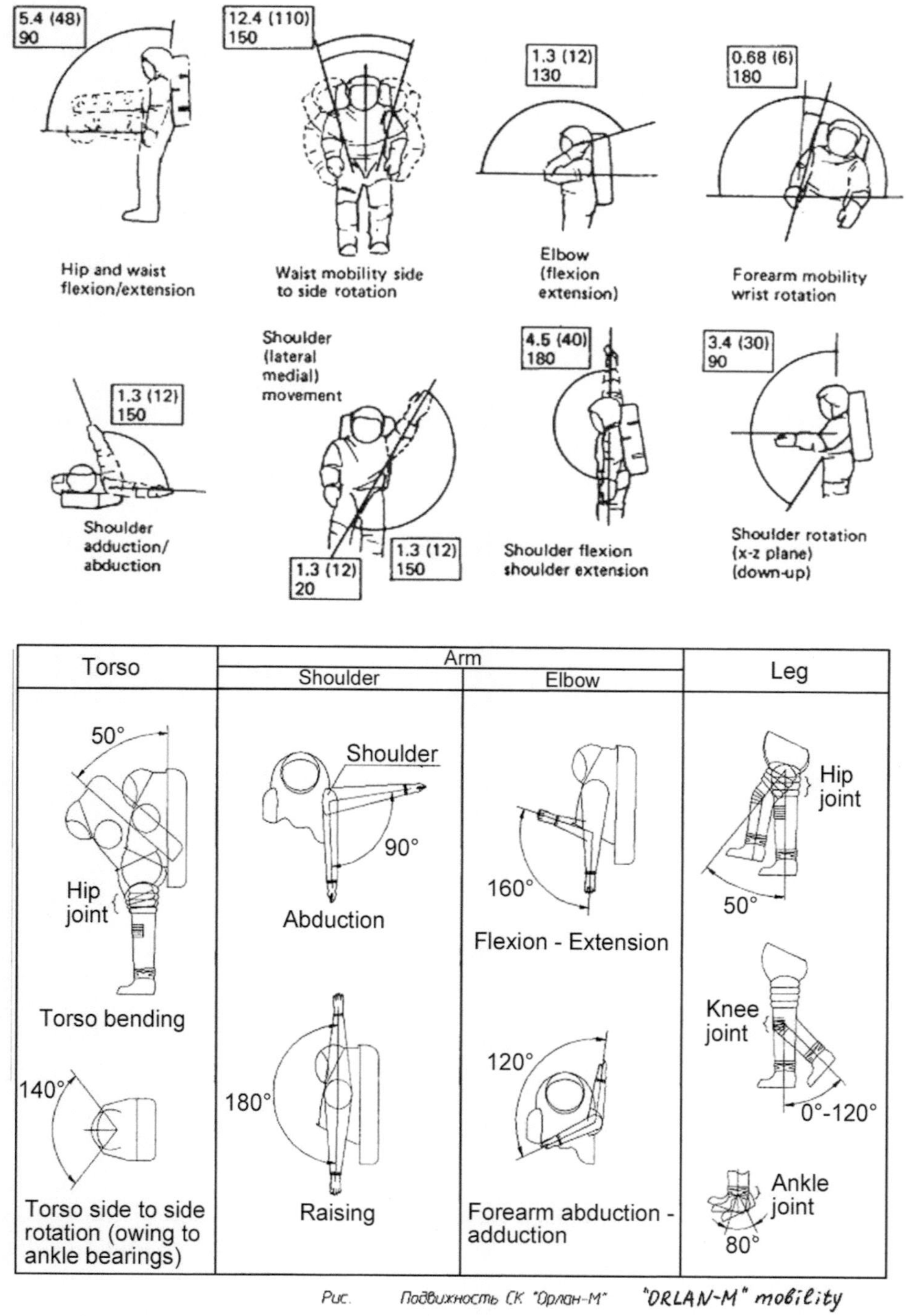

Fig. 6.10. Typical U.S. (top) and Russian (bottom) body movement requirements (sources: NASA and Zvezda).

a low suit pressure and pure oxygen, decompression must also be taken into consideration. Entering an oxygen suit atmosphere with a pressure below 500 hPa from a "normal" cabin pressure can result in decompression sickness (DCS) with symptoms similar to a rapid ascent whilst diving, and so prebreathing of pure oxygen is employed to reduce this risk. The actual risk, defined as the R-factor, depends on the actual suit pressure, the cabin pressure, and the time for pressure reduction. If pressure reduction is too rapid, nitrogen in the tissue forms bubbles in the joints which cause pain commonly described as "the bends". The R-factor is an experimentally determined quantity that defines the likelihood of DCS occurring. Assessment of the risk of DCS is defined slightly differently in the U.S. and Russian standards. In the U.S., the R-factor for normal EVA operations is set at $R = 1.4$, which is a 5% risk of DCS, so with the selected pressure for the American ISS space suit of 296 hPa it is necessary to prebreathe pure oxygen for four hours before an EVA. An alternative used by the Space Shuttle was to reduce the cabin pressure to 703 hPa with a 28% oxygen content for some 22 hours prior to making an EVA (Fig. 6.11a). On the Russian side, the R-factor is set at $R = 2$ with a prebreathing time of 30 minutes at a suit pressure of 500 hPa (Fig. 6.11b). Despite the different approaches, no instances of DCS have been reported during actual EVA operations.

A number of safety requirements must be considered during the design of a space suit system

- A quantitative safety target of 0.9995 for the probability of safely returning from an EVA.
- Fail operational/fail safe
  - No single EVA system failure shall lead to catastrophic or critical consequences or loss of mission.
  - No second EVA system failure independent of the first shall lead to catastrophic or critical consequences.

To satisfy these requirements redundancies and backup systems are used.

Furthermore the space suit shall provide sufficient protection to limit exposure to radiation to a low level (e.g., for bone marrow 25 rem at 5 cm depth). No EVA shall occur while the spacecraft is traveling through the South Atlantic Anomaly, or during periods of predicted high solar activity.

Operational requirements

Today's space suits are designed not only to be used by a variety of different people, but also to have an operational life of seven years during which they might be used in up to some 30 EVAs. With a permanent space station in orbit, suit maintenance is also an important design requirement to reduce cost.

Depending on the mission scenario, space station, or planetary use, consideration must be given to the gravitational environment. The mass of a space station suit for use in microgravity is less critical than for a lunar suit (g/6) or a

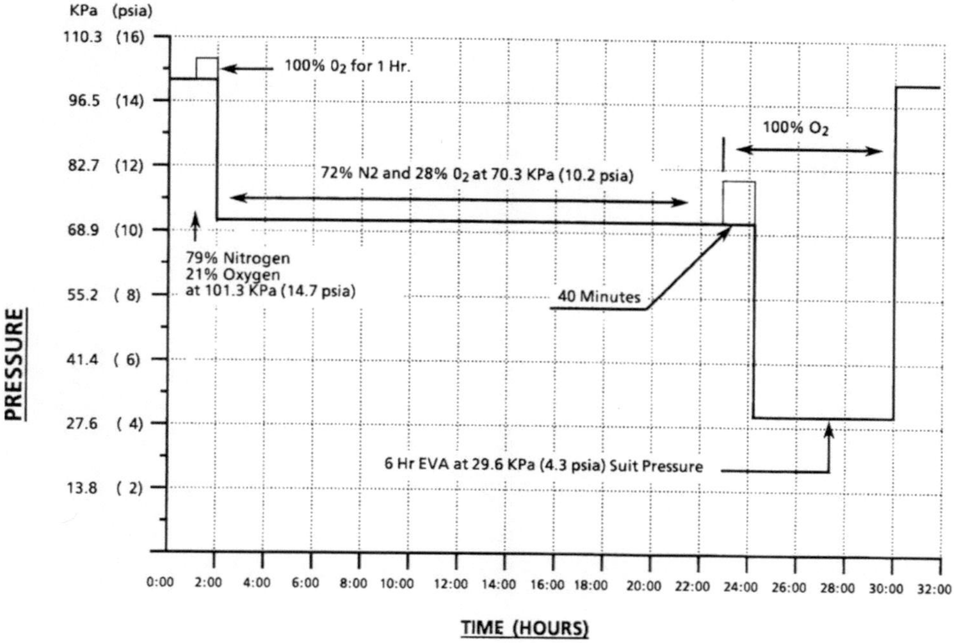

Fig. 6.11a. The U.S. prebreathing protocol for the STS (source: NASA).

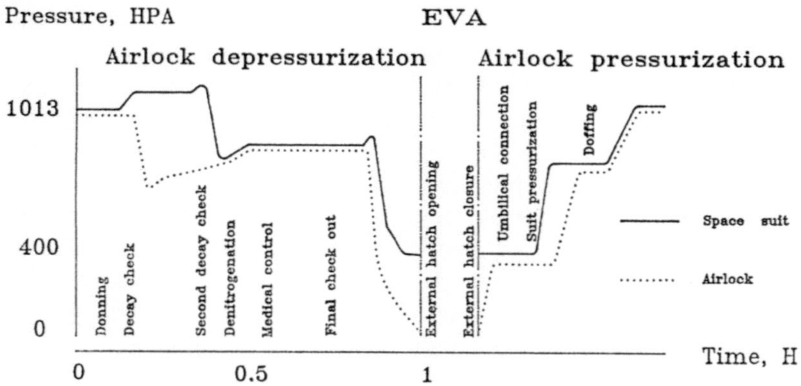

Fig. 6.11b. The Russian prebreathing protocol (source: A. Barer, Zvezda).

Mars suit (3g/8). A suit for use on a celestial body must provide protection against the materials present in that environment and enable its user to walk efficiently.

The operating capabilities in a space suit impose a set of design requirements on the tools and equipment to be used, the layout of the work sites, and the design for in-orbit maintenance and repair (e.g., Hubble Space Telescope, Mir, and the ISS). Tools, both mechanical and power tools, must be designed for use with space suit gloves. Examples are adequately sized handles, the need for a gloved hand to be able to grip a piece of equipment, the provision of rails to enable a spacewalker to move around efficiently, adequate illumination at the work site, room for working with tools, the elimination of protruding sharp items, the absence of hazardous fluids, protection against electrical voltage shocks, and limited touch temperatures. There must also be a supply of water, possibly food, hygiene precautions, and comprehensive biomedical monitoring and alert systems.

6.3 SPACE SUIT DESIGN

The 50 years of using EVA suits and rescue suits have resulted in a large variety of types, because for each new planned project a new suit concept has been developed, sometimes even in competition between rival concepts. In America in excess of 120 experimental, prototype, and development suit projects have been introduced, from which 15 operational suit systems emerged, mainly due to industrial competition for each new technology demonstrator or operational system. In the U.S.S.R./Russia 19 suit development projects resulted in 14 operational suit systems, due to their single-source policy and philosophy of progressive improvement. The only space suit manufacturer in the former Soviet Union and today's Russian Federation is RD & PE Zvezda. In America, Hamilton Sundstrand is prime contractor of the suits for the Shuttle and the ISS. Previous prime contractors were B.F. Goodrich (Mercury suit), David Clark (Gemini, Shuttle reentry suit), NASA with the International Latex Corporation (ILC) and Hamilton Sundstrand (Apollo and Skylab suits). The selected prime contractor for the future Constellation Space Suit System (CSSS) is Oceaneering Space Systems.

Early spacecraft were very small, and the space suits had to be highly optimized for mass and volume. Hence for the first 15 years the operational suits were of the all-soft type, using only soft materials. Of course, this had the drawback of requiring a low suit pressure and necessitated prebreathing or reduced cabin pressure. When the Soviet Union's Salyut space stations and the U.S. Shuttle entered service, improved operational procedures resulted in the introduction of semi-rigid suits with hard upper torsos. From the last decade of the 20th century, the trend towards all-hard suits for use on celestial bodies was the main focus of space suit technology research. But this philosophy will most likely be changed for the next generation of multipurpose suit systems.

6.3.1 The space suit system

A modern space suit system for EVA has four major subsystems: the suit enclosure forming the pressure-retaining protective "shell"; the life support system that provides a breathable atmosphere, removes toxic products and secures thermal comfort; the monitoring, control, communication, and safety system that provides commands, gives operational instructions, and controls the suit system and the astronaut from a safety point of view; and the maneuvering unit that secures the position and provides autonomous movement. Furthermore, when designing these subsystems consideration must be given to how they will support operational procedures and interface with the EVA equipment and tools.

At the present time two operational EVA systems are in use on the ISS: the U.S. ISS EMU and the Russian Orlan-MK. The U.S. suit is based on the Shuttle EMU and the Russian suit is an improvement of the Salyut/Mir suit. These systems differ in many ways but are both supported, independently, by the ISS facilities (Fig. 6.12).

Rescue suits are simpler in design and less autonomous. The main operation is inside the spacecraft, so most of the life support system functions are provided using umbilical(s) and the period of autonomous operation is limited to about 30 minutes.

- Operating Pressure (Nominal): 296 hPa (4.3 psi)
- SSA Weight @ 1-G: 55.5 kg (122 lbs) nominally for both Shuttle & ISS
- LSS Weight @ 1-G: 69 kg (153 lbs) Shuttle, 85kg (187 lbs) ISS
- EMU STS @ 1-G: 124 kg (275 lbs)
- EMU ISS @ 1-G: 140 kg (309 lbs)
- EMU ISS + SAFER @ 1-G: 169 kg (374 lbs)
- Minimum Hatch Size: 559 mm x 762 mm (22 in. X 30 in.)
- LSS, EVA, Primary: 8 hours nominally, longest used 8.93 hours
- LSS, EVA, Backup: 30 minutes certified, never used in flight

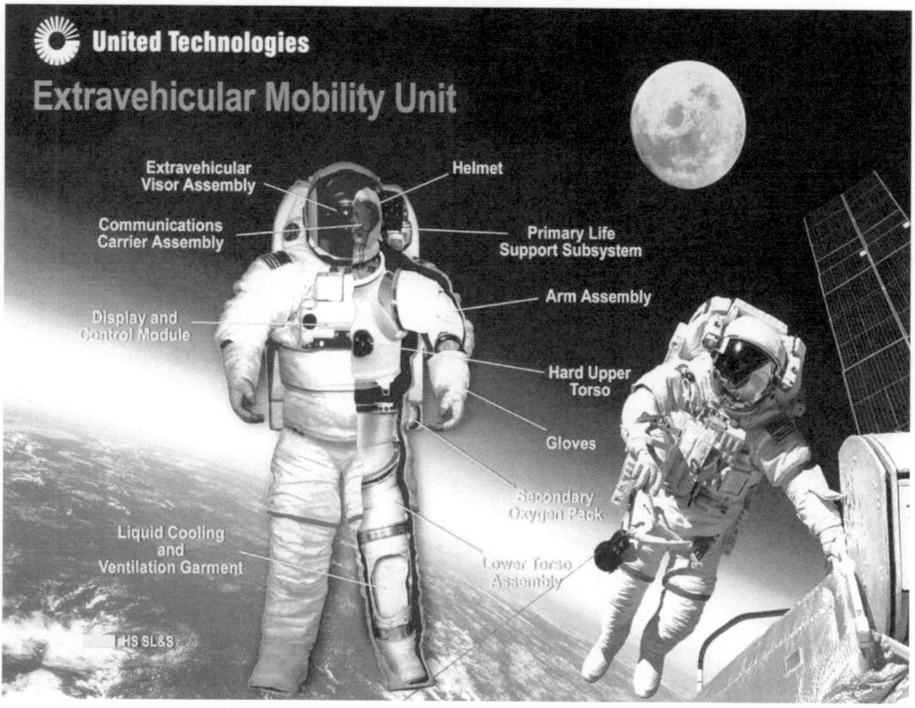

Fig. 6.12a. U.S. ISS EMU (source: Hamilton Sundstrand).

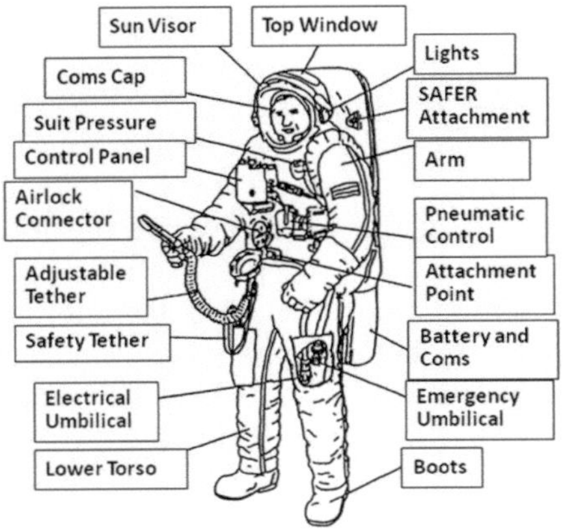

- Nominal duration of the autonomous mode — 7 hours
- CO_2 absorption cartridge operating time (with air locking time included) — 9 hours
- Suit positive pressure:
  - nominal mode — 392 hPa
  - emergency mode — 270 hPa
- Oxygen available (main and back-up) — 1 kg each
- Cooling water available — 3.6 kg
- Assured heat removal:
  - average — 350 W
  - maximum — up to 600 W
- Consumed power (totally by the suit systems) — up to 54 W
- Quantity of telemetry measured parameters — 29
- Space suit mass (wet) — ≤ 112 kg
- Service life — up to 30 EVAs during 7 years (no return to the Earth)

Fig. 6.12b. Russian Orlan-MK (source: Zvezda).

6.3.2 The suit enclosure

Suit enclosure general design

The suit enclosure design is governed by three factors: selecting the materials for its components, the envisaged method of entering the suit, and the sizing requirements. If the overall suit enclosure must be light and permit adequate mobility, soft materials are best. However, soft materials change the volume and stiffness with increased suit pressure, and are mainly used for lower suit pressures. At the other end of the scale an all-hard suit has a fixed, suit pressure independent, volume, but is heavier and it is very complicated to achieve mobility with adequate bearings. A semi-rigid suit combining soft and hard materials gives good mobility in the extremities and a constant volume upper torso and helmet part. This type of suit design has been the prevailing concept for the last 30 years. It consists, in principle, of two suits: an inner, partly soft and partly hard pressure retention suit, and a soft passive outer thermal and mechanical protective garment.

The all-soft suit concept is primarily used in rescue suits, where the spacefarer is pressurized for a limited time and requires the capability to be seated, standing, and walking. In the last decade, additional types of soft biosuits and mechanical counter pressure suits have been tested, but this research is experimental because the design of the new exotic materials is complex. The major characteristics of the different suit types are listed in Table 6.2.

The suit entry concept is governed by the operational mission requirements and also by the crew size, since some concepts oblige the spacefarer donning/doffing the suit to seek the assistance of colleagues who are not wearing suits. An all-soft suit can be designed for either rear entry, which requires support, or front entry, which does not require support. Semi-rigid suits have either an unassisted rear entry design or a waist entry, which requires considerable assistance. Thus far, all-hard suit designs have all been rear entry (Fig. 6.13).

The user population size range (Table 6.1) requires careful design measures. By use of sizing straps on the inside and outside of the arms and legs, and in the

Table 6.2. Characteristics of different basic suit concepts.

	All-soft suits	*Semi-rigid suits*	*All-hard suits*
Mass	Low	Medium	Medium to high
Mobility	High	High	Low
Sizing	Flexible	Flexible	Parts replacement
Suit pressure	Low	Medium	Normal atmosphere
Deformation by pressure	High	Limited	None
Life support system attachment	Complex	Integrated part of torso	Integrated part of torso
Lifetime	Short	Medium (soft parts)	Long
Maintenance	Difficult	Easy, also in orbit	Easy, also in orbit
Present usage	Sokol	Orlan-MK and STS/ISS EMU	None. Only test suits
Past usage	Apollo A1C/A7L and Shuttle Launch Entry Suits	Krechet, never flew although flight-ready	None

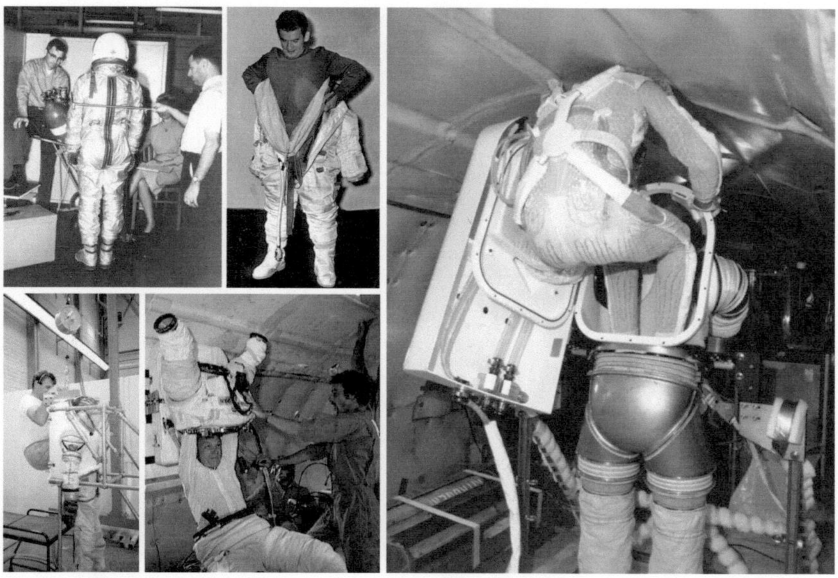

Fig. 6.13. Suit entry donning/doffing concepts: soft, rear-entry Gemini, U.S. (top left); soft, front-entry Sokol, RF (top center); semi-rigid, rear-entry EVA 2000 (lower left); semi-rigid, waist-entry STS/ISS EMU (lower center); all-hard, rear-entry U.S. MkII suit (right) (sources: NASA, Zvezda, ESA).

crotch, an all-soft suit can be made so that one size can accommodate the entire population range (e.g., the Russian Sokol). Given the relative low cost, individual sizing is also an option for multiple use. On the other hand an all-hard suit can only accommodate a range of body sizes by means of exchanging different elements of the suit enclosure. This type of suit is not practical for permanent in-orbit (space station) and planetary use if the same suit is to be used by several crew members. Semi-rigid suits combine the advantages of a soft suit with the disadvantages of a hard suit. The Russian Orlan suits are of a single size and have strap adjustments on the legs, arms, and crotch, making them very appropriate for permanent in-orbit use such as on Mir and the ISS. On the other hand, the U.S. EMU has up to five elements of different sizes for each component of the suit enclosure. As long as the suit was used on board the Shuttle, all maintenance (and sizing) was done on Earth, and thus did not complicate the logistics of a mission. However, for permanent use on the ISS the EMU requires a large set of sizing elements to be either stored on board the station or transported up and down for each new crew member.

Torso

The upper torso of the suit enclosure is the central element of the suit, since it is the main interface to other subsystems (i.e. life support system, chest pack, and airlock) and thus becomes the load-bearing element. It also integrates the suit entry, either by a separable waist attachment or by a rear door. The arms are attached using bearings and, depending on the required mobility, the lower torso may or may not have a large bearing. The upper torso is manufactured out of glass fiber (STS EMU) or aluminum (Orlan and ISS EMU) (Fig. 6.14).

Mechanically attached to the torso are the helmet, the life support system, any umbilicals, and a chest pack. For storage and donning/doffing, the whole suit is attached to the wall of the airlock by means of attachment points in the front (Orlan) or the rear (EMU).

A soft upper torso has no load-bearing function, so the life support system must have straps to enable it to be worn as a backpack (Apollo) or on the midriff

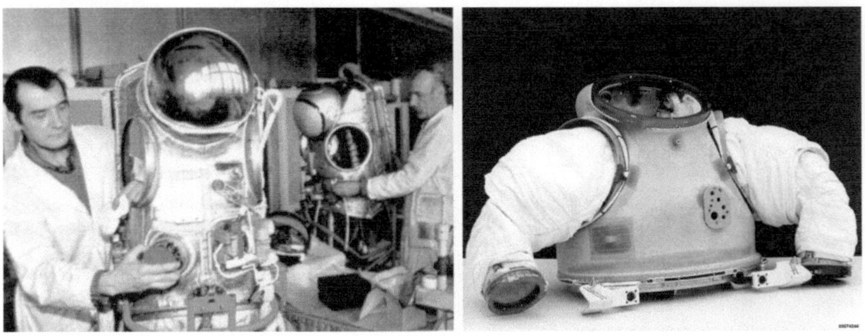

Fig. 6.14. Orlan-M torso and STS EMU torso (sources: Zvezda and ILC Dover).

(Skylab). Today the soft suit concept is used for rescue suits, for which no life support system is integrated into the suit system because life support functions are provided by the spacecraft through umbilicals.

In the case of a suit for use on a celestial body, the inside of the upper torso must be designed such that the overall suit mass is transferred to the shoulders and back of the user.

Arms

The genuine movement pattern of the arm when performing a specific task that includes applying a given force is extremely difficult for a pressurized space suit to replicate. For shoulder, elbow, and wrist movements the suit must facilitate rotating, bending, stretching, and twisting with as few restrictions as possible. Soft

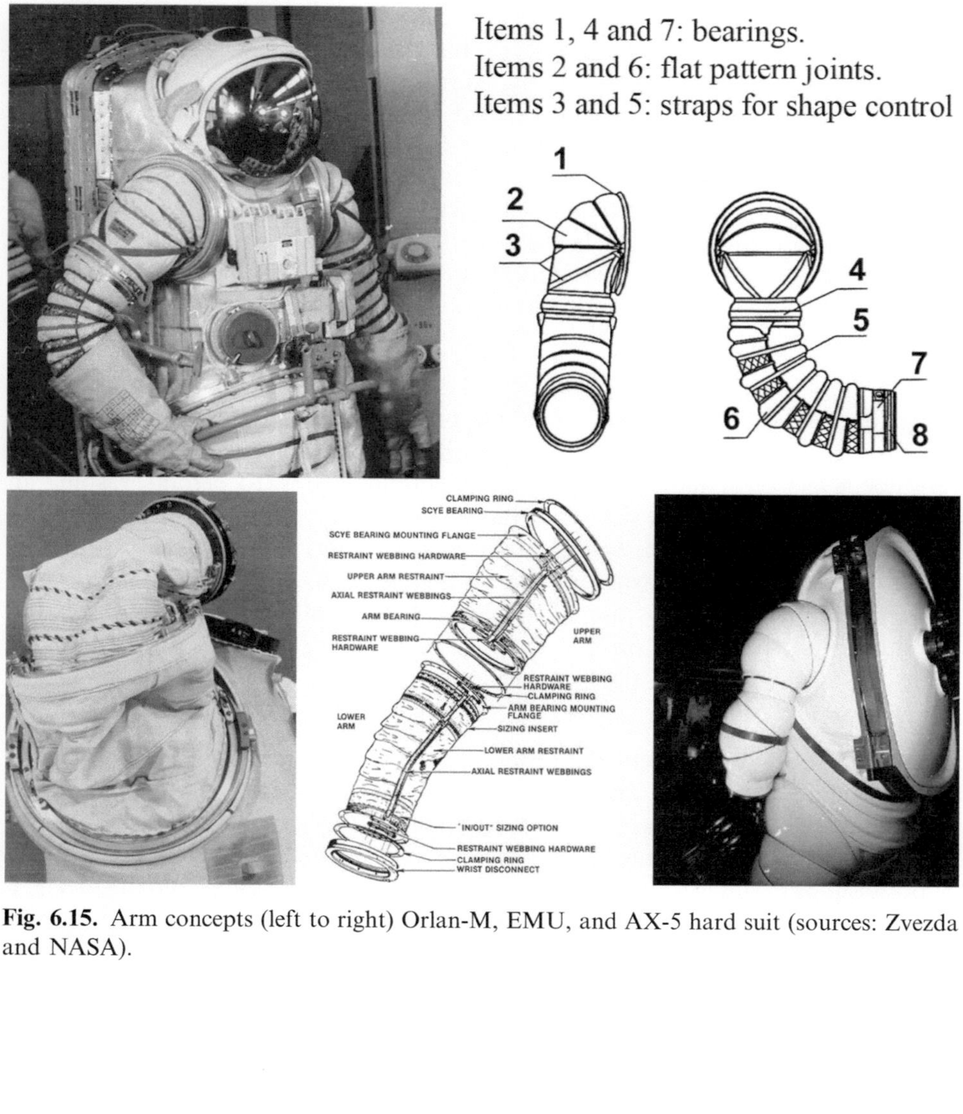

Fig. 6.15. Arm concepts (left to right) Orlan-M, EMU, and AX-5 hard suit (sources: Zvezda and NASA).

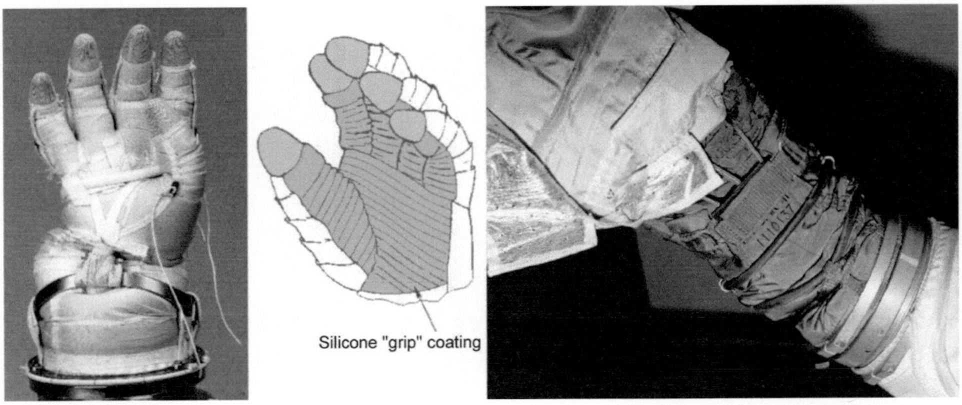

Fig. 6.16. Inside of glove (STS EMU) without and with thermal cover (left), and backside (Orlan-DM) with partly removed thermal cover (right) (sources: ILC Dover and Zvezda).

materials permit most of these movements but they require bearings, and restraints are needed to counteract the ballooning effect of the suit pressure. Many concepts have been tested using, for example, guided metal rings and steel wires to control the volume of the upper arm in particular. Modern suits (Fig. 6.15) use the all-soft flat pattern concept for shoulder and arm, utilizing three bearings (scye, arm, and wrist). Hard suits have no ballooning characteristics but require a large number of bearings in different planes of the arm to facilitate near-unrestricted movement. This is at the cost of higher mass and increased volume due to the many bearings (Fig. 6.15, final illustration).

Gloves

The pair of gloves for a space suit is the only element that is manufactured for each and every user individually. The tactility requirements imply a design with very few layers of material, in particular at the finger tips, to permit grasping an object of some 8–12 mm in size when the suit is pressurized to 300–500 hPa. Also, the hand contour must be well reproduced and ballooning prevented. For this purpose the fingers are sewn together by a large number of textile pieces, and metal rings and soft straps are often used, all this without hampering the grip functions of the hand (Fig. 6.16).

Lower torso and boots

In the semi-rigid suit concept the lower torso (i.e., waist and legs) is a soft part, and depending on mobility requirements it may include one or more bearings. For use on the ISS there is essentially no requirement for lower-torso mobility; in a free-floating situation the lower torso serves more as a means of balancing to stay in position. But some activities, in particular transport and positioning using a robot arm, require the spacewalker to be affixed to a foot restraint (anchor) and in this configuration it must be possible to bend and twist using the lower body,

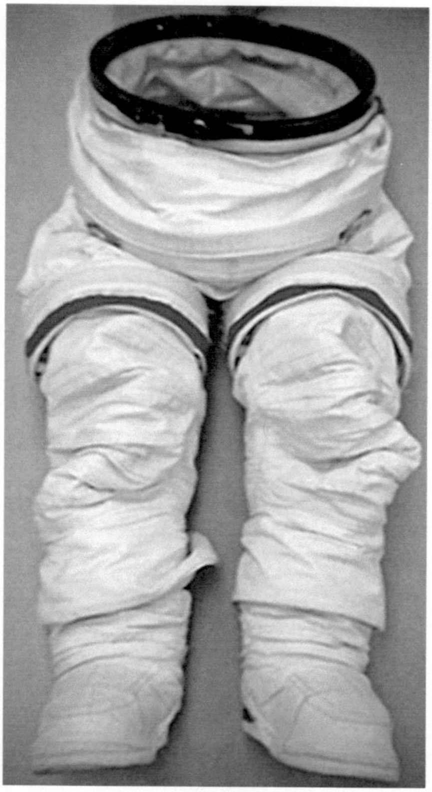

Fig. 6.17. EMU lower torso (source: NASA).

for which a waist bearing and/or ankle bearings are of great advantage. To permit kneeling, the knee design is often of the same type as the elbow, namely a flat pattern concept (Fig. 6.17).

A space suit for use on a celestial body introduces new requirements for the lower torso, because the user must be able to descend/ascend a ladder, walk not only on flat terrain but also on slopes, and climb on/off a rover vehicle in low gravity. For these actions ankle, hip, or thigh bearings, and possibly also a waist bearing, could become necessary.

Helmet

Helmets use a dual-layer polycarbonate material for the major pressure shell and an outer protective carbon fiber half-shell. When working in very bright sunlight, sun visors that slide between the protective transparent shell and the outer shell are used. Helmets also serve as the attachment points for lights and video cameras, which can be applied on both sides of the helmet. In the U.S. design the helmet is detachable using a metal ring to simplify the somewhat longer donning/doffing process, but the Russian design integrates the helmet into the upper torso.

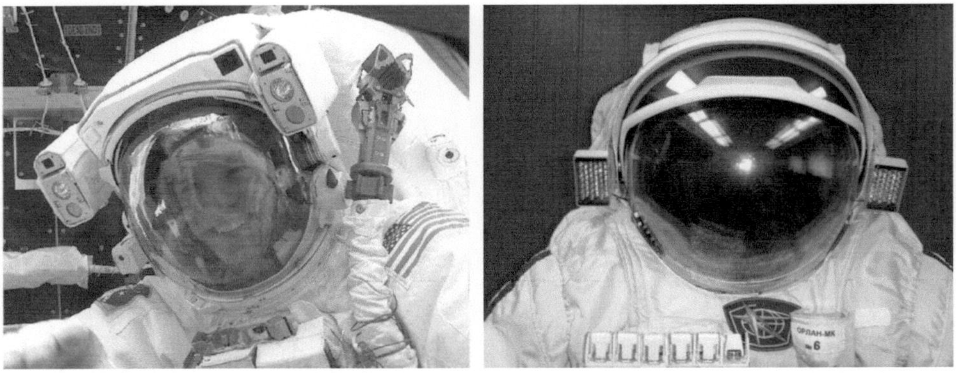

Fig. 6.18. EMU (left) and Orlan (right) helmets (sources: NASA and Zvezda).

Suit materials

The suit enclosure has two major elements: the inner pressure-retaining suit and the outer thermal and mechanical protective garment (Fig. 6.19). The inner suit has two layers: the bladder, which is a rubber or polyurethane material, and a Dacron or nylon cover restraint to provide mechanical strength and prevent ballooning. The thermal and mechanical protective garment outer covering has an inner nylon layer, five to seven layers of aluminized Mylar as thermal insulation and an outer mechanical protective layer of a fabric of polyamide threads (Phenilon-type material for the Russian suit) or Teflon-coated Nomex with Kevlar reinforcement (Ortho fabric for the U.S. suit).

In addition, the user has a liquid cooling and ventilation garment as underwear to give the appropriate thermal comfort during operations. It is a thin one-piece garment that is donned separately from the rest of the suit, and it consists of a liner and a web of nylon mesh with fine plastic tubes interwoven in

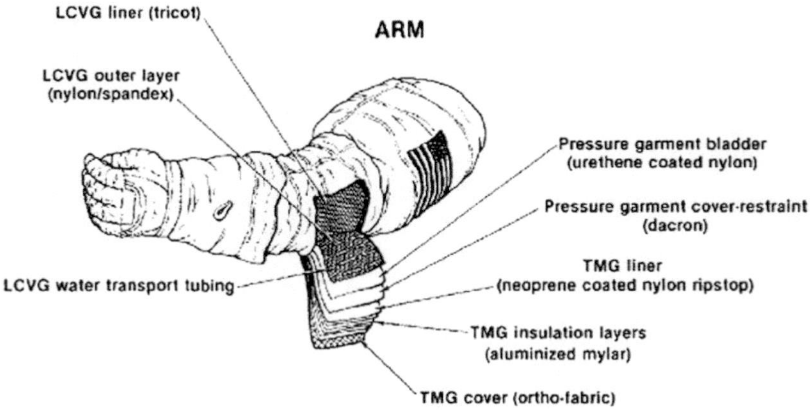

Fig. 6.19. The materials used to build up the suit layers (source: NASA).

well-defined patterns over the arms, chest, back, and legs in order to circulate cooling water over most of the body surface.

6.3.3 The life support system

Life support system general design

The EVA suit life support system must operate in a fully autonomous mode for up to 8–10 hours implementing the basic survival functions, such as oxygen supply and CO_2 removal. It is also essential to provide optimal working conditions inside the suit at varying workloads and across the full range of environmental conditions. For example, the external temperature in LEO or on the Moon spans the range of some $-140°C$ to $+140°C$. The in-suit temperature may vary in the range 10–27°C with a maximum humidity of 70%. Ventilation gas cooling and water cooling must be combined to achieve an optimal in-suit environment. For operations lasting 8 hours or more, some basic supply of water, possibly food, and certainly hygiene provisions are necessary.

Rescue suits are designed to provide basic survival functions, such as brief cabin depressurization or emergency evacuation, so both the workload and the operational time are limited. Due to the seating/walking capability, required low weight, and short emergency operation time, rescue suits mainly rely on the life support system of the spacecraft and an umbilical for the "nominal" seated phases. For rescue phases a small oxygen bottle is sufficient to support the brief operation time.

Over the years the required performance for a space suit life support system has given rise to a number of different concepts. The rescue suits (i.e., Sokol and Shuttle Launch/Escape Suit) use an open ventilation loop which vents rest oxygen plus CO_2 overboard. This results in high gas consumption at a limited cooling capability (less than 100 watts for a maximum of 1 hour) but also low weight and high mobility when disconnected from the umbilical in the emergency mode. A closed ventilation loop with a temperature control and CO_2 removal via the umbilical considerably reduces gas consumption and permits use over a period of several hours at limited workloads, but the operational range is restricted because of the umbilical. Closed ventilation loop systems have in the past been used for the Soyuz 4/5 suits (270 watts for 2.5 hours) and for the Mercury, Gemini, Apollo Command Module, and Skylab suits.

To achieve high cooling capabilities the LSS must include a water cooling loop. This is the case for current EVA suits. In these self-contained systems, oxygen and water are consumables and CO_2 is a waste product whose capture involves the use of another consumable. For occasional use on a space station or on a celestial body, it might be acceptable not to recover oxygen and water (Orlan, EMU, and Apollo suits) but this requires the replenishment of oxygen and cooling water, and replacing CO_2 removal cartridges. A first step to reduce resupply is to introduce regenerative methods for CO_2 removal, as in the new generation of EMU for the ISS.

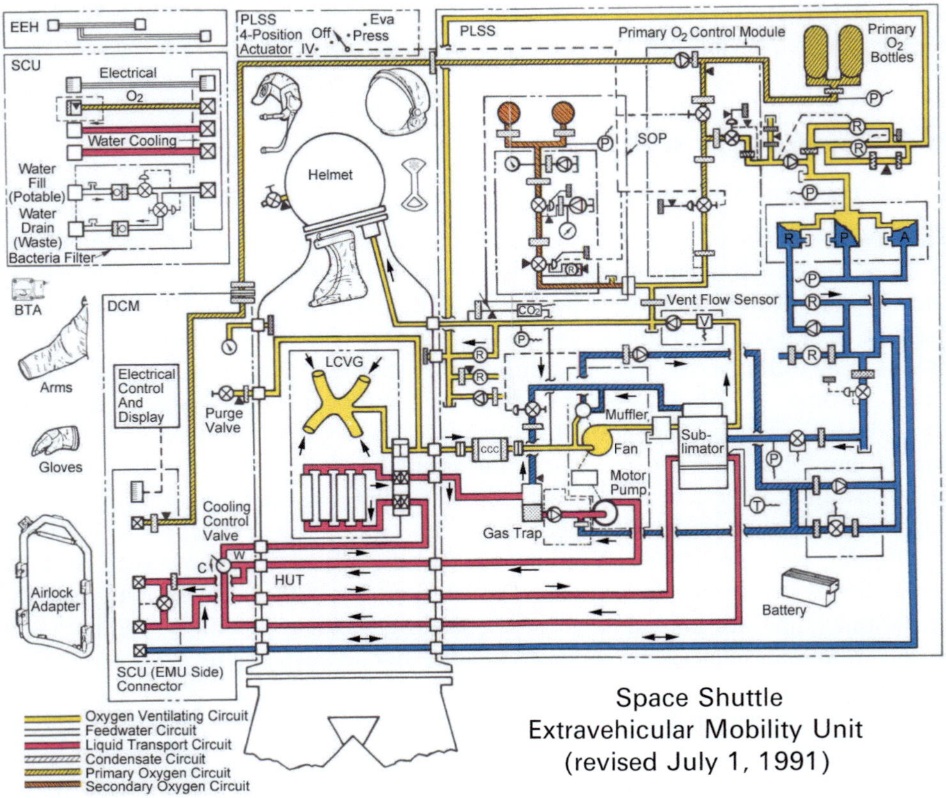

Fig. 6.20. Schematic of Shuttle EMU Life Support System (source: NASA).

Present life support systems for EVA suits use three basic subsystems, or loops, to provide relevant functions (Fig. 6.20): a ventilation loop including oxygen supply, oxygen circulation, CO_2 removal, and air cooling; a liquid cooling loop; and a thermal heat rejection loop. In addition, they contain components like an in-suit drinking bag, a urine collection device, and possibly some snack bar.

Pressurized versus unpressurized LSS

There exists a major difference in space suit design between East and West in terms of the concept of integrating the LSS into a backpack (Fig. 6.21). The U.S.S.R./Russian Orlan suit exploits the inside of the rear-entry door as the carrying structure for the LSS. Thus the LSS of the Orlan operates in the pure oxygen atmosphere of the suit itself, which allows easy maintenance in orbit and low manufacturing cost because the components need not be certified for use in vacuum. The drawback of operating all the electrical components in an oxygen atmosphere is the need for major design and safety precautions. The LSS of the U.S. EMU suit is contained in a backpack carried on the outside of the hard

Fig. 6.21. Pressurized backpack Orlan (left) and unpressurized backpack EMU (right) (sources: Zvezda and NASA).

upper torso. It requires structural protection and attachment points, and all the components must be certified for vacuum. The very dense packaging means complicated maintenance procedures, and major components can only be replaced on Earth, thereby increasing the overall system cost.

Ventilation loop

The primary functions of the ventilation loop are to provide a breathable gas, remove CO_2, control humidity, and impart a certain cooling effect (Fig. 6.22). In today's suits the breathable atmosphere is pure oxygen due to the required oxygen partial pressure and the complexity of a two-gas system, but at the cost of potential prebreathing for a duration depending on the final overall suit pressure.

The oxygen is blown into the back of the helmet and flows down over the front of the helmet in order to remove exhaled CO_2 and humidity from the face area. It is forced to flow down over the body of the user to yield a certain cooling effect, as the inlets of the return ducts (part of the liquid cooling and ventilation garment) are located at the ankles and wrists. A ventilation fan sucks (U.S.) or pushes (Russian) it through a carbon dioxide and contamination control cartridge, and into the oxygen cooling part of the heat exchanger/sublimator unit, then it is returned to the suit with addition for lost oxygen due to metabolic consumption and nominal suit leakage. Humidity is controlled by a water separator that draws condensate from the exhaled gas.

The primary oxygen system contains some 0.6 to 1.0 kg of pure oxygen at high pressure in one or two bottles, which is sufficient, depending on the suit pressure,

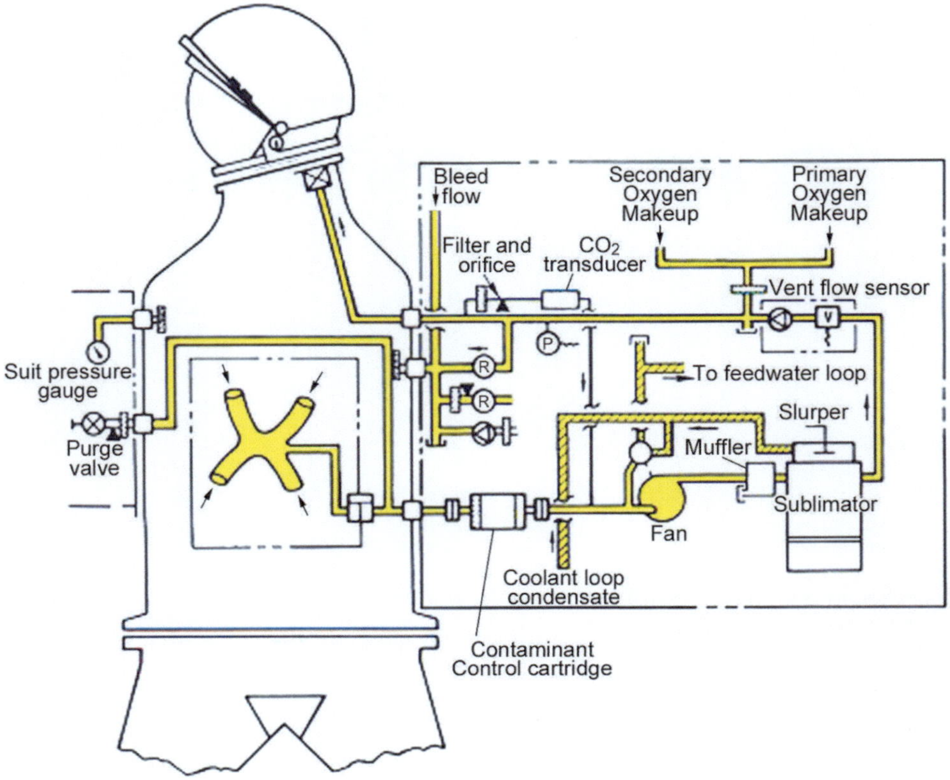

Fig. 6.22. Ventilation loop schematic (in this case the STS EMU) (source: NASA).

for 8–10 hours of nominal operations. If the primary oxygen system fails, the secondary oxygen system is automatically activated. It will also start if the life support system fails to maintain the suit pressure either due to a component failure or to off-nominal suit leakage, fails to remove CO_2, fails to remove humidity, or fails to remove excess heat. With about 1 kg of oxygen, the backup can provide a breathable atmosphere for 30 minutes at an average workload, which is sufficient to finish the task in hand and return to the airlock.

Carbon dioxide removal

Since the 1960s the standard method of removing CO_2 in manned spacecraft has been by means of lithium hydroxide. This is a very efficient and secure method, but it is a one-way, non-regenerable process and the number of LiOH cartridges increases with the duration of the mission. About 1.2 kg of LiOH is required for a nominal 8 hours of EVA. Normally a small amount of active charcoal (~100 g) is also included in the cartridge to remove other contaminants and odors. The cartridge has an overall mass of some 3 kg (Fig. 6.23). The drawback with this

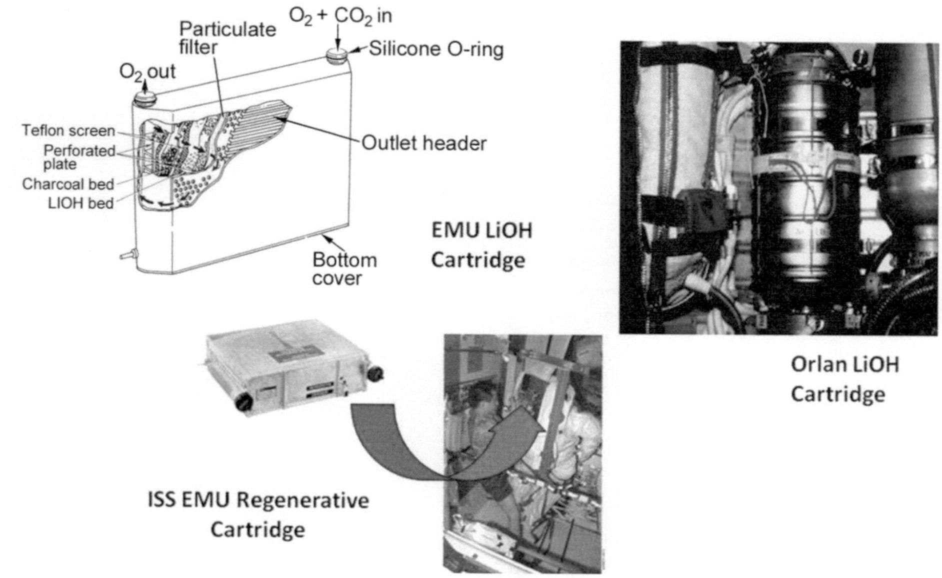

Fig. 6.23. Different methods for carbon dioxide removal in space suits (source: Hamilton Sundstrand, NASA, and Zvezda).

non-regenerative process is the transportation cost, in particular when the space suits are based in orbit or used for a lunar or Mars mission.

In the meantime, a new regenerative system based on the absorption of CO_2 by a metal oxide has been introduced for the space suits on the ISS. This canister consists of a set of metal oxide absorbent sheets and an active charcoal bed for control of trace gases. They are regenerated (in pairs) in a special unit by flowing hot air (~200°C) through them. The overall system is much heavier than the non-regenerable one, but after about 12 sorties with two suits (spacewalks are made by two-person teams), the mass penalty has been equalized. For future space stations and bases on celestial bodies, regenerable CO_2 systems for space suits are necessary to maintain an overall oxygen balance by recovering oxygen from carbon dioxide.

Liquid cooling loop

Major metabolic heat removal is performed by a water cooling loop using a web of fine tubes integrated into the liquid cooling and ventilation garment. The desired temperature is set by a control valve on the chest of the suit, which regulates the bypass of water through the undergarment or the liquid cooling section of the heat exchanger/sublimator of the LSS (Fig. 6.24).

Heat removal

The overall metabolic and component heat produced in the space suit during an EVA lasting 7 hours is about 2.5 to 3 kWh. The standard method for heat

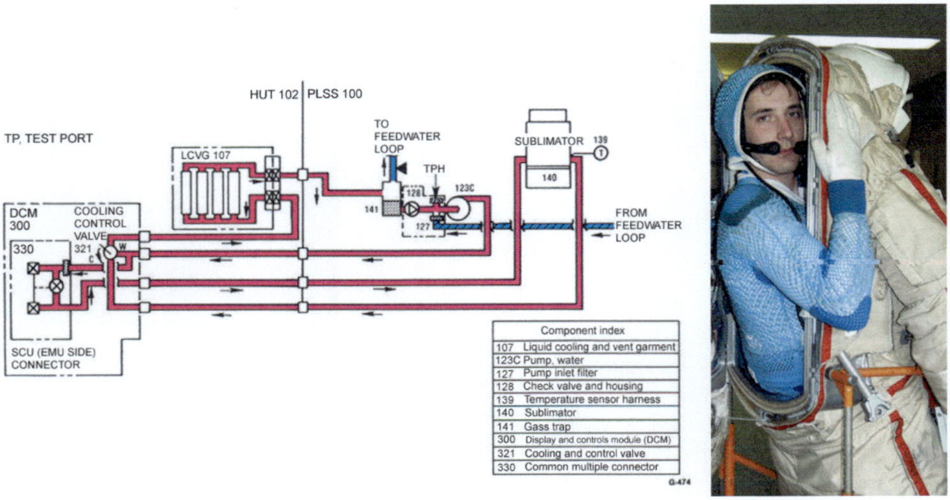

Fig. 6.24. Liquid cooling loop schematic of the EMU (left) and LCVG of the Orlan suit (right, in blue) (sources: NASA and Zvezda).

dissipation in space is the use of radiators. Present radiator technology requires an area much larger than is available on the only practical part of the suit (the backpack) to dissipate the total amount of heat from the suit. This situation became evident in the 1960s when the development of lunar suits got under way. Thus for both the U.S. and U.S.S.R. the only practical solution at that time, a sublimator, was developed. This concept is still used in today's EVA suits.

The operation of a sublimator (Fig. 6.25) is based on the sublimation process (i.e., the transfer from solid ice to water vapor without an intermediate liquid phase). This process requires a large amount of heat drawn from the liquid cooling water and the gas in the ventilation loop in the combined heat exchanger

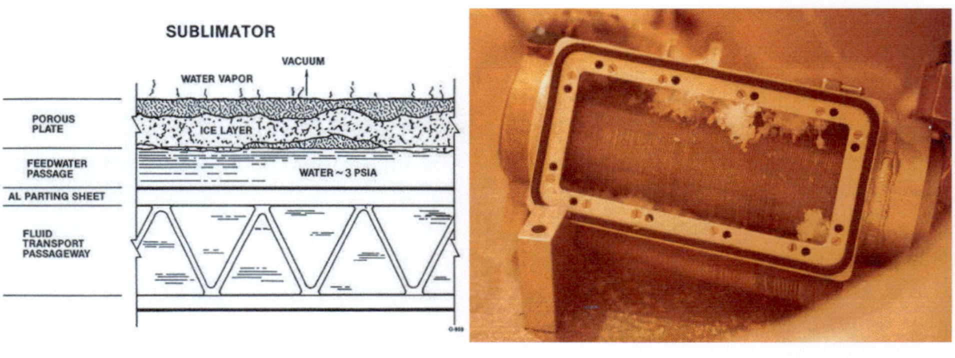

Fig. 6.25. Schematic of sublimator function (left) and operating Orlan sublimator with ice crystals on the outside (right) (sources: Hamilton Sundstrand and Zvezda).

and sublimator. A special water loop, the feed water loop with a pressure of ~200 hPa, is on the sublimator side exposed to a porous plate where water freezes upon penetrating the plate because of the temperature drop caused by the outer layer of ice sublimating in the vacuum. This reduces the water temperature in the feed water passage on the heat exchanger side, exposed to the liquid cooling and ventilation gas loops. The required amount of feed water for the sublimation process, about 4 kg, is a consumable.

Since the sublimation process requires a vacuum, it is usable only in space and on the Moon. For a Mars planetary suit, which will have to operate in an atmosphere at a pressure of 6–11 hPa, other concepts will have to be developed.

Drinking and hygiene

The relative high workload results in a considerable loss of water during an EVA, and so a bag is installed on the inside of the hard torso to enable the user to drink up to roughly 0.6 kg of water by sucking on a mouthpiece. Sometimes there is also a food stick. Equally important are the hygiene provisions. For male users a urine collection device and pampers for solid waste are used, and for female users larger pampers are applied.

6.3.4 Monitoring and control, communication, and safety

System monitoring and control

The operation of the space suit requires the monitoring of suit functions, biomedical functions of the user, and a range of operational functions. The space suit caution and warning system monitors a number of parameters to assure the safe function of the suit system and to support safe return of the crew in case of an emergency situation. These parameters include primary oxygen pressure, secondary oxygen pressure, suit pressure, ventilation flow in the suit, CO_2 level, fan operation, water pump operation, feed water pressure, cooling water temperature, battery voltage and current, and the status of the communication system. The major safety-related parameters (e.g., suit pressure, battery charge) are displayed on the control panel at the chest (Fig. 6.26). In addition, all of the monitored parameters are transmitted as telemetry to Earth via the spacecraft or the ISS. Safety information is also displayed inside the helmet by means of light-emitting diodes on the helmet rim and by audio information.

The control panel and all operational switches for emergency oxygen, fans, suit temperature, feed water, power, communication, audio volume, and lights are located at the front of the suit, and because many of the switches are not directly visible from the inside of the helmet a mirror on one arm is used to set any control (with the text on the suit in reverse). On the top of the control module is located an alphanumerical liquid crystal display for relevant safety and operational information.

Fig. 6.26. Displays and controls in the chest area, Orlan-M (left) and STS EMU (right) (sources: Zvezda and NASA).

The space suit directly monitors the heart rate and body temperature of its occupant continuously. Workload is monitored indirectly by observing oxygen consumption, CO_2 level, and cooling water temperature.

Communication

To exploit the "buddy system", no EVA is performed with fewer than two people. The space suit provides voice and telemetry communication between all spacewalkers and the spacecraft or space station and with Mission Control via the spacecraft by an open-loop system. All relevant safety and biomedical information is transmitted to Earth, and all caution and warning information is transmitted to the other spacewalkers for instant support.

Safety

Safety considerations are key drivers of the design of a space suit system. There must be redundancy of critical components (e.g., fans, pumps, oxygen tanks, pressure regulators, sensors, dual bladders), backup systems (emergency oxygen, injector, purge valve, arm cuffs), and at least two crew members per EVA.

A quantitative safety target of 0.9995 for the probability of crew safe return from an EVA must be guaranteed. This can be accomplished by implementing a fail operational/fail safe approach

- No single EVA system failure shall lead to catastrophic or critical consequences or loss of mission
- No second EVA system failure independent of the first shall lead to catastrophic or critical consequences.

Fail-operational means that in the event of a failure such as a fan, pump, or sensor, the EVA is able to be continued by activating a redundant component or function. A second failure shall permit the EVA crew member to at once complete the work in a predefined manner and then return to the airlock. Only in the event of an additional failure or safety-critical situation, an emergency such as suit leakage or failure of the primary oxygen supply, shall the EVA be terminated and the crew member return to the airlock in less than 30 minutes, the operations limit for the emergency/secondary oxygen supply.

6.3.5 Maneuvering systems

Safety during crew member transfer in the Shuttle cargo bay or across the surface of a space station is provided using two safety tethers. Additionally, when a spacewalker disengages one of these safety tethers, the other hand maintains a grasp on a handrail. This method has been successfully used for all EVA activities on Russian and U.S. spacecraft and space stations.

It was envisaged early on that maneuvering devices would be needed to address the possibility of crew errors or "double failures". Such a device would also enable an EVA crew member to depart without tethers out to a distance of about 100 m from a spacecraft or space station in order to capture or repair another freely floating object. In the 1980s both the U.S.A. and U.S.S.R. tested such devices (Fig. 6.4 and 6.27), but they were very heavy ($\sim$150 kg) and expensive. The very extensive use of EVA in the construction of the ISS, and its large dimensions, reintroduced the requirement for a maneuvering device that would minimize the consequences of tether failure, and in the mid-1990s the U.S. introduced a lightweight ($\sim$30 kg) Simplified Aid For EVA Rescue (SAFER). This unit

Fig. 6.27. Russian Orlan MMU in 1990 and U.S. EMU SAFER in 1994 (sources: Zvezda and NASA).

(Fig. 6.27) is attached to the backpack of the space suit and movements are by means of cold-gas jets. This system was also soon introduced in a Russian development for the Orlan ISS suits.

6.4 WORKING IN THE SPACE SUIT

6.4.1 Preparation and execution of an EVA

The execution of an EVA in Earth orbit or on another celestial body is preceded by a long sequence of preparatory activities, including detailed planning, tool preparation, and training for the task in hand.

EVA activities can be defined into two classes (from NASA-STD-3000, para. 14.1.2)

(1) Planned EVA—conducted to accomplish tasks included in the nominally scheduled timeline to support selected mission operations.
(2) Unplanned EVA—conducted to accomplish tasks not included in the nominal timeline, but which may be required to achieve mission success, mission enhancement, or to repair or override failed systems.

Each planned EVA is broken down into a series of steps, the equipment prepared, routes specified for moving to and from the work site, along with how the work site will be set up, the sequence in which the tasks will be tackled and the tools required, and the time required. Such EVAs are practiced several times on Earth prior to a flight in order to make sure that all elements (tools, sequence, and time) are sufficient for a single EVA and, if not, then the work will be assigned to several EVAs, with secure intermediate stops.

An unplanned EVA is an activity decided upon during a mission in response to unexpected events, such as if a planned EVA is unable to be accomplished in a given time because of unexpected problems, or if an equipment failure on the spacecraft or station necessitates urgent repair. In these cases, planning is done on Earth and, after training by other astronauts confirms the activities and tools required, the results are transmitted to the in-orbit crew for implementation. General training is given prior to any mission in preparation for this type of activity, and focuses mainly on the use of standard tools.

Three types of facility are available on Earth for EVA training (Fig. 6.28): underwater simulated microgravity training in a Neutral Buoyancy Facility, vacuum chambers, or special work site setups. Other modern training methods are by means of virtual reality facilities.

The conduct of an EVA requires a wide variety of equipment, starting with the airlock. Outside the airlock the spacewalker needs foot restraints, handrails, tethers, tools, and sometimes assistance from a robot arm controlled by a colleague inside the station (Fig. 6.29). In addition to providing a variable-pressure chamber, the airlock is the storage and servicing facility for the space suits. It also

Fig. 6.28. (Left to right) Neutral Buoyancy Facility in EAC, Cologne; Russian thermal vacuum chamber; and NASA work site setup (sources: ESA/DLR, Zvezda, and NASA).

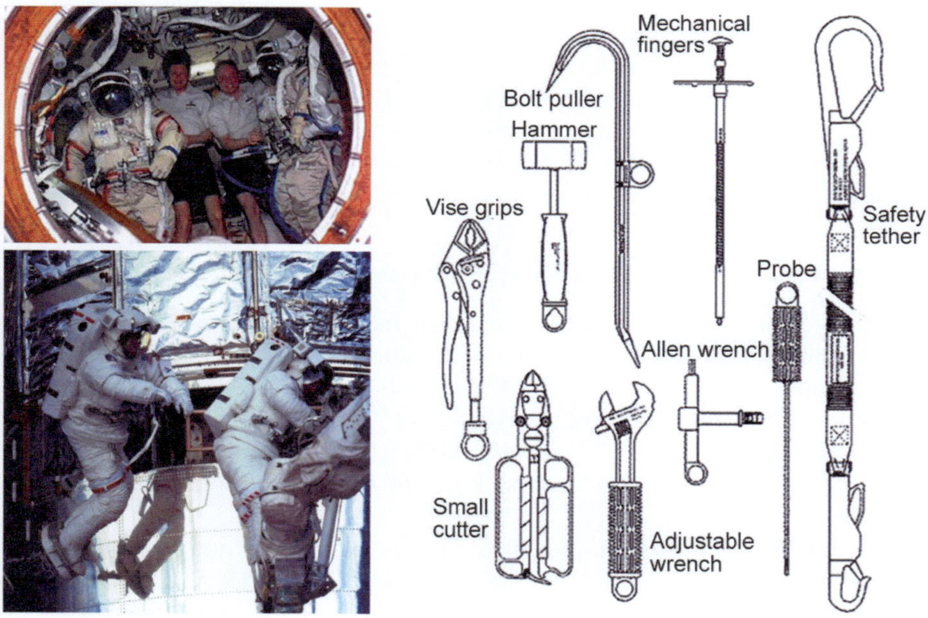

Fig. 6.29. Servicing Orlan in the ISS Pirs airlock (top left), use of foot restraint and robot arm for Hubble servicing (lower left), and set of NASA standard tools (right) (source: NASA).

houses the equipment for suit check-out. The replenishment of oxygen, cooling and feed water, battery recharge, and replacement or regeneration of carbon dioxide cartridges is performed with a suit attached to the wall by fixtures. After each EVA the suits are cleaned by wiping the interior with a cloth moistened by a disinfectant, and then they are dried using a blower to avoid microorganisms and odor. Any repair or changes of malfunctioning parts are also done in the airlock.

For the weightless exterior work the EVA crew members travel along a series of handrails, and use two safety tethers fixed to either the handrails or other suitable parts of the structure. For some positioning, a robot arm can rapidly move the person and their equipment into a location that would be difficult to reach by using handrails. A foot restraint on the robotic arm can hold a spacewalker stable, leaving both hands free to work.

Tools are either from a standard set, or specifically created for a certain task. All tools are stored and transported in bags that are loaded during in-orbit preparation for the EVA, and attached by Velcro and straps to prevent them from drifting away in space.

An EVA emergency is defined as a situation where the EVA has to be terminated in as short a time as possible. In the event of one crew member suffering a loss of suit pressure, a critical CO_2 level, or overheating, the recovery procedures are to activate the emergency oxygen and injector (Russia) and suit purge (U.S.), immediately return to the airlock, and switch to an umbilical system. If a crew member is incapacitated, the other spacewalker will affix a buddy umbilical and transport the first person to the airlock. If a spacewalker becomes physically disconnected from the vehicle, he must activate the SAFER to attempt to regain contact. If necessary, the other crew member will attempt to attach a safety line (STS and the ISS). In the case of the Shuttle, the drifting crew member could be captured by the robotic arm.

6.4.2 Personal reflections on an EVA by ESA astronaut Christer Fuglesang

During my years as an ESA astronaut I had fairly extensive experience of EVA. First, I trained with the Russian Orlan suit in Star City and later I did (much more) training at the Johnson Space Center with NASA's EMU suit. I performed five EVAs during two Space Shuttle flights to the ISS, in 2006 and 2009. They were all done using the EMU and the airlock on the U.S. section of the ISS. What I describe below relates to these kinds of EVAs, unless specially mentioned otherwise.

Preparation starts a day in advance. The suit is checked out and prepared; all tools to be used are gathered and put in place. About 15 hours prior to the planned start of the EVA—defined by the opening of the outer hatch—the two EVA crew members enter the airlock (A/L) and close the hatch behind them to start the so-called "camp-out" where they sleep in a lower pressure (about 2/3 atm) with a somewhat elevated oxygen fraction (27–28%). There are also other methods to prepare the body for the upcoming low-pressure environment, but this method is now the most common. The shortest method consists of staying in the suit for four hours and breathing pure O_2. For the Orlan, the body preparation protocol is shorter because the suit has a higher pressure when outside the spacecraft (some 0.4 atm, versus about 1/3 atm; 4.5 psi to be precise). Finally an aspirin is taken, supposedly to lower the risk of some medical complications during an EVA.

In the morning, the EVA crew members don oxygen masks, repressurize the crew lock, and open the hatch to the rest of the station. It is very important that they keep the masks tight and on all the time, otherwise fresh nitrogen will enter the body and break the prebreathing protocol. The masks have long hoses to the O_2 supply so that the EVA crew members can, for example, go to the toilet. Many take a Sudafed pill to decongest the airways of the head in preparation for the forthcoming pressure changes.

After about an hour, the hatch is closed again and the pressure lowered. This time two or three other crew members are also in the A/L to help the EVA participants to don their suits. The EMU is very difficult to put on by yourself, although a test was performed that showed it to be possible. The Orlan, on the other hand, is much easier and a crew member can don that without assistance.

Once back at 2/3 atm pressure, the O_2 masks can be taken off and the donning of the suit begun. To ensure that no mistake is made, typically one of the helpers will do the donning while another checks off each step in the protocol list to ensure that the tasks are done in the right order and that nothing is missed out. The EVA participants have already put on the nappies (very seldom needed, but just in case), the underwear (soft, white long-johns), and the liquid cooling/warming garment.

First the "trousers" of the space suit are donned, with the boots already connected. Due to the problems some people have had with the bladder in the boots, a helper will often take off the boot and pull the bladder into a good position.

The hardest part is to enter the HUT (Hard Upper Torso) with the arms, but not gloves, already connected. This is fixed to a mounting on a wall and by pushing with the legs against the floor, and with some twisting and rotating, one gets into it. This is a somewhat painful process for many. Often some help is needed to keep the interior parts of the HUT free from catching a shoulder or the back. The water bag, which already has been placed in the chest area, might have to be pulled back into position.

Thin comfort gloves come next, then the EMU gloves, the comm-cap, and helmet. Communication tests are made to ensure good radio contact. Pressure tests are carried out to verify that there are no leaks. After this, the EVA astronauts start a period of pure O_2 prebreathing. In the meantime the helpers assist the EVA crew members put on wrist mirrors, the cuff checklist (a small checklist, mainly with procedures to handle contingencies with the suit, but crew members can also make notes on it), a light and TV assembly on the helmet, all tools and tethers needed for the EVA, and finally the SAFER on the back.

The helpers then float the suited crew members into the so-called Crew Lock. The U.S. A/L is made of two parts, called the Crew Lock (from which the crew egress to space) and the Equipment Lock (larger, in which the suits are donned). The call signs for the EVA participants are EV1 and EV2; normally EV1 is the lead person. EV1 is placed facing outwards and "down", in order to operate the outer hatch. EV2 is placed in the opposite direction to operate the valves and some other systems. The helpers close the hatch between the Equipment Lock and the Crew Lock. EV2 opens the valve and starts the pump to depress the Crew Lock. In order to save as much air as possible, it is not vented to space but pumped over to the rest of the station. However, when the pressure becomes really low the pump becomes inefficient and is switched off. Then EV1 opens the valve on the outer hatch. During the depressurization there are often sounds in the suit due to the overpressure valve opening to let O_2 out of the suit. I experienced different sounds in different suits. While in the Crew Lock the EVA crew members also verify the status of their suits by checking the monitoring display for suit pressure, oxygen supplies, cooling water temperature, power of the battery, CO_2 pressure, etc. The final check before opening the hatch is to ensure that the tethers are in place between the suits and hooks on the wall; it is imperative that at all times each EVA crew member has at least one tether attached to the structure.

Finally EV1 can open the hatch and, on receiving a GO from the EVA-conductor, move out into space. Someone either inside the ISS, the Shuttle, or on Earth leads an EVA. When the Shuttle was docked to the ISS it was always a crew member in the Shuttle that did this, and this person was known as IV (Intra Vehicular). Lately, with no Shuttle at the ISS, someone in Mission Control Center, Houston, in the case of the EMU, and Mission Control Center, Moscow, for Orlan, has played this role. For me it was always an IV in the Shuttle. The Mission Control Centers are not always on line due to radio outages, but they follow as much as they can and often notice mistakes by the crew and call for corrections. For any off-nominal situation, the crew will also consult with the ground on the best actions to take.

The EVA crew will normally have trained several times in the pool to perform the tasks, but is not required to remember the details by heart. The IV has the checklist and tells the spacewalkers what to do, step by step. For example, an instruction might be: "Translate to the Columbus module, BRT to Handrail 301, and take a picture of the experiment box beside it."

When one exits the airlock one can get the feeling of falling, or vertigo. Actually the feeling of falling may also occur later, upon observing the Earth passing by many hundred kilometers below. It then helps to just look at the station. It is also easy to unconsciously grip very hard onto the handrails. One astronaut realized that after an hour his hand muscles were exhausted. Being conscious of everything—having good situational awareness—is very important for a successful EVA. Very little force is needed to hold onto a handrail, since one is weightless and the 100+ kg suit does not matter.

Another important factor is to know where each part of your suit, and any cargo that you may be carrying, is. The view from inside the suit is limited and many items protrude from the station, some sufficiently sensitive to be designated as "no-touch". There are also "keep-out zones", mainly due to radiation from antennas.

Although most concentration goes to the work, it is unavoidable now and then to feel the pleasure of the fabulous view of the Earth. Until humans return to the Moon, spacewalks are the ultimate space experience.

Fig. 6.30. Christer Fuglesang participates in the third STS-128 spacewalk outside the ISS (source: NASA).

The tethers for the EMUs are on reels and 85 feet long, so one doesn't have to switch them very often. In contrast, the Orlan protocol uses two fairly short tethers and one moves along with them from handrail to handrail. With the EMU reel, one can give a small push and float alongside the station for a long while; a wonderful feeling, although I never had my hands far from a handrail. Moving from one place to another is called 'translating' in EVA lingo. This brings up another important issue: language, nomenclature, and clear communication. Establishing clear communication protocols that cannot be misunderstood and repeating every key piece of information is a must. For a non-native English speaker (or Russian for Orlan) this is not always easy, even after many years in the country.

At a work site, one normally fixes oneself using so-called 'local tethering'. This can be a short tether or a device known as a BRT (Body Restraint Tether). The BRT is metallic, but flexible, with a claw that can be locked on a handrail. The flexibility can be fixed by rotating a ring on the BRT. It is carried on the space suit at the hip. It is also often used to transport not-so-large items. Another common device is a foot restraint that can be moved around. There are plenty of interfaces where they can be mounted all over the station. When a task requires a lot of force, a foot restraint will normally be used. Often if a large cargo has to be moved, an EVA crew member on a foot restraint at the end of the robotic arm holds it while another person operates the arm from inside. The arm is also used if one has to work in a place which would be difficult (if not impossible) to reach simply by translating (e.g., anywhere on the solar arrays).

The day–night cycles of 90 minutes present certain challenges. For example, one may want to regulate the suit cooling as the temperature varies, although workload is a bigger factor for body temperature. During my first EVA I didn't change setting at all. During the second, I first started to sweat when I was fighting some very hard electrical connectors, and later got cold when I did nothing except hold a large piece of equipment at the end of the robotic arm in night time. I had forgotten to turn down the cooling after increasing it when I was warm. One should avoid sweating, since it could cause condensation on the visor and other problems. The water that circulates through the liquid cooling/warming garment can only provide cooling, not warming. However, there are battery-driven heaters in the gloves that can be activated to warm the hands. Another challenge is the darkness on the night side of an orbit. There are good headlights on the helmet, so locally one sees well (although they tend to burn out now and then during EVAs) but one can easily lose orientation. This can happen even in daylight. I once realized that I had taken a wrong turn somewhere and had to backtrack. When training in the pool there is always light and the bottom of the pool is easily seen, so one is always aware of where one is and how one is oriented. In addition, one feels in the suit where down is. Orientation is an aspect that is easier in the pool, but otherwise it is generally harder to work in the pool than in space. In the pool one has to fight turning moments (there is only one orientation of the suit that is stable in the water), whereas that is not the case in space. Many tools are also heavier in water and it can be tiring to work with them. Tether manipulation is tricky both in the pool and in space. Every piece of equipment must always be secured either to the suit or to the station. This means you always move around with many small tethers that can very easily become entangled. One always has to make sure the hook is properly closed when one puts a tether on something, and do a pull test to verify it is secured. Once I missed that and lost a tool to space. It happens now and then, despite rigorous training and close attention during an EVA.

When the EVA approaches its end after some six hours, one's hands and arms are normally fairly tired, but it is essential to maintain concentration in order to be careful and attentive and to retain one's situational awareness.

The final part is the "clean up" when all the tools that have been used are gathered and inventoried. EV2 ingresses first, followed by EV1, who then closes the hatch. A helper inside the Equipment Lock repressurizes the Crew Lock by opening the vent valve. To avoid ear and sinus problems as the pressure is restored, it is important that the helper and the spacewalkers coordinate properly. In one instance, due to some misunderstanding, the repress was too rapid and caused a crew member's eardrums to burst. The helpers reinstall the EVA astronauts onto the wall mounts and help them to doff the suits. This normally goes quite fast. While the EVA crew members get some well-deserved food after a full day's work, the helpers clean and dry the suits ready for reuse.

6.5 FUTURE DEVELOPMENT

6.5.1 Background

Apart from the Apollo suits used in the lunar environment some 40 years ago, all subsequent EVA suits have been for Shuttle and space station operations in Earth orbit. Over the years, the major design parameters (gravity, atmosphere, temperature environment, and required mobility) all show a progressive increase in the design constraints from suits for use in Earth orbit to lunar suits and finally the Mars suits (Table 6.3). Today's EVA suits for the ISS are in many respects similar to what is needed for a lunar suit, but in terms of mass they would hardly be acceptable for lunar applications, and they fall far short of the mobility required for walking on a celestial body.

EVA suits for use in a microgravity environment are less sensitive to overall mass. For these suits the selected technologies and the specified mass are more driven by a combination of development and production cost versus higher

Table 6.3. Major design requirements for future space suits.

	LEO requirements	*Lunar requirements*	*Mars requirements*
Gravity	0 g	1/6 g	3/8 g
Environment		High dust content	Severe dust storms (~5 m/s). Dust is corrosive, skin irritating, and hygroscopic
Atmosphere	Vacuum	Vacuum	6–11 hPa. 95% CO_2, 2.7% N_2, 1.6% Ar, 0.13% O_2
Temperature	−150°C to +140°C	−170°C to +150°C	−140°C to +20°C
Mobility	Mainly upper body. Lower body limited	Walking capability needed	Walking capability needed
Place of maintenance	Space station or ground	On lunar surface	On Martian surface
Interfaces	Space station and crew transport spacecraft	Lunar Lander, Shelter, Rover	Mars Lander, Shelter, Rover
Other remarks			Suit operation after ~1 year in microgravity for the astronauts!

transportation cost for heavier suits. It is also important that suits intended for use in Earth orbit can be maintained on the ground if necessary.

6.5.2 Lunar requirements

The U.S. Apollo A7LB and the Soviet Krechet suits for use on the lunar surface were in the mass range 96–106 kg, yielding an effective suit weight on the Moon of 16–18 kg for a duration of 7–10 hours. The Apollo suits were all-soft to keep the mass down, but this also required a low suit pressure of 255 hPa. The slightly heavier Soviet suit had a hard upper torso and a higher suit pressure of 400 hPa. The lunar missions were limited in number, so the suits could be manufactured for each individual astronaut. Six of seven U.S. missions assigned to land on the Moon succeeded and a total of 14 astronauts required such suits. Towards the end of the program crews conducted three EVAs with no maintenance in between sorties except for oxygen replenishment, battery recharge, and replacement of water and CO_2 removal cartridges. In effect, these missions took the suit to its operational limit.

The next generation of lunar EVA suits will be designed for a higher suit pressure to avoid time-consuming and oxygen-intensive prebreathing. Life (maintainability) and garment sizing will be key drivers for frequent, long-term use by a larger crew population. Ease of repair and maintenance in a lunar base will be mandatory, and regenerative life support system concepts must be implemented in order to reduce consumables in the long term. On the other hand, the thermal control system concept with sublimators might be maintained if a ready source of water is available.

The crews of the early Apollo lunar missions were quarantined upon returning to Earth and tested for contamination by alien biology. Once this was established not to be a concern, the quarantine was discontinued. The worst aspect of operating on the lunar surface proved to be the dust, which is very fine grained, adhesive, and abrasive. Particular attention will have to be given to the suit enclosure design for future longer lunar missions in order to avoid the accumulation or entrapment of dust that could be introduced into the habitat areas where it will damage equipment.

6.5.3 Martian requirements

The major challenge in designing a suit for Mars will be the atmospheric conditions, because although the pressure is low there are high winds with strong gusts. Also the dust on Mars is much more severe than on the Moon, and its effects more complex. It is corrosive, skin irritating, and hygroscopic—a hazardous combination for the crew and their equipment. Therefore for longer durations and potential exposure of crew and material to this dust requires unique methods for preventing dust entry into habitats and contact with the crew. Although the suits must be kept outside the habitable area, the fact that they will need maintenance will require there to be complex air locking and rover interfaces.

The use of suits in the Martian environment results in challenges to solve several contradictory requirements

- The partial atmosphere prevents the use of sublimators for cooling.
- The low temperatures in many cases will require heating—not cooling.
- The 3g/8 environment requires new materials and designs (the Apollo suit would feel like about 40 kg on Mars). Walking requires "normal" leg and foot mobility.
- The crew will have gained 10–12 months of microgravity experience during the trip to Mars, and will then be required to adapt to working in a 3g/8 environment wearing a Martian suit.
- Prebreathing should be avoided since consumables are expensive, but a higher suit pressure increases suit mass.
- Each member of a small crew will probably possess a tailored suit, but for a larger crew the sizing of the suits and logistics of parts will be critical.
- The distance of Mars from Earth requires that all spares be carried on board, with a sophisticated maintenance system.
- Real-time communications in safety-critical situations will not be possible, so advanced built-in diagnostics and preventive actions systems are required.

6.5.4 Prototype testing

For more than a decade various prototypes for elements of lunar and Mars suits have been tested. The mobility and walking capability of different soft, part-hard and all-hard suits have been tested, and excursions on simulated lunar surface landscapes have been undertaken, with and without rovers.

Not only the suit enclosure, but also the life support system is continuously being tested for new technologies to make the system totally regenerative, free of additional consumables.

Fig. 6.31. (Left) Testing of a part-hard suit (in this case the ILC H-suit). (Middle) Walking test in a simulated Moon landscape with a NASA Mark III suit. (Right) Mobility test of a Russian suit (source: NASA).

6.5.5 Future technologies

New suit materials are investigated in search of a soft material that will not change its shape and volume at higher suit pressures (700–1,000 hPa), yet still provide a much better mobility than hard suits. This is essential to increase pressure in order to reduce or totally eliminate prebreathing, with loss of time and consumables. Also, new sizing concepts must be developed for larger crews on long-term missions to other celestial bodies. This is to limit the number of suits that must be carried. When resupply from Earth is impractical, new spare parts concepts must be evaluated to keep the logistics at an acceptable level.

The main design improvements for the life support system will be non-venting heat sinks, and interfacing CO_2 removal to the base life support system in order to recover the oxygen in an overall closed system. If the suit pressure can be increased to the base habitat cabin pressure, then the suit atmosphere will have to be changed to an oxygen/nitrogen one.

New technologies will also be required for monitoring and control, and for safety, in particular when real-time contact with a Mission Control Center is impossible. Prolonged EVA operations time will also require new means of supporting EVA crew members. Voice control/voice recognition, larger graphic displays, and helmet head-up displays are some of the potential options.

6.5.6 Next-generation U.S. suits

The U.S. Space Shuttle was retired after some 30 years of excellent operation. In the new Constellation space transportation system the Orion manned vehicle is to be based on a capsule configuration of the Apollo type. This implies a new operational scenario and new vehicle interfaces that are not able to be fulfilled with the present STS Safety and Rescue Suit and STS/ISS EMU EVA systems.

The envisaged mission phases (Fig. 6.32) for the new suit system, known as the Constellation Space Suit System (CSSS), will be implemented in a two-step approach with launch and entry plus in-space contingencies available for the initial operation of the Orion spacecraft, and a second step providing the lunar surface configuration.[1]

To deal with the constraints of using a capsule with limited space, and to limit the overall cost for the development and operations of the various configurations for the three mission phases, NASA has chosen a modular approach (Fig. 6.33)

- A single spacesuit system with two configurations which share many components
- Common, evolvable infrastructure
- Minimum set of hardware to meet all mission phase requirements
- Modular, reconfigurable, component-based open architecture.

[1] By mid-2011 the development of the Constellation Space Transportation System was put on hold for redefinition, and this also applies to CSSS activities.

Fig. 6.32. Mission phases for the use of CSSS (source: NASA).

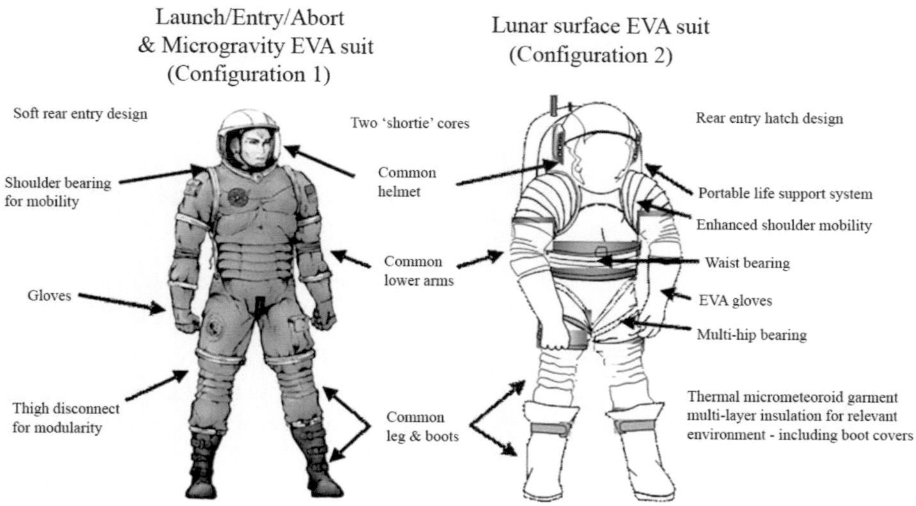

Fig. 6.33. The NASA Constellation Space Suit System (source: NASA).

The NASA mission description for the CSSS can be summarized as follows

- ISS/Constellation Exploration Vehicle—provide launch, entry, and abort (LEA), limited-duration pressurized survival, and contingency microgravity EVA capability for missions to ISS for up to six crew members
- Lunar sortie—provide LEA, extended-duration pressurized survival (up to 120 hours), microgravity EVA capability, and surface (g/6) EVA capability for a lunar mission (~2 weeks, 1 week on surface) for up to four crew members
- Lunar outpost—in addition to the above, provide surface EVA capability for a lunar mission duration of up to 6 months
- Mars—surface operation EVA capability on Mars for extended duration (as a future development option).

The LEA and in-space contingency operations will be performed by a "Configuration 1" suit that does not have an autonomous life support system. An umbilical concept will be used for an EVA from a capsule in Earth orbit. Thus the

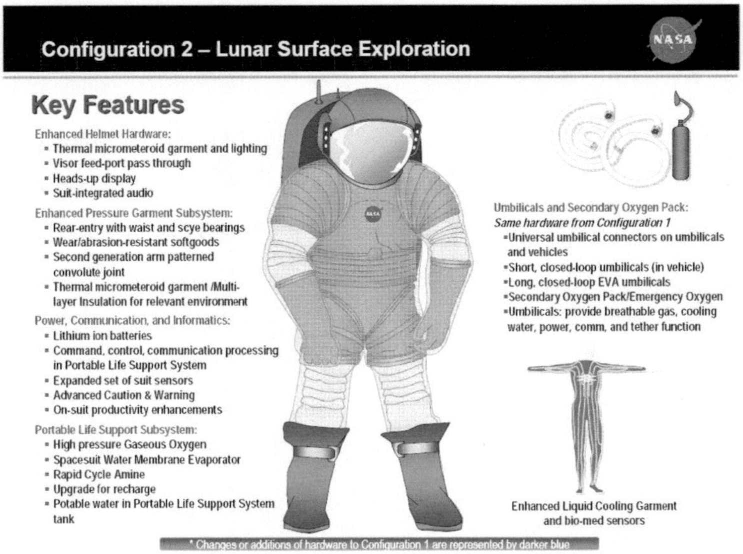

Fig. 6.34. CSSS Lunar Surface Exploration space suit (source: NASA).

CSSS is basically only operational from the Orion vehicle, and ongoing maintenance and repair of the ISS will have to be performed by the ISS EMU or the Orlan suits for the remaining life of the station, which implies parallel use of several suit systems for the next decade at least.

The Lunar Surface Exploration suit (Fig. 6.34) will use a number of components from the first configuration (Fig. 6.33), but with a hard upper torso and a portable life support system. Compared with the present EMU, a number of new components and materials will be introduced and suit entry will be via the rear.

Technical and schedule challenges have resulted in a phased or block approach implementation

- LEA and microgravity EVA capability to be available for the first flight of the Orion spacecraft no later than 2014.
- Surface EVA capability is required for the first lunar sortie missions by 2018.

NASA perceives that a single-suit *system* providing LEA, microgravity EVA, and lunar surface EVA capabilities is a feasible approach, with the potential payoff of reduced up-mass, reduced logistics and spares, and reduced life cycle costs.

6.5.7 Ultimate goals

The advancement of technologies for future space suits is running in parallel with present program-oriented system developments. New improved materials

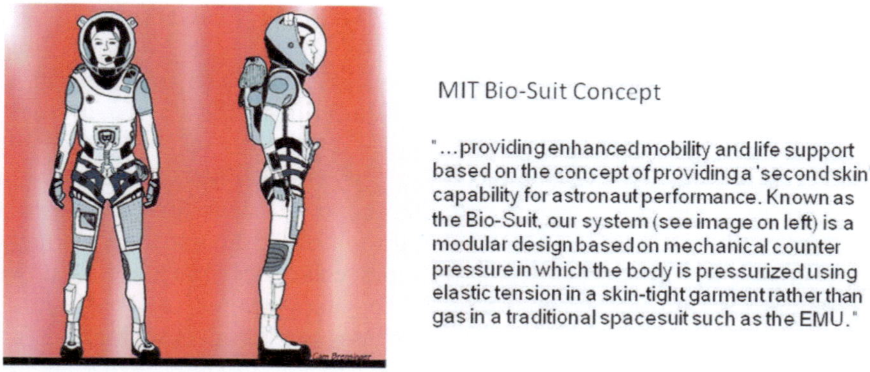

Fig. 6.35. A skin-like Bio-Suit concept from MIT (source: MIT MVL).

Fig. 6.36. A challenging view of future space suit potentials (source: NASA).

(Fig. 6.35), regenerable life support functions, information management, and control systems will all find their way into future space suit systems. Maybe one day planetary exploration in an entirely new form might become reality (Fig. 6.36).

6.6 SPACE SUITS FOR SUBORBITAL "TOURIST FLIGHTS"

In the near future, commercial enterprises will offer flights to the fringe of space (i.e., ballistic flights to altitudes of 100–110 km) to enable "space tourists" to experience microgravity for several minutes.

The mission-implied requirements for the protection of passengers in a suborbital spacecraft are a cabin and life support system for passengers with limited mobility for a maximum mission duration of 2.5 hours. Even during such a short flight there is the potential for rapid cabin decompression that would result in suffocation after several seconds. To guard against this, the pilots and passengers should wear safety suits for all phases of the mission (Fig. 6.37). The metabolic loads to be expected are limited due to low mobility in the cabin and the short mission duration. Thus a simple air-cooling system which blows cabin air through the suit via an umbilical would suffice. If so required, the umbilical could be disconnected for short zero-gravity exercises in the cabin. Such tourists will be inexperienced passengers with little or no pre-flight training, and so the major feature of a protective suit must be an automatically closing helmet visor at the onset of decompression similar to the system realized in the Russian Sokol suit. The suit must be equipped with an oxygen emergency system by a separate bottle or umbilical supplying breathing oxygen to the suit helmet and a suit pressure of some 200 hPa, and exhaled gases must be vented by a bleed valve in the suit.

The suit enclosure can be very much simplified compared with today's rescue suits. With a large population of space tourists, sizing can be implemented by having a number of standard sizes that do not require individual adjustment, because the time to wear the suit is short and hardly any mobility is required. From a hygiene point of view no special precautions others than a pamper would be necessary. Gloves would be necessary, but no special tactility is required and thus a simple pressure-tight and form-fixed shape should suffice.

Fig. 6.37. Animations of space tourists (sources: Virgin Galactic on the left and Space Adventures on the right).

6.7 REFERENCES/RECOMMENDED LITERATURE

Abramov, I.P. and Skoog, A.I.: *Russian Spacesuits*, Springer/Praxis (2003).

Abramov, I.P., Severin, G.I., Stoklitsky, A.Yu., and Sharipov, R.Sh.: *Space Suits and Systems for EVA*, Mashinostroeinie (1984) [in Russian].

Harris, G.L.: *The Origins and Technology of the Advanced Extravehicular Space Suit*, Univelt (2001).

NASA Man-System Integration Standard, NASA-STD-3000, NASA (1995).

Thomas, K.S. and McMann, H.J.: *US Spacesuits*, Springer/Praxis (2005).

7

Astronaut selection and training

Hansulrich Steimle and Carol Norberg

ABSTRACT

In this chapter a review is given of astronaut selection campaigns carried out to date. The initial selections were undertaken in the United States and the Soviet Union. In the beginning the candidates were predominantly men having military backgrounds. Male engineers and doctors followed. With the exception of one occasion in the early 1960s, it was not until around 1980 that women were admitted as astronaut candidates. They have remained a minority. Nowadays Europe, Canada, Japan, and China also have corps of astronauts. The selection processes are long and thorough to ensure that the healthiest and most psychologically stable individuals are selected, with the potential to work well in a team while confined in a small space for extended periods of time. Having been selected, the candidates have a long and diverse training regime which requires flexibility and adaptability of both them and their families. The ultimate prize of a trip into space is not guaranteed, and the waiting period can be long. Intricate planning of space missions, which starts long before a mission occurs, produces detailed work schemes for the astronauts to follow when they are finally in space.

7.1 ASTRONAUT SELECTION

7.1.1 General requirements

A human spaceflight program, whether national or international, is the ultimate prerequisite for a spacefaring nation to engage astronauts. Once this has been politically established, there must be a specific desire to create an astronaut team. A viable space system—a transportation vehicle and perhaps a space station—must either exist or be in development to realize human spaceflight. The programmatic working environment should be objective oriented (e.g., to undertake scientific research or technology development), and the required number of people with appropriate expertise must be recruited. The principal programmatic boundaries could be

- assignment to a specific mission or into a newly built or existing team
- selection of a generalist versus specialist
- permanent employment or for participation in a specific mission increment of a partner's program
- implementation of a national program (e.g., U.S., Russia, China, Europe)
- opportunities to engage astronauts within operational systems of the U.S., Russia, or China
- task assignment: commander, pilot, mission specialist, payload specialist, flight engineer, medical specialist, educator astronaut.

In making the selection, persons with professions such as test pilot, engineer, physicist, geologist, biologist, chemist, and medical doctor are generally of interest. The complementary training required might also play a role in the selection process, as well as consideration of the complementary skills that an applicant may bring to the existing astronaut corps. A medical and psychological screening process must be organized to ensure the recruitment of healthy and stable individuals who will be able to cope both physically and mentally with the stresses of being a career astronaut. Furthermore, suitability for participation in public relations activities, such as having good communications skills, is an asset in the modern media-dominated world.

To be an astronaut is an extremely demanding profession. An astronaut is a public figure and receives high public recognition. A person with a military or scientific background can suddenly become popular on a scale more typical of a movie star. This was certainly the case for the very first astronauts of the Soviet Union and the United States. Even today, the first astronauts of smaller nations can attain that status in their home countries after making their first flight into space. It requires the highest professional competence, and an astronaut requires to be aware that he/she must be presentable at all times.

Desirable prerequisites such as entry profession could either be a pilot in the capacity of a test pilot or a scientist with proven competence (e.g., a doctorate). The ability to master multi-tasking capabilities in a complex operational environment is of paramount importance. The ability to communicate well in a number of languages is a necessary asset. The family environment must be highly stress tolerant, since time for private life is minimal. Any objectionable or peculiar characteristics, habit, behavior, or political standing would reduce the chance of selection. The ideal candidate should be intelligent, agreeable, charismatic, likable, a good speaker, and a team player.

Thus far, the United States has been the nation that has sent the largest number of its citizens into space, followed by the U.S.S.R./Russia, and then Europe. The first astronaut programs of the U.S.A. and U.S.S.R. almost exclusively employed test pilots. Having had the experience of flying high-performance aircraft, these men were perceived to have the "right stuff" required to successfully complete a space mission. Their ability to withstand g-forces and operate a parachute in an emergency were also assets. Initially spaceflight was focused on testing what could be done, and how a human being could operate in a weightless environment. As spaceflight became more focused on science, it became necessary to recruit scientists and engineers as astronauts.

Most astronaut selection today is carried out by the partners of the ISS. The Multilateral Crew Operations Panel staffed by representatives of each of the ISS partners has agreed "principles regarding processes and criteria for selection, assignment, training, and certification of ISS (expedition and visiting) crew members." As is to be expected, the selection processes are rather similar for the participating nations. The astronaut selections of the ISS partners are treated in turn in the following sections.

7.1.2 United States of America

The first astronauts selected in the United States were the Mercury Seven (Fig. 2.5). President Dwight Eisenhower had decided that the best candidates to become America's first astronauts were to be found amongst its military pilots. As there was limited room in the Mercury capsule, a maximum height of 5 feet and 11 inches was imposed. About 500 candidates were screened. They went through a battery of elaborate tests that included medical checkups, psychiatric studies, and endurance tests. Finding the right tests for the selection process was a major part of the task for the selection team. The number of candidates reduced step by step in the selection process.

The first NASA astronaut classes contained only military test pilots, but once the United States had successfully carried out a number of space missions and confidence in its ability to launch a person into space was established, other professions became of interest. This is illustrated in Table 7.1, which shows NASA astronaut recruitment from 1959 until 2009. At the end of the Apollo era there was a long pause in astronaut recruitment from 1969 until 1978. New astronauts were then needed for the upcoming Space Shuttle missions, leading to 35 astronauts being recruited in 1978 including the first female American astronauts: Anna Fisher, Shannon Lucid, Judith Resnik, Sally Ride, Rhea Seddon, and Kathryn Sullivan. After the 20th astronaut class was recruited in 2009 the need for further astronaut recruitment in the United States became uncertain.

The historical development of the astronaut corps is illustrated in Fig. 7.1. At its peak in the year 2000 NASA had about 150 active astronauts. In mid-2011 there were around 60 and the intention was to maintain at least that number through 2016. The current plan is to select small groups of astronaut candidates periodically as required. There is a need to carry out recruitment campaigns regularly in order to maintain the corps, since astronauts leave for a number of reasons. They may choose to leave for personal reasons or they can be obliged to quit (e.g., for medical reasons). With the retirement of the Shuttle, astronauts are

Table 7.1. NASA astronaut recruitment 1959–2009.

Class	*Year*	*Recruits*	*Pilots*	*Scientists/ Mission Specialists*
1	1959	7	7	
2	1962	9	9	
3	1963	14	14	
4	1965	6		6
5	1966	19	19	
6	1967	11		11
7	1969	7	7	
8	1978	35	13	22
9	1980	19	8	11
10	1984	17	7	10
11	1985	14	6	8
12	1987	15	7	8
13	1990	23	7	16
14	1992	19	4	15
15	1995	19	10	9
16	1996	35	10	25
17	1998	25	8	17
18	2000	17	7	10
19	2004	11	2	9
20	2009	9	3	6

for the time being only able to participate in long-duration missions to the ISS. This increases the likelihood of reaching their career radiation exposure limit, compared with participation in shorter flights. Recently it has been discovered that astronauts returning from space missions can suffer from a swelling of the optic disk, a condition called "papilledema" that in some cases results in vision impairment that would preclude assignment to another mission for as long as this persists. The condition may be caused by redistribution of body fluids due to the

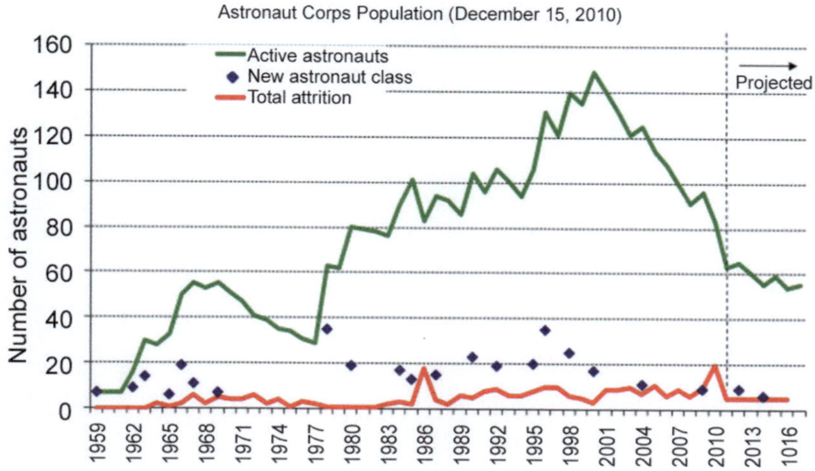

Fig. 7.1. Historic development of the NASA Astronaut Corps population (source: *Ensuring the Readiness of the Astronaut Corps*, A White Paper, NASA Astronaut Office, March 25, 2011).

microgravity environment. There is an anthropometric size limitation for astronauts flying in the Soyuz spacecraft to ensure that they can fit into the confined space of the capsule. This type of limit did not exist for the Shuttle, but the selection criteria for future recruitments will have to be modified to take size into account as long as the Soyuz is the only vehicle available to transport NASA astronauts to and from the ISS.

The need for astronaut recruitment can be assessed utilizing an analytical model introduced in 2008:

$$\begin{matrix}\text{Number of}\\ \text{astronauts in}\\ \text{post-flight activities}\end{matrix} + \begin{matrix}\text{Number}\\ \text{in orbit}\end{matrix} + \begin{matrix}\text{Number required}\\ \text{in space in the}\\ \text{following 5 years}\end{matrix} + \text{Margin} \rightarrow \begin{matrix}\text{Minimum}\\ \text{manifest}\\ \text{requirement}\end{matrix}$$

By this, the minimum manifest requirement is obtained by summing the number of astronauts occupied with post-flight activities and recovery, the number of astronauts in space, the number required for the assignments planned for the coming five years, plus a margin to take into account unpredictable constraints such as astronauts being unavailable, attrition, and mission constraints. This margin was initially 50% of the sum of the first three numbers but in 2010 was decreased to 25%. Whereas there were about 28 flight opportunities to the ISS per year with the Shuttle, there are only about six using the Soyuz, so there has been a considerable reduction in the number of astronauts required for missions.

The switch to long-duration spaceflights has led to modifications to the astronaut selection process, including a more stringent medical checkup. During the selection process, applicants spend two weeks in medical screening sessions and interviews. All new recruits to the astronaut corps must have skills that will

Table 7.2. NASA astronaut candidate basic qualification requirements (source: NASA).

<table>
<tr><th>Non-piloting background</th><th>Piloting background</th></tr>
<tr><td colspan="2">Bachelor's degree from an accredited institution in engineering, biological science, physical science, or mathematics. An advanced degree is desirable</td></tr>
<tr><td>Degree must be followed by at least 3 years of related, progressively responsible, professional experience. An advanced degree may be substituted for experience according to a predefined scale. Teaching experience is considered to be qualifying experience</td><td>At least 1,000 hours of pilot-in-command time in jet aircraft. Test pilot experience is an asset</td></tr>
<tr><td colspan="2">The candidate must pass the NASA long-duration spaceflight physical, which includes the following:
• distance visual acuity must be correctable to 20/20, each eye;
• near visual acuity must be correctable to 20/20, each eye;
• blood pressure must not exceed 140/90 measured in the sitting position;
• standing height between 62 and 75 inches (157.5–190.5 cm)</td></tr>
</table>

enable them to work with robotics, scientific experiments, and EVA tasks. They must also have an aptitude to learn the Russian language if they are not already familiar with it.

Astronaut recruitment today begins with an advert in the Federal Government's Official Jobs Website, USAJOBS. Active military personnel apply through their own military service. Only citizens of the United States are eligible for the positions. There are two sets of basic qualification requirements, one for a piloting background and the other for a non-piloting background (as described in Table 7.2). After applications have been screened, the candidates of interest are invited for two rounds of medical evaluation and interviews. In the case of class 20, the advert appeared in August 2007 and the successful applicants were announced in June 2009. The number of applicants received in response to an advert varies between about 2,500 and 3,500 persons each time, but the call for applicants in November 2011 to form the class of 2013 attracted 6,300 applicants, which was the second highest number ever received; the maximum being 8,000 in 1978 concluding a long period without recruitment.

7.1.3 Soviet Union/Russia

In the Soviet Union/Russia the selection of cosmonauts has never been a public process. Information on cosmonaut selection is still somewhat difficult to access, and there are inconsistencies in the available material. However, an effort is made herein to give a brief and accurate summary of what is known. Initially the Air Force had the monopoly on recruiting and training cosmonauts. Table 7.3 is a summary of Air Force recruitment up to 2010. In 1967 a resolution ended the

Table 7.3. Cosmonaut recruitment to Air Force groups 1960–2010.

Group	*Year*	*Recruits*	*Pilots*	*Engineers*	*Other*
1	1960	20	20		
Women's group	1962	5			5 parachutists
2	1963	15	6	8	1 navigator
Additional enrollment	1964	1	1		
3	1965	22	12	8	1 doctor and 1 engineer
Additional enrollment	1966	1	1		doctor
4	1967	12	6	3	3 navigators
5	1970	9	7	2	
6	1976	9	9		
7	1978	2	2		
8	1987	5	5		
9	1988	3	3		
10	1989	3	3		
11	1990	3	3		
Additional enrollment	1991	1	1		
Additional enrollment	1996	2	1		1 doctor
12	1997	7	7		
Additional enrollment	1998	3	1	1	1 politician
13	2003	5	3	1	1 civilian engineer
14	2006	3	3		
Additional enrollment	2007	2	2		
15	2010	3	2	1	

military monopoly on cosmonaut recruitment and opened the door for civilian organizations to promote their own candidates for space missions.

Sergei Korolev was an advocate for the recruitment of civilian cosmonauts, and it was his design bureau that was first to recruit new groups of cosmonauts.

Table 7.4. Cosmonaut recruitment by RSC Energia 1966–2010.

Group	*Year*	*Recruits*
1	1966	8
Additional enrollment	1967	2
2	1967	3
3	1972	3
4	1973	4
Additional enrollment	1977	1
5	1978	7
6 (Female selection)	1980	2
Additional female enrolment	1983	1
7	1984	2
8	1985	2
9	1987	1
10	1989	4
11	1992	3
12	1994	2
13	1996	3
14	1997	2
Additional enrollment	1998	1
Transferred	1999	1
15	2003	3
16	2006	2
17	2010	4

Recruitment by RSC Energia is summarized in Table 7.4. It was still mainly male cosmonauts that were recruited, but there were a few females. Elena Kondakova (Energia group 10) made a 169-day visit to the Mir space station from October 1994 to March 1995. She was the third (and at of the time of writing the last)

Table 7.5. Cosmonaut recruitment by the Institute for Medical and Biological Problems 1972–2003.

Group	*Year*	*Recruits*
1	1972	3
2	1978	3
3 (Female selection)	1980	4
4	1985	1
5	1989	3
6	2003	1

female Russian cosmonaut to fly into space, excluding those that were privately funded. Among the currently active cosmonauts there is one woman, Yelena Serova of Energia group 16.

In 1972 the Institute for Medical and Biological Problems in Moscow started to select groups of medical doctors as prospective cosmonauts. It was thought at the time that doctors would be required to support the crews on long-duration missions, and to carry out medical research in space. Only 2 of the 15 doctors recruited in this way have flown in space to date (Table 7.5). One was Valeri Polyakov, who holds the world record for the longest single spaceflight: 438 days. Four recruits were women, but none of them ever flew in space. The female recruitments in 1980 were related to NASA recruitment of female mission specialists for the Shuttle.

There have been additional cosmonaut selection categories. Vladimir Chelomey's design bureau selected its own engineers to train as cosmonauts to fly on the Almaz military space stations that the company was building, but none of them flew and the group was disbanded in the 1980s. Other institutes and organizations within the space industry have selected cosmonaut candidates, but many did not fly before moving on to other duties. On several occasions Soviet journalists were selected for training, but again none of them flew into space.

The Interkosmos program (Fig. 7.2) was set up to coordinate the space activities of the Soviet Union in cooperation with other Eastern bloc countries.

Fig. 7.2. The logo of the Soviet Interkosmos program.

Founded in 1966, it was expanded in 1976 to include manned spaceflight. During late 1976 military pilots from Czechoslovakia, Poland, and East Germany were screened, and two from each country were selected for cosmonaut training. In a second selection round in 1978, 10 more pilots were selected from Bulgaria, Cuba, Hungary, Mongolia, and Romania. In 1979 Vietnam joined the program. Between 1978 and 1981 there were nine manned Soyuz missions to Salyut 6 in support of the Interkosmos program. On each flight two cosmonauts from the same nation served as the prime and backup crews. In 1979 the Soviet Union expanded their guest cosmonaut program beyond Interkosmos. This led to Soyuz flights by cosmonauts from France, India, Syria, and Afghanistan and later to the sale of seats commercially.

When the military monopoly on recruitment ended, responsibility for running the medical selection of cosmonauts was given to the Institute for Medical and Biological Problems. The medical screening of both military and civilian cosmonaut candidates culminates in a decision by the Chief Medical Commission (GMK) about who should receive basic space-related training. Aspiring cosmonauts must then pass inspection by the Joint State Commission (GMVK), which is essentially a political committee. If successful, they are designated cosmonaut candidates and can begin mission training.

In Russia cosmonaut training takes place at the Yuri Gagarin Training Center in Star City, Moscow. Cosmonauts have been trained there since 1961. By November 2010 it had trained a total of 422 cosmonauts, of whom roughly half had gone on to participate in a spaceflight. The majority of them were from the U.S.S.R./Russia or U.S.A., with the remainder coming from 30 other countries. In 2009 Star City was transferred from the Russian Air Force to the Russian Space Agency, signifying a transition from military to civilian control. The merger of the military group of cosmonauts and two civilian groups from RSC Energia and from the Institute for Medical and Biological Problems has formed a single cosmonaut corps under one agency. Since the start of 2011 all active cosmonauts belong to a single Cosmonaut Space Agency.

7.1.4 European Space Agency

The European Space Agency carried out its first astronaut selection in the late 1970s in conjunction with an agreement with NASA to provide the first Spacelab reusable science laboratory for the Space Shuttle in exchange for flight opportunities for ESA astronauts. The first astronauts selected were Claude Nicollier (Switzerland), Wubbo Ockels (The Netherlands), and Ulf Merbold (West Germany). In the 1980s several European countries ran their own national recruitment campaigns for astronauts who went on to fly on either the Russian Soyuz or the U.S. Shuttle. It was agreed in 1990 to establish a European Astronaut Centre (EAC) in Cologne, Germany, on the same site as the German Space Agency's astronaut training center. It became the home base for European astronauts. In later years, a new European astronaut recruitment campaign was initiated to have

Fig. 7.3. The European Astronaut Corps, 1998: (back row left to right) Paolo Nespoli (Italy), Thomas Reiter (Germany), Christer Fuglesang (Sweden), Frank de Winne (Belgium), Michel Tognini (France), Hans Schlegel (Germany), Philippe Perrin (France), André Kuipers (The Netherlands); (front row left to right) Pedro Duque (Spain), Gerhard Thiele (Germany), Jean-François Clervoy (France), Umberto Guidoni (Italy), Léopold Eyharts (France), Reinhold Ewald (Germany), Roberto Vittori (Italy), Claude Nicollier (Switzerland) (source: ESA and J-L. Atteleyn).

sufficient crew available for the planned Columbus and Hermes spaceplane projects. Each ESA member state could propose up to five of its nationals as candidates, with ESA making the final selection of six new astronauts in 1992 who were trained for Shuttle missions and missions to Mir using the Soyuz spacecraft. In the end the Hermes project did not result in flight opportunities.

In the 1990s Europeans gained substantial experience of spaceflight, in particular with Thomas Reiter of Germany, who spent 6 months on board Mir, becoming the first European to undertake a long-duration mission. However, it was not until 1998 that a single European Astronaut Corps was established. This was formed by amalgamating existing ESA astronauts and astronauts selected by the national space agencies of the individual member states. From then on, EAC was responsible for the selection of ESA astronauts. Fig. 7.3 shows the corps after this amalgamation. Note that at that time there were no female members. With time, astronauts moved on to new functions or retired until the corps had only eight members remaining in 2008. In response, a new astronaut selection was initiated in order to meet Europe's commitments to the ISS program.

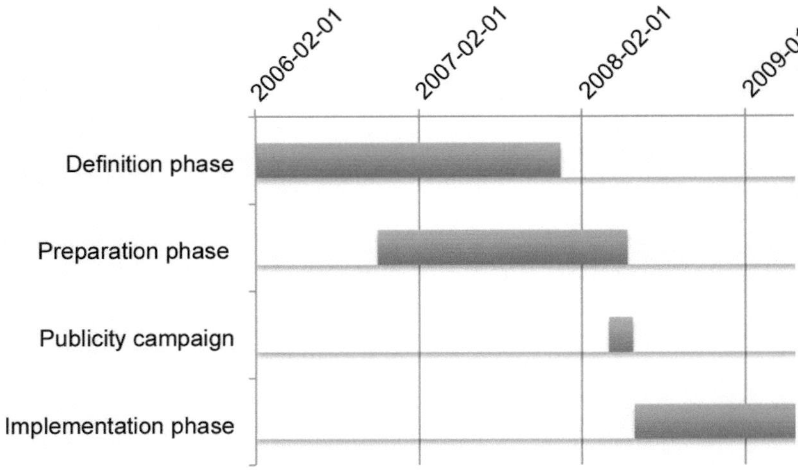

Fig. 7.4. Timeline for ESA astronaut selection 2008/2009.

The procedure recently followed to recruit astronauts in Europe was adapted from previous European selections, and from those of NASA and the U.S.S.R. It was divided into three phases: the definition phase, the preparation phase, and the implementation phase. Fig. 7.4 illustrates the timeline for these phases and the associated publicity campaign. The process began in February 2006, and ended on May 20, 2009 with the announcement of six new ESA astronauts.

A key question for the definition phase was whether an ESA-wide selection should be made, or whether national preselections should be accepted. The prospect of an ESA-wide selection was somewhat daunting, since it would involve the processing of applications by potentially more than 10,000 individuals. On the other hand, national preselections would inevitably be less fair to individual candidates, and would lead to a small and potentially inhomogeneous group from which ESA would have to select. In addition, some countries stated that they would not be able to carry out a national preselection to the required standard. The only option was therefore to proceed with an ESA-wide selection. During the definition phase, the requirements for an astronaut candidate were driven by the fact that the successful candidates would ultimately have to participate in long-duration missions to the ISS. In November 2007 the Human Spaceflight, Microgravity and Exploration Programme Board of ESA approved the planning and procedures for the selection.

The formal process started by establishing a selection board composed of members representing politics, programmatics, medical aspects, operations, science, technology, and, last but not least, public relations. An application form was drafted, followed by iterations and refinements. It was decided to accept applications only via the Internet. (It was the first time that an astronaut selection was carried out in which applications were only accepted via the Internet.) The agency also elected to go public using multiple media. An Announcement of Opportunity seeking astronaut candidates was made in major publications in Europe and on television. People from all member states were eligible to apply, even those countries that did not contribute financially to ESA's optional human spaceflight program. Applicants were required to possess a university degree in natural science, engineering, or medicine, and be either a pilot or have at least 3 years of relevant postgraduate experience. The preferred age range was 27 to 37. In addition, the applicants were expected to have language skills, and in particular a high level of linguistic skill in English. With regard to medical and psychological health the following requirements were defined:

- the applicant must be free from any disease
- the applicant must be free from any dependency on drugs, alcohol, or tobacco
- the applicant must have the normal range of motion and functionality in all joints
- the applicant must have visual acuity in both eyes of 100% (20/20) either uncorrected or corrected with lenses or contact lenses
- the applicant must be free from any psychiatric disorders
- the applicant must demonstrate cognitive, mental, and personality capabilities to enable him/her to work efficiently in an intellectually and socially highly demanding environment.

To ensure that only applications from serious and healthy applicants would require processing, all applicants had first to submit proof of their identity and a certificate of their medical status: either a JAR-FCL 3, Class 2 medical certificate from an Aviation Medical Examiner who was certified by his/her national Aviation Medical Authority, or an equivalent certificate. Only after providing this information could the applicants access the on-line application form. The final version of the application form had over 40 questions. Predefined answers could be selected to most of the questions, but there were three open questions at the end of the form. In total 9,199 applications were submitted, of which 90% were screened by a computer-based analysis support tool that applied a set of elimination criteria and assigned a score to those applications that it did not reject. Test pilots were sorted into a special group, and those that were not eliminated were assessed individually by a group of experts. The performance of the tool was verified periodically by comparing its scoring with that obtained by a group of experts from EAC and ESA Human Resources.

The selection process continued with the implementation phase which included the following steps:

1. Initial selection according to basic criteria based on the written application
2. Psychological tests focused on cognitive skills for selected candidates at the DLR Institute for Aviation and Space Psychology, Hamburg

3.. A second round of psychological tests and interviews focused on behavioral assessment at EAC, Cologne
4. One full week of medical tests at either the Institute for Space Medicine and Physiology (MEDES) in Toulouse or the DLR Flight Clinic in Cologne
5. A professional interview
6. An interview with the Director-General of ESA.

The purpose of the medical tests was to find candidates that were currently healthy and had every chance of remaining so in the future. The examinations were based on established standards in the following medical systems:

- general medicine
- ears, nose, and throat
- ophthalmology
- pulmonology
- cardiovascular system
- hematology
- abdomen and digestive system
- endocrine and metabolic
- genitourinary
- musculoskeletal and orthopedics
- dermatology
- neurology
- psychiatry and human behavior
- obstetrics and gynecology
- dental
- infectious diseases
- anthropometry
- radiation exposure
- nutrition
- physical fitness.

Once all of the candidates had been medically evaluated, their confidential medical files were sent to the ESA Medical Board which was staffed by numerous European medical experts representing the various medical disciplines tested during the medical screening process. Each candidate's medical test results were reviewed. At this stage the main reasons for rejection were related to a candidate's eyes and cardiovascular system. The final selection factors were personal history, family status, sex, religion, political activities, professional history, personal appearance, representation, ability to communicate, language bandwidth, and acceptance in the community of the European and international space programs. The board then conducted interviews preparatory to drawing up the list of finalists for the agency's Director-General, for further interview and final selection.

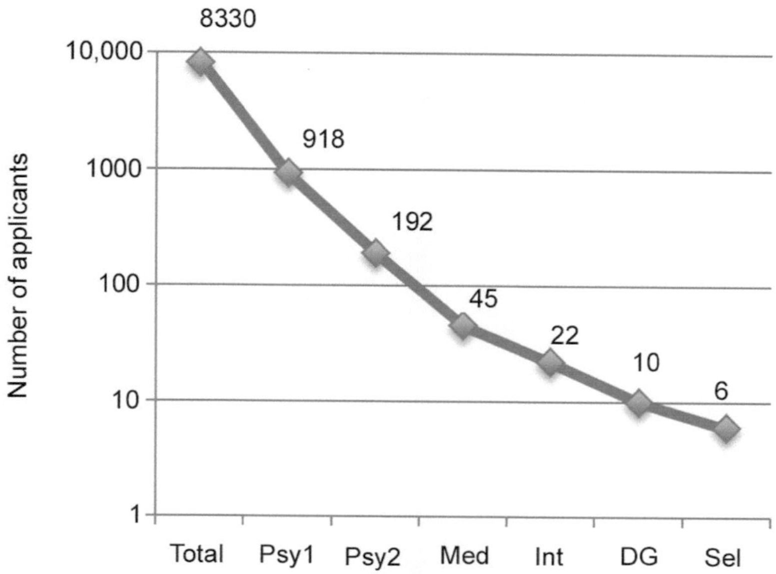

Fig. 7.5. Applicants admitted to each phase of the 2008/2009 selection process.

Fig. 7.5 shows how the number of applicants reduced at each stage in the selection process from 8,330 valid applications to 6 selected candidates. The abbreviations on the horizontal axis have the following meanings: "Total" is the total number of applications, "Psy 1" and "Psy 2" are the first and second rounds of psychological tests, "Med" is the medical testing, "Int" refers to the professional interviews, "DG" means interviews by the Director-General, and "Sel" is a successful candidate. In Fig. 7.6 the progress of two groups, one with scientists and engineers and the other with test pilots, is tracked through the selection process. It was found that both groups performed similarly, with the final selection including four scientists/engineers and two test pilots. There were 1,287 valid applications from female candidates; 15.5% of the total. Fig 7.7 shows that the ratio between female and male applicants remained close to the initial value through to the six finalists; five men and one woman: Alexander Gerst (Germany), Andreas Mogensen (Denmark), Luca Parmitano (Italy), Timothy Peake (United Kingdom), Thomas Pesquet (France), and Samantha Cristoforetti (Italy).

It is normal practice for an astronaut candidate and their family members to receive a training course in making public appearances. In addition, contractual arrangements must be made in order to enable the new recruit to transfer into the new assignment as a trainee astronaut.

To date about 50 Europeans have flown in space; about 10% of the total number of astronauts. Table 7.6 lists their flights. In addition, a small number of astronauts born in Europe have become American citizens and gone on to become NASA astronauts.

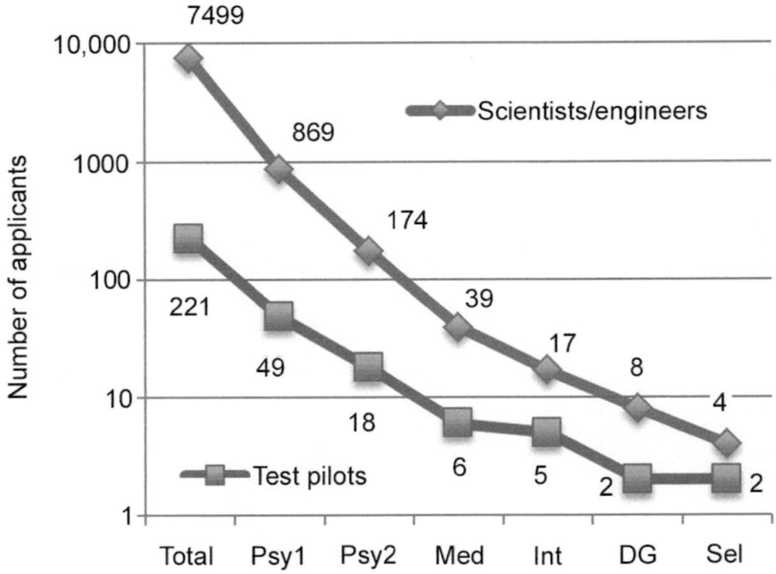

Fig. 7.6. The historical development for applications from scientists/engineers and test pilots during the 2008/2009 ESA astronaut selection process.

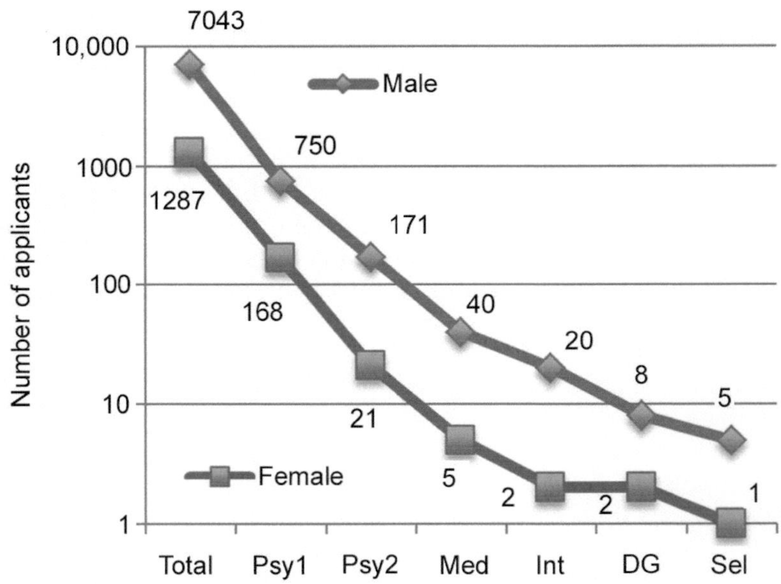

Fig. 7.7. The historical development of male and female applications during the 2008/2009 ESA astronaut selection process.

Table 7.6. European astronauts and their flights presented in chronological order (adapted from an article by Jörg Feustel-Buechl, ESA Bulletin 116, 2003).

Astronaut	*Home country*	*Space agency*	*Mission*
Vladimir Remek	Czechoslovakia	U.S.S.R.	Soyuz flight to Salyut 6, 1978
Miroslav Hermaszewski	Poland	U.S.S.R.	Soyuz flight to Salyut 6, 1978
Sigmund Jähn	GDR	IKF	Soyuz flight to Salyut 6, 1978
Georgi Ivanov	Bulgaria	U.S.S.R.	Soyuz flight, 1979
Bertalan Farkas	Hungary	U.S.S.R.	Soyuz flight to Salyut 6, 1980
Dumitru Prunariu	Romania	U.S.S.R.	Soyuz flight to Salyut 6, 1981
Jean-Loup Chretien	France	CNES	Soyuz flight to Salyut 7, 1982
Ulf Merbold	Germany	ESA	Spacelab-1, Shuttle, 1983
Patrick Baudry	France	CNES	Shuttle, 1985
Reinhard Furrer	Germany	DFVLR[a]	Spacelab-D1, Shuttle, 1985
Ernst Messerschmid	Germany	DFVLR	Spacelab-D1, Shuttle 1985
Wubbo Ockels	The Netherlands	ESA	Spacelab-D1, Shuttle, 1985
Aleksandr Aleksandrov	Bulgaria	U.S.S.R.	Soyuz flight to Mir, 1988
Jean-Loup Chrétien	France	CNES	Soyuz flight to Mir, 1988
Helen Sharman	U.K.	Private funding	Juno, Soyuz flight to Mir, 1991
Franz Viehböck	Austria	Austria	Astromir 91, Soyuz flight to Mir, 1991
Ulf Merbold	Germany	ESA	Spacelab IML-1, Shuttle, 1992
Klaus-Dietrich Flade	Germany	DARA/DLR	Mir 92, Soyuz flight to Mir, 1992
Dirk Frimout	Belgium	Belgium	Shuttle, 1992
Michel Tognini	France	CNES	Soyuz flight to Mir, 1992
Franco Malerba	Italy	ASI	Shuttle, 1992
Claude Nicollier	Switzerland	ESA	Shuttle, 1992

(*continued*)

Table 7.6. European astronauts and their flights presented in chronological order (adapted from an article by Jörg Feustel-Buechl, ESA Bulletin 116, 2003) (*cont.*)

Astronaut	*Home country*	*Space agency*	*Mission*
Hans Schlegel	Germany	DARA/DLR	Spacelab-D2, Shuttle, 1993
Ulrich Walter	Germany	DARA/DLR	Spacelab-D2, Shuttle, 1993
Jean-Pierre Haigneré	France	CNES	Soyuz flight to Mir, 1993
Claude Nicollier	Switzerland	ESA	Shuttle, 1993
Ulf Merbold	Germany	ESA	Euromir 94, Soyuz flight to Mir, 1994
Jean-François Clervoy	France	ESA	Shuttle, 1994
Thomas Reiter	Germany	ESA	Euromir 95, Soyuz flight to Mir, 1995
Claude Nicollier	Switzerland	ESA	Shuttle, 1996
Maurizio Cheli	Italy	ESA	Shuttle, 1996
Umberto Guidoni	Italy	ASI	Shuttle, 1996
Jean-Jacques Favier	France	CNES	Spacelab LMS-1, STS-78, 1996
Claudie André-Deshays	France	CNES	Soyuz flight to Mir, 1996
Reinhold Ewald	Germany	DARA/DLR	Mir 97, Soyuz flight to Mir, 1997
Jean-François Clervoy	France	ESA	Shuttle flight to Mir, 1997
Jean-Loup Chrétien	France	CNES	Shuttle flight to Mir, 1997
Léopold Eyharts	France	CNES	Soyuz flight to Mir, 1998
Pedro Duque	Spain	ESA	Spacehab, Shuttle, 1998
Jean-Pierre Haigneré	France	CNES, ESA	Soyuz flight to Mir, 1999
Michel Tognini	France	CNES	Shuttle, 1999
Claude Nicollier	Switzerland	ESA	HST service, Shuttle, 1999
Jean-François Clervoy	France	ESA	HST service, Shuttle, 1999
Gerhard Thiele	Germany	DLR, ESA	Shuttle, 2000
Umberto Guidoni	Italy	ESA	Shuttle, 2000

Astronaut	*Home country*	*Space agency*	*Mission*
Claudie Haigneré	France	CNES, ESA	Soyuz flight to ISS, 2001
Roberto Vittori	Italy	ASI, ESA	Soyuz flight to ISS, 2002
Philippe Perrin	France	CNES	Shuttle flight to ISS, 2002
Frank De Winne	Belgium	Belgium, ESA	Soyuz flight to ISS, 2002
Pedro Duque	Spain	Spain, ESA	Soyuz flight to ISS, 2003
André Kuipers	The Netherlands	ESA	Soyuz flight to ISS, 2004
Roberto Vittori	Italy	Italy, ESA	Soyuz flight to ISS, 2005
Thomas Reiter	Germany	Roscosmos, ESA	Shuttle flight to ISS, ISS crew, 2006
Christer Fuglesang	Sweden	NASA/ESA	Shuttle flight to ISS, 2006
Paolo Nespoli	Italy	NASA/ESA	Shuttle flight to ISS, 2007
Hans Schlegel	Germany	NASA/ESA	Shuttle flight to ISS, 2008
Léopold Eyharts	France	NASA/ESA	Shuttle flight to ISS, ISS crew, 2008
Frank De Winne	Belgium	NASA/ESA	Soyuz flight to ISS, ISS crew, 2009
Christer Fuglesang	Sweden	NASA/ESA	Shuttle flight to ISS, 2009
Paolo Nespoli	Italy	NASA/ESA	Soyuz flight to ISS, ISS crew, 2010-2011
Roberto Vittori	Italy	NASA/ASI/ ESA	Shuttle flight to ISS, 2011
André Kuipers	The Netherlands	ESA	Soyuz flight to ISS, 2011

[a] DFVLR was a precursor of DLR, the German Space Agency.

7.1.5 Canada

The Canadian astronaut corps was created in 1983 by selecting six people to fly on Shuttle missions. In 1992 four more astronauts were recruited. Over time, members moved on to other activities. Fig 7.8 shows the members of the corps who were active in 2002.

By 2008 the corps was down to three active astronauts, so in March of that year a new campaign with the slogan "Do you have what it takes?" was initiated to recruit two new astronauts. An on-line application generated 5,351 applicants.

Fig. 7.8. Canadian Space Agency Astronaut Corps, 2002: (standing, from left to right) Chris Hadfield, Dave Williams, and Bjarni Tryggvason; (seated, from left to right) Bob Thirsk, Julie Payette, and Steve MacLean (source: Canadian Space Agency).

Fig. 7.9. Survival test (source: Canadian Space Agency).

The selection process included the following steps:

1. Initial screening of the applications
2. Preliminary interviews
3. Robotics, flight operations, and physical tests held in Quebec
4. Survival tests, Nova Scotia
5. Medical and psychological exams, Ontario.

To test their robotic skills the candidates participated in a simulated manipulation of Canadarm2.[1] The survival tests included exercises in fighting floods and fires, and having to escape several times from a mock-up of a helicopter under water (Fig. 7.9). The process ended in May 2009 with the selection of Jeremy Hansen, a Canadian Air Force pilot, and David Saint-Jacques, a medical doctor. An overview of the number of applicants at each stage in the process is shown in Fig. 7.10.

To date, eight Canadians have flown in space; one of them living on the ISS for six months. The flights of Canadian astronauts are summarized in Table 7.7.

[1] Canadarm2 is a 17 m long robotic arm on the ISS contributed by Canada.

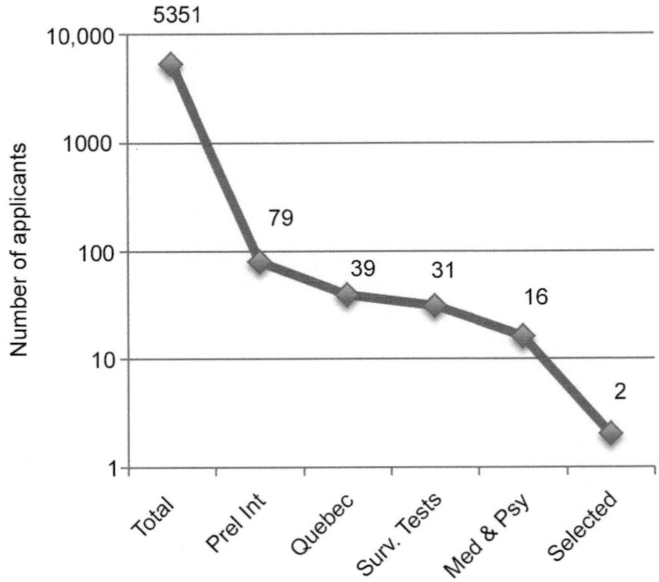

Fig. 7.10. Applicants admitted to each phase of the 2008/2009 Canadian astronaut selection process.

Table 7.7. Canadian astronauts and their flights presented in chronological order.

Astronaut	*Mission*
Marc Garneau	Shuttle, 1984
Roberta Bondar	Shuttle, 1992
Steve MacLean	Shuttle, 1992
Chris Hadfield	Shuttle flight to Mir, 1995
Marc Garneau	Shuttle, 1996
Bob Thirsk	Shuttle, 1996
Bjarni Tryggvason	Shuttle, 1997
Dave Williams	Shuttle, 1998
Julie Payette	Shuttle flight to ISS, 1999
Marc Garneau	Shuttle flight to ISS, 2000
Chris Hadfield	Shuttle flight to ISS, installation of the Canadarm2 on ISS, 2001
Steve MacLean	Shuttle flight to ISS, 2006
Dave Williams	Shuttle flight to ISS, 2007
Julie Payette	Shuttle flight to ISS, 2009
Bob Thirsk	Soyuz flight to ISS, ISS crew 2009

7.1.6 Japan

Japan selected its first astronauts in 1985 and made a further three selection rounds in the 1990s. Fig. 7.11 shows the active members of the Japanese astronaut corps in 2004. In the fifth selection in 2009 the Japanese Space Agency, JAXA, selected Kimiya Yui, Takuya Ohnishi, and Norishige Kanai from 839 male and 124 female applicants in a multi-step process (Fig. 7.12). According to JAXA, "astronauts have been selected based on the screening of application documents; a series of written exams on the English language, general and specialized knowledge of natural science, interviews, and medical and psychological examinations." It also used an isolation chamber in its 1998/1999 and 2008/2009 astronaut selections. Candidates were observed around the clock for a week inside an isolation chamber with a floor area of 77 m^2 to assess the ability of the group to undertake assigned tasks while in confinement. One such task was to follow detailed instructions to create a large number of origami cranes. Table 7.8 summarizes the flights of Japanese astronauts to date.

Fig. 7.11. The members of the Japanese Astronaut Corps, 2004: (standing, from left to right) Soichi Noguchi, Satoshi Furukawa, Akihiko Hoshide, Koichi Wakata, Mamoru Mohri, Takao Doi; (seated, from left to right) Chiaki Mukai, Naoko Yamazaki (courtesy of JAXA).

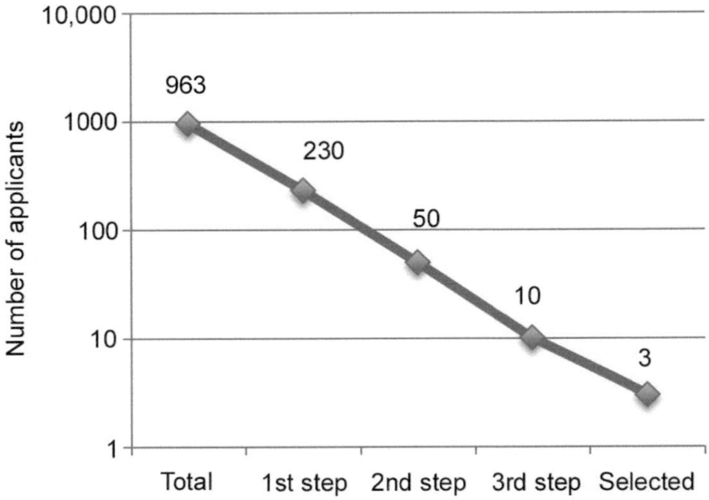

Fig. 7.12. Applicants admitted to each phase of the 2008/2009 Japanese astronaut selection process.

Table 7.8. Japanese astronauts and their flights presented in chronological order.

Astronaut	*Mission*
Mamoru Mohri	Shuttle, 1992
Chiaki Mukai	Shuttle, 1994
Koichi Wakata	Shuttle, 1996
Takao Doi	Shuttle, 1997
Chiaki Mukai	Shuttle, 1998
Mamoru Mohri	Shuttle, 2000
Koichi Wakata	Shuttle flight to ISS, 2000
Soichi Noguchi	Shuttle flight to ISS, 2005
Takao Doi	Shuttle flight to ISS, 2008
Akihiko Hoshide	Shuttle flight to ISS, 2008
Koichi Wakata	Shuttle flight to ISS, ISS crew 2009
Soichi Noguchi	Soyuz flight to ISS, ISS crew 2009/2010
Naoko Yamazaki	Shuttle flight to ISS, 2010
Satoshi Furukawa	Soyuz flight to ISS, ISS crew 2011
Akihiko Hoshide	Soyuz flight to ISS, ISS crew 2012

7.1.7 China

The first recruitment to the Shenzhou program took place in the 1990s, resulting in 14 astronaut candidates. In 2010 China selected another 5 men and 2 women from a pool of 45 military pilots. A prerequisite for the female candidates was that they be married, in the belief that this would ensure their physical and psychological maturity. Table 2.9 summarizes the Shenzhou crewing.

7.2 ASTRONAUT TRAINING

Upon selection, astronaut candidates face a complex training program to qualify them for a range of mission assignments. The principal objective of the training process is for the participant to become competent as a professional astronaut and to improve on this for further assignments as commander, pilot, mission specialist (systems operator, data, video, etc.), payload specialist (payload system operator, science coordinator), or science instrument operator working on behalf of scientists on the ground.

With the exception of China, all career astronauts are trained to participate in the ISS program according to requirements agreed by the partners. Training centers exist in the various ISS partner countries

- NASA Johnson Space Center (JSC), Houston, Texas, U.S.A.
- Yuri Gagarin Training Center, Star City, Moscow, Russia
- Tsukuba Space Center, Tsukuba Science City, Japan
- Canadian Space Agency (CSA), Saint-Hubert, Quebec, Canada
- European Astronaut Center (EAC), Cologne, Germany.

There are three stages to the ISS astronaut training program

- basic training, 1 to 1.5 years
- advanced training, 1 year
- increment-specific training, 1.5 to 2.5 years.

The training schedule is illustrated schematically in Fig. 7.13. In parallel with their training, astronauts have duties such as participating in public relations activities and supporting miscellaneous special projects. Training continues even during the spaceflight. For a new astronaut, there is the need to adapt to the new environment. In-orbit training can be necessary, for example, to refamiliarize an astronaut with procedures or to prepare for an unplanned task.

Until recent times, both a prime and a backup crew were trained for each mission to avoid delays if a crew member could not fly for some reason. This practice has now ceased, and today the next crew in line serves as the backup crew until 6 months prior to launch. The transition to a six-person crew on board the ISS, remaining in space for months at a time, has necessitated a change in the crew training philosophy. Since it is not possible to predict in advance all of the tasks that they will have to perform, each member of the crew must possess a

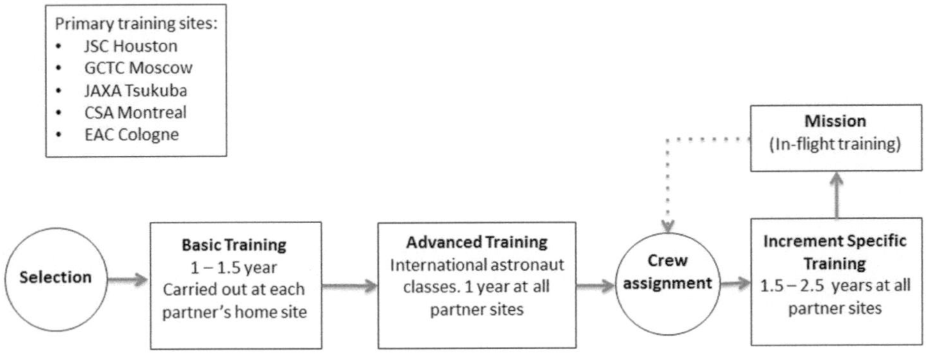

Fig. 7.13. ISS partner astronaut training schedule.

broad knowledge base, including ISS systems and maintenance, robotics, payload science, EVA, and the Soyuz. There is enough cross-training capability in the crew now to provide backup in the case of incapacitation of a crew member; this was lacking in the three-person crews.

Whereas previously basic training was mainly obtained in Houston and Star City, the partner countries are now providing most of the initial training at home. The goal of basic training is to give the participants the knowledge and skills required to follow mission training after they have received an assignment. This part of the training is largely classroom based. As an example, the basic training at ESA is divided into four blocks, the contents of which are detailed in Table 7.9.

The blocks in the NASA training program from astronaut candidate (ASCAN) to flight are shown in Fig. 7.14. As the only transfer vehicle between the ground and the ISS is currently the Soyuz, it is essential that all astronauts be competent in Russian. The Soyuz system is operated entirely in Russian, including all communication with the Moscow Ground Control Center. The Soyuz commander is always a Russian Space Agency cosmonaut. Hence, Russian is introduced in the basic training program for non-Russians. NOLS is an acronym for National Outdoor Leadership School. The education provided to astronauts focuses on leadership and tolerance. NEEMO is an underwater habitat off the Florida coast in which astronauts can spend several weeks as part of their training. A T-38 Talon is a high-performance aircraft used to provide spaceflight readiness training. Maneuvering this jet places the crew under stress in a rapidly changing situation that is physically uncomfortable; it requires multi-tasking and crew coordination. In addition, there is an element of real danger—over the years several astronauts have been killed in crashes. At NASA this training is considered an invaluable preparation for spaceflight.

Once a crew has been assigned to a mission, the mission-specific training begins. Each ISS partner provides the relevant training for their contributions to the ISS. Soyuz training occurs at this stage, and has two forms depending on the composition of the three-person crew. Each crew consists of at least one Russian Space Agency cosmonaut and one NASA astronaut. There are two forms of train-

Table 7.9. Basic training for astronauts provided by the European Space Agency.

Block	*Contents*
1. Introduction	Introduction to ESA and European space industry Introduction to global space activities Introduction to space technologies
2. Fundamentals of human spaceflight	Spaceflight and aerospace engineering Electrical engineering Computer engineering Life science Material science Fluid physics Earth science Space science
3. Space systems and operations	ISS European contributions to ISS Soyuz systems, infrastructure, and operations Human physiology in space Health care in space
4. Special skills and miscellaneous	Presentation and media skills Diving and EVA Fitness training Russian Photography and video skills training Robotics Human behavior and performance training Survival training Microgravity environment including participation in a parabolic flight campaign Flight training, JAR-FCL[a] theory, practice, and certification Tours of the ISS partner facilities

[a] Joint Aviation Requirements–Flight Crew Licensing.

ing that depend on whether the third member of the crew is Russian or is from NASA or one of the other international partners. A Russian cosmonaut takes the role of mission commander and occupies the center seat in the Soyuz capsule. To the left is flight engineer no. 1, who has a similar role to a co-pilot. To the right is flight engineer no. 2. (The role of flight engineer no. 1 requires a higher level of ability in Russian than flight engineer no. 2.). Fig. 7.15 shows cosmonaut Yuri Onufrienko occupying the commander's position and NASA astronaut Carl Walz in the left-hand seat of a Soyuz that was docked to the ISS; it clearly illustrates the cramped conditions inside the capsule.

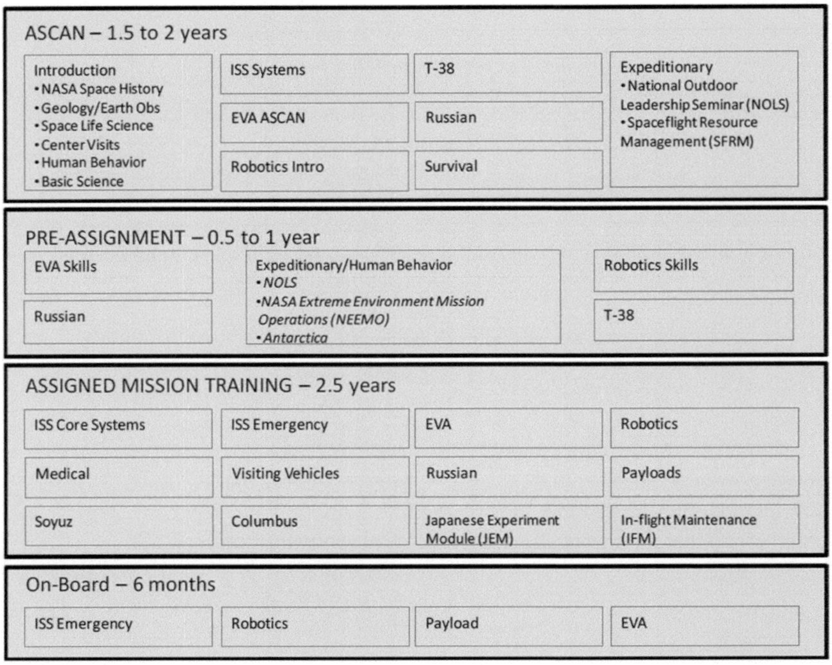

Fig. 7.14. Summary of NASA training for astronauts to prepare them for a visit to the ISS (source: NASA).

Fig. 7.15. Cosmonaut Yuri Onufrienko and NASA astronaut Carl Walz in the Soyuz TM-33 spacecraft docked to the ISS, 20 April 2002. The crew moved the Soyuz to a new docking port on the station in preparation for the arrival of a new crew (source: NASA).

There are minimum training requirements for a six-person ISS crew included in the agreements between the ISS partners. Each crew member must be qualified for tasks under the following planning guidelines:

A. At least one United States On-orbit Segment (USOS) and one Russian Segment (RS) ISS core system specialist are required to be on orbit at all times in order to perform nominal and off-nominal system operations.[2] To meet this requirement, there must be one USOS and one RS core system specialist on each Soyuz as a precaution against a return to three-person crew operations. As a result, there will be two USOS core system specialists and two RS core system specialists on orbit.
B. At least one Space Station Remote Manipulator System (SSRMS) robotics specialist is required on orbit at all times. There will be two such specialists on orbit to support complex robotics operations during nominal six-person crew operations.
C. At least two Crew Medical Officers (CMOs) are required on orbit at all times, resulting in two CMOs on each Soyuz as a precaution against a return to three-person crew operations.
D. At least one specialist per payload on orbit at all times for a predetermined set of USOS payloads operated during an increment.
E. At least one specialist per payload on orbit at all times for a predetermined set of RS payloads operated during an increment.
F. At least two USOS Extravehicular Maneuvering Unit EVA specialists and two RS Orlan EVA specialists on orbit during nominal six-person crew operations.
G. At least one USOS Japanese Experiment Module (JEM) Remote Manipulator System (RMS) robotics specialist and one USOS JEM RMS operator on orbit during nominal six-person crew operations.
H. The number of Special Purpose Dexterous Manipulator specialists on orbit will be determined based on the complexity of operations requirements.
I. At least two ATV specialists for any increment that includes an ATV docking or undocking, and at least one ATV specialist for any increment that includes an attached ATV.
J. At least two SSRMS qualified crew members will be assigned to perform HTV rendezvous, track, and capture operations for an increment that could include an HTV arrival or departure, at least one of whom must be qualified as an HTV system specialist. Additionally, there will be one HTV system specialist for any increment that includes an attached HTV. The HTV operators/specialists will be USOS crew members.
K. At least one Progress vehicle Teleoperator Control System (TORU) specialist is required on orbit at all times. There will be two such specialists on orbit

[2] The term United States On-orbit Segment applies to the on-orbit elements of the National Aeronautics and Space Administration (NASA), Japan Aerospace Exploration Agency (JAXA), European Space Agency (ESA), and Canadian Space Agency (CSA); the term Russian Segment (RS) applies to the on-orbit elements of Roscosmos.

during nominal six-person crew operations. The TORU specialists will be RS crew members.

These crew requirements impose limitations on mission assignments. It is clear that the flight opportunities for astronauts from countries other than the U.S. and Russia are more limited than for NASA astronauts and Roscosmos cosmonauts.

7.3 ASTRONAUT CODE

It is common practice in many professions to follow an ethical code specially written to describe the responsibilities of the members of the profession. European astronauts established the Charta of the European Astronaut Corps for the following reasons:

- to maintain an ethical guideline for team members
- to announce the moral boundaries of the corps when employing their competence and expertise
- to set the standards for training, spaceflight employment, and conduct of space experimentation on behalf of users.

The charta (Fig. 7.16) is signed by the ESA astronauts in 2001. In 2000 a code of conduct was drawn up to govern the conduct of ISS crew members. NASA published its own astronaut code of conduct in 2008 (Fig. 7.17).

7.4 HUMAN SPACEFLIGHT OPERATIONS

In the case of the ISS, a network of control stations participate in on-board operations. The U.S. Payload Operation and Integration Center is located at the Marshall Space Flight Center in Huntsville, Alabama. The Mission Control Center at the Johnson Space Center in Houston, Texas is in charge of the U.S. segment. It is also the lead control center for the entire station. The Mission Control Center (TSuP) in Korolev, formerly Kaliningrad, near Moscow, is in charge of the Russian segment. Partner centers are the Canadian Space Agency Mission Control Center in Quebec, the JAXA Manned Spaceflight Control Center in Tsukuba and the Columbus Control Center near Munich. There is coordination at least twice per day between all task-sharing operations centers of the ISS consortium.

Planning for an "increment" on board the ISS starts years before it takes place. The high-level planning is put into a document called the On-Orbit Operations Summary. Detailed scheduling for each week of the increment is then put into the Short Term Plan. This is available via the Onboard Short Term Plan

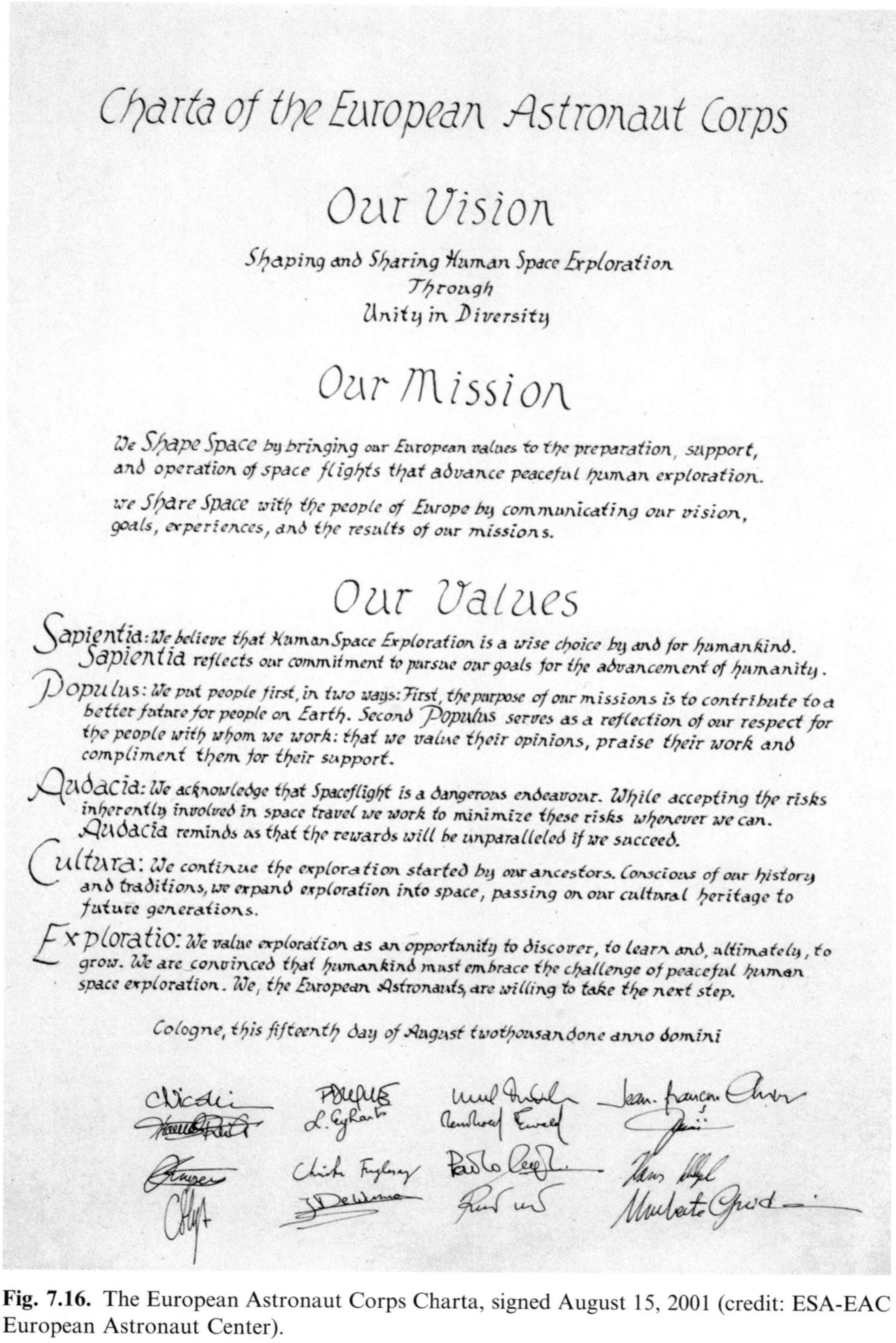

Charta of the European Astronaut Corps

Our Vision

Shaping and Sharing Human Space Exploration
Through
Unity in Diversity

Our Mission

We Shape Space by bringing our European values to the preparation, support, and operation of space flights that advance peaceful human exploration.

we Share Space with the people of Europe by communicating our vision, goals, experiences, and the results of our missions.

Our Values

Sapientia: We believe that Human Space Exploration is a wise choice by and for humankind. Sapientia reflects our commitment to pursue our goals for the advancement of humanity.

Populus: We put people first, in two ways: First, the purpose of our missions is to contribute to a better future for people on Earth. Second Populus serves as a reflection of our respect for the people with whom we work: that we value their opinions, praise their work and compliment them for their support.

Audacia: We acknowledge that Spaceflight is a dangerous endeavour. While accepting the risks inherently involved in space travel we work to minimize these risks whenever we can. Audacia reminds us that the rewards will be unparalleled if we succeed.

Cultura: We continue the exploration started by our ancestors. Conscious of our history and traditions, we expand exploration into space, passing on our cultural heritage to future generations.

Exploratio: We value exploration as an opportunity to discover, to learn and, ultimately, to grow. We are convinced that humankind must embrace the challenge of peaceful human space exploration. We, the European Astronauts, are willing to take the next step.

Cologne, this fifteenth day of August twothousandone anno domini

Fig. 7.16. The European Astronaut Corps Charta, signed August 15, 2001 (credit: ESA-EAC European Astronaut Center).

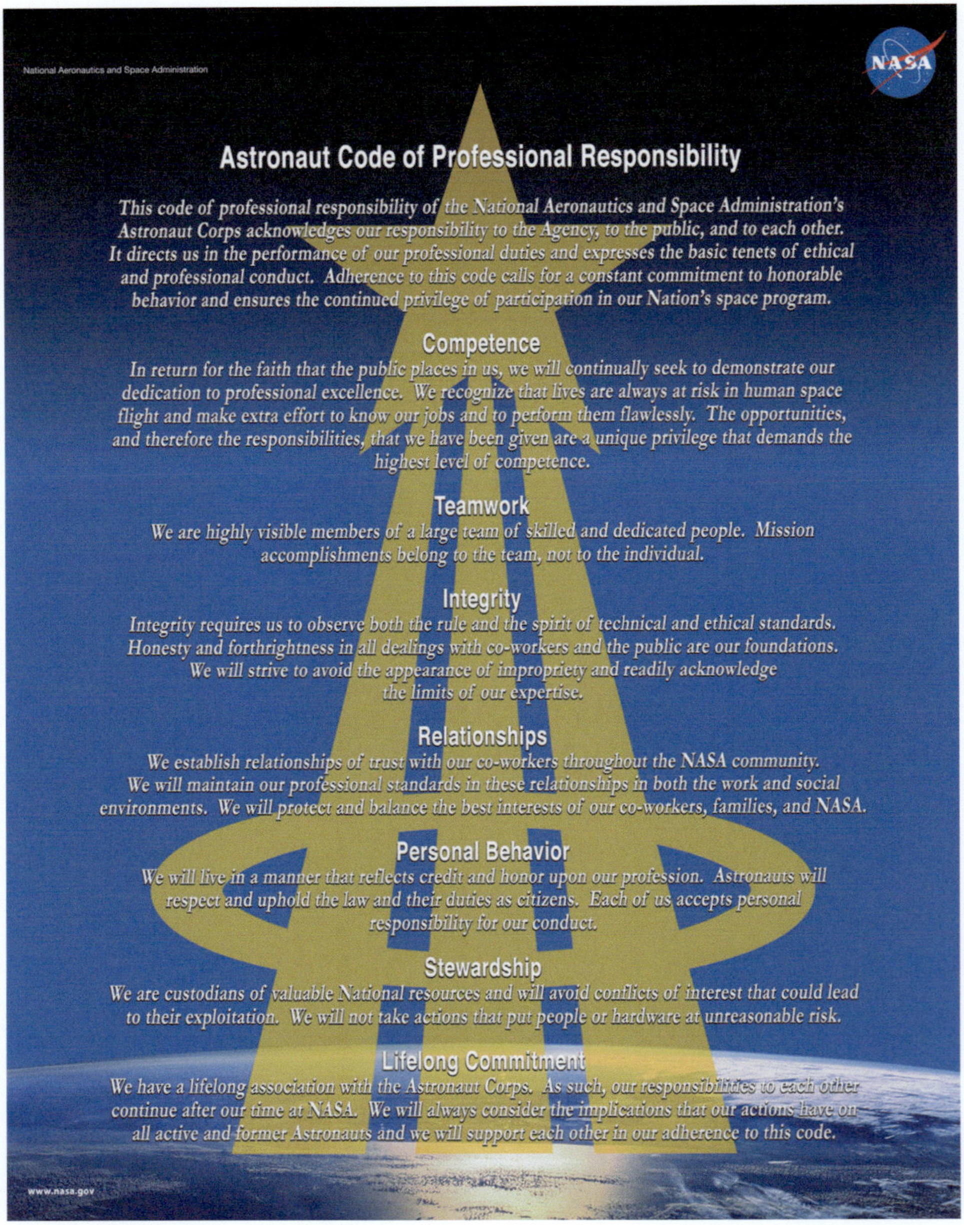

Fig. 7.17. NASA Astronaut Code of Conduct (source: NASA).

Viewer (OSTPV), an Internet-based viewer that shows activities both on the ISS and on the ground in an integrated timeline. Fig. 7.18 shows a portion of an OSTP for March 15, 2012. The timeline is continuously updated on the ground and uploaded to the crew. In addition, a detailed schedule for each day is sent to astronauts on the

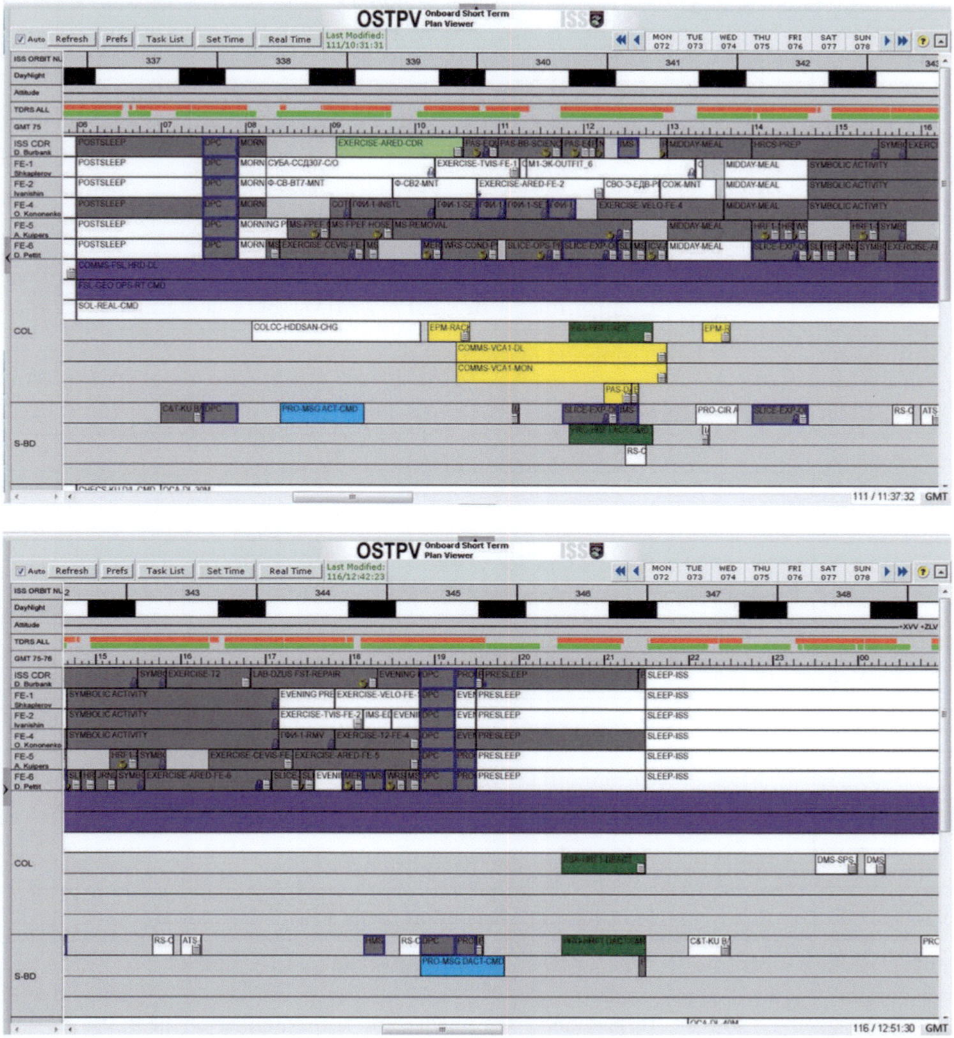

Fig. 7.18. Screenshots of the portions of the OSTPV for March 15, 2012 that describe the astronauts' planned activities for the day (source: ESA Columbus Control Center, Munich).

ISS from the Russian Mission Control Center in the form of a radiogram called a Form 24, an example of which is presented in Table 7.10. It is rather detailed, and a remnant of the former Soviet and Russian means of controlling on-board operations. The abbreviations "CDR" and "FE" stand for mission commander and flight engineer, respectively. The commander is elected among the six crew members in residence at any time, and acts as coordinator and ultimately head of decision making on board the station.

Table 7.10. Radiogram No. 8683u, Form 24 for March 15, 2012 (source: Mission Control Center, Moscow for the International Space Station, TsUP, original version in Russian).

Time (GMT)	*Crew*	*Activity*
06:00–06:10	CDR, FE-1 FE-4, FE-5, FE-6	Morning inspection of the station
06:00–06:10	FE-2	Inspection station PSA Testing SM/DCS-2 section 3.3 page 3 Control Fuse-4 BPP, BPP-7 MIM1/BOT paragraph 8.2 p. 8-2
06:10–06:40		The morning toilet
06:40–07:30		Breakfast
07:30–07:55		Daily Planning Conference (S-band)
07:55–08:30	FE-5	Preparation for work
07:55–08:15	CDR, FE-1, FE-2, FE-4, FE-6	Preparation for work
08:15–08:25	FE-6	Marangoni—Set up WB
08:15–10:15	FE-1	Verifying reportage lamp SS 307. Tagup (S-band)
08:15–09:45	FE-2	Cleaning BT-7 on GZHT4/ES FGB paragraph 4.3.2 III.4 p. 4-7
08:25–09:25	FE-6	Physical exercise (CEVIS)
08:30–09:00	FE-5	Marangoni—Disconnect cable FPEF
09:00–09:45	FE-5	Marangoni—Removal of the filter hose FPEF
09:00–09:15	FE-4	Inspection of the replaceable half-coupling of the 4GB4 hydraulic unit of cooling loop KOB-2
09:05–10:35	CDR	Physical exercise (ARED)
09:15–10:15	FE-4	RELAKSATSIYA—Hardware setup. Tagup (if required)
09:25–09:35	FE-6	Glovebox (MSG)—Activation
09:45–10:45	FE-2	Cleaning ventilation screens on the interior panels of FGB (pp. 116, 316, 231, 431)/ES FGB sh.6 paragraph 4.3.2 p. 4-7
09:45–12:35	FE-5	Marangoni—Dismantling FPEF
10:05–10:20	FE-6	MERLIN—Replacing the desiccant

Time (GMT)	*Crew*	*Activity*
10:15–10:45	FE-4	RELAKSATSIYA—Operation mode setup. Tagup (if required)
10:15–11:15	FE-1	Physical exercise (TVIS)—Day 4
10:20–11:00	FE-6	WRM—Start pumping condensate
10:35–11:00	CDR	PASSAGES—Installation of equipment for the experiment
10:45–11:05	FE-4	RELAKSATSIYA—First measurement session
10:45–12:15	FE-2	Physical exercise (ARED)
12:45–13:00	FE-6	ICV—Change of procedure at the midpoint
11:00–11:45	CDR	PASSAGES—Collection of scientific data. Session 1
11:05–11:45	FE-6	SLICE—Prep. operations
11:05–11:35	FE-4	RELAKSATSIYA—Re-installing the hardware mode. Tagup (if required)
11:15–11:20	FE-1	MRM1 Comm configuration/RTC CM p. 8.8 III.2 p. 8-26 (2,813 SMEs III.2)
11:20–13:20	FE-1	Installation of a sliding shelf in the glove box MRM1
11:35–11:55	FE-4	RELAKSATSIYA—Second measurement session
11:45–12:05	CDR	PASSAGES—Equipment stowage
11:45–12:25	FE-6	SLICE—Operations
12:05–12:10	CDR	EPM—Laptop activation and PCMCIA memory card installation
12:10–13:40	FE-4	Physical exercise (VELO)—Day 4
12:10–12:15	CDR	NEUROSPAT—Renaming questionnaire
12:15–12:55	FE-2	Refilling (separation), EDV (KOV) for the "Electron" (Fill EDV (KOV) #1000 from CWC #1072, 1039)/SM SOGS Section 3.2 p. 3-3, 3-4
12:25–12:40	CDR	IMS and conference on stowage of goods
12:25–12:35	FE-6	SLICE—Replacement of burner

(*continued*)

Table 7.10. Radiogram No 8683u, Form 24 for March 15, 2012 (source: Mission Control Center, Moscow for the International Space Station, TsUP, original version in Russian) (*cont.*)

Time (GMT)	*Crew*	*Activity*
12:35–12:45	FE-6	Marangoni—Dismantling of WB
12:55–13:00	CDR	PASSAGES—PCMCIA stow
12:55–13:35	FE-2	TO—Coolant/Coolant point 4.4 CM sh.1 p. 4-26
13:00–14:00	CDR, FE-5, FE-6	Lunch
13:20–13:25	FE-1	MRM1—Reconfiguration of the communication system/RTC CM p. 8.8 III.4 p. 8-26 (2,813 SMEs III.4)
13:40–14:40	FE-1, FE-2, FE-4	Lunch
14:00–15:30	CDR	HRCS—Preparation for the laying of cable
14:00–14:20	FE-5	Set up equipment for the measurement of body mass
14:00–14:40	FE-6	SLICE—Operations
14:20–14:30	FE-5	HRF1—Measurement of body mass
14:30–14:40	FE-5	Swap the CWC 1062 for CWC-I 1016 to pump the condensate to the tank WPA WW
14:40–17:10	FE-1, FE-2, FE-4	Symbolic activity
14:40–14:50	FE-6	SLICE—Flash card replacement
14:50–15:00	FE-6	HRF1—Measurement of body mass
15:00–15:15	FE-6	Logging
15:10–15:30	FE-5	SLAMMD—Stow equipment
15:15–15:35	FE-6	Symbolic activity
15:30–15:50	CDR, FE-5	Symbolic activity
15:35–17:05	FE-6	Physical exercise (ARED)
15:50–16:50	CDR	Physical exercise (T2)
16:20–17:20	FE-5	Physical exercise (CEVIS)
16:50–18:20	CDR	DZUS—Repair fastener

Time (GMT)	*Crew*	*Activity*
17:05–17:25	FE-6	SLICE—Check ventilation
17:10–17:50	FE-4	RELAKSATSIYA—Final operation and dismantling of equipment. Tagup (if required)
17:10–17:50	FE-1	Evening work preparations
17:10–18:10	FE-2	Physical exercise (TVIS)—Day 4
17:25–17:35	FE-6	SLICE—Replacing videotape
17:35–17:55	FE-6	Evening work preparations
17:50–18:50	FE-4	Physical exercise (T2)
17:50–18:50	FE-1	Physical exercise (VELO)—Day 4
17:55–18:10	FE-6	MERLIN—Removal of desiccant
18:10–18:30	FE-2	Data editing of IMS/IMS pp. 2.2.1, 2.2.2, 2.3, pp. 2-2, 2-4
18:10–18:25	FE-6	Private psychological conference
18:20–18:50	CDR	Evening work preparations complementary to work
18:25–18:40	FE-6	End pumping condensate WRM
18:30–18:50	FE-2	Evening work preparations
18:40–18:50	FE-6	Microgravity Science Glovebox (MSG)—Power off
18:50–19:15		Daily Planning Conference (S-band)
19:15–19:30	CDR, FE-5, FE-6	Conference on the scientific program
19:15–19:30	FE-1, FE-2, FE-4	Evening work preparations
19:30–19:35	CDR	Start playback of photos/videos
19:30–21:30	FE-1, FE-2, FE-4, FE-5, FE-6	Pre sleep
19:35–21:25	CDR	Pre sleep

(*continued*)

Table 7.10. Radiogram No 8683u, Form 24 for March 15, 2012 (source: Mission Control Center, Moscow for the International Space Station, TsUP, original version in Russian) (*cont.*)

Time (GMT)	*Crew*	*Activity*
21:25–21:30	CDR	End of data playback photos/video
21:30–06:00		Sleep
Task list	FE-1, FE-2, FE-4	Preparation of reports for the Roscosmos site URAGAN (hurricane)—Observations and photography EKON—Observations and photography

Notes:
1. SM Window #9 shutter opening is at crew discretion w/Report to MCC.
2. See OSTP for references to U.S. activities.
3. Pre-sleep ops.: daily food prep., dinner, pre sleep.

End of radiogram

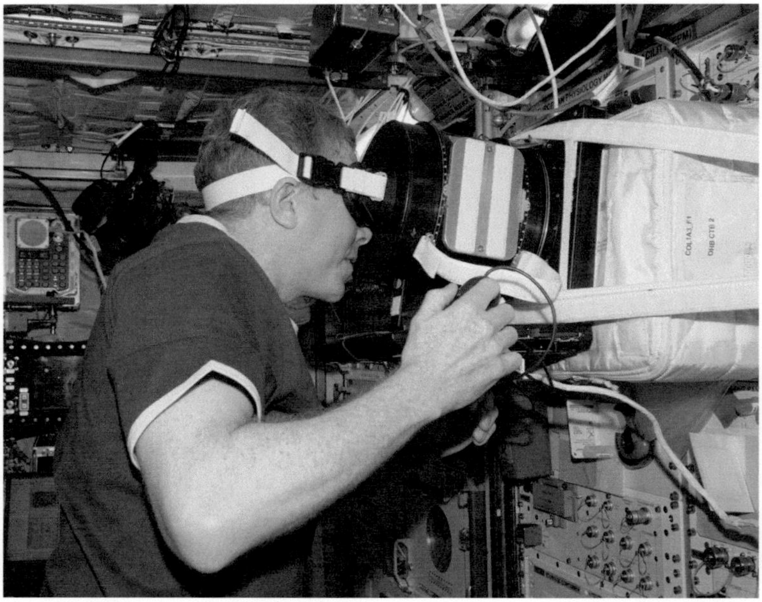

Fig. 7.19. NASA astronaut Mike Fossum working with the PASSAGES experiment in the Columbus Laboratory, October 2011 (source: NASA).

The Form 24 shown is for March 15, 2012, which was the day when the European PASSAGES experiment was run for its final time. The experiment was to study how the brain processes visual information gathered in space to manage motion in relation to obstacles (e.g., passing through a restricted opening). Fig. 7.19 shows Mike Fossum participating in the experiment on a previous occasion. Other experiments that the astronauts worked on on March 15 were

- ICV, Integrated Cardiovascular Alternate Experiment
- Marangoni, surface fluid physics experiment
- NEUROSPAT, an experiment to understand the brain's electrical activity in microgravity
- RELAKSATSIYA (Relaxation), Earth observation experiment
- SLICE, Structure and Liftoff in Combustion Experiment
- SLAMMD, Space Linear Acceleration Mass Measurement Device.

They also carried out maintenance tasks. Throughout the day astronauts undertook physical exercise using one of the available training devices

- CEVIS, cycle ergometer
- ARED, advanced resistive exerciser
- TVIS, treadmill
- VELO, ergometer bike with load trainer
- T2/COLBERT, advanced treadmill.

In the afternoon the crew spent some time on "symbolic activities" which involved signing commemorative memorabilia.

The cyclogram has traditionally been used by the U.S.S.R./Russia as the principal planning tool for all operational steps from launch to landing of flight crew, including contingencies. A cyclogram is still used for up–down transfers of the Soyuz system. Handwritten cyclograms were produced during early Soviet space missions in order to detail the operations of the spacecraft. A particularly famous cyclogram is a detailed time chart of the 96-day Salyut 6 mission. Cosmonauts Grechko and Romanenko drafted it in advance of their mission in order to have a detailed summary of the flight on paper, and refined it in space to track the passage of time using eight methods. Their cyclogram provides a unique record of their flight, and in 2010 was sold at auction in the U.S. for $50,000.

7.5 REFERENCES/RECOMMENDED LITERATURE

Astronaut Fact Book, NASA NP-2005-01-001JSC 2005, NASA (*http://spaceflight.nasa.gov/spacenews/factsheets/pdfs/astro.pdf*)

Chambers, M.J. and Chambers, R.: *Getting Off the Planet: Training Astronauts*, Apogee Books (2005).

Farand, A.: *The Code of Conduct for International Space Station Crews*, ESA Bulletin 105, ESA (2001).

Feustel-Büechl, J.: *25 Years of European Human Spaceflight*, ESA Bulletin 116, ESA (2003).

Hall, R., Shayler, D., and Vis, B.: *Russia's Cosmonauts, Inside the Yuri Gagarin Training Center*, Springer/Praxis (2010).

International Space Station 6-Crew Strategic Planning Document, SSP 50826 (Draft), NASA Johnson Space Center (2008).

Ley, W., Wittmann, K., and Hallmann, W.: *Handbook of Space Technology*, Wiley–Blackwell (2009).

Preparing for the High Frontier: The Role and Training of NASA Astronauts in the Post-Space Shuttle Era, National Academy of Sciences of the U.S.A., The National Academies Press (2011).

Roach, M.: *Packing for Mars*, Norton (2010).

Robinson, S.K., Forrester, P.G., McDonald, H., and Whitson, P.: *Ensuring the Readiness of the Astronaut Corps*, A White Paper, Astronaut Office, NASA Johnson Space Center (2011).

Seedhouse, E.: *Prepare for Launch: The Astronaut Training Process*, Springer/Praxis (2010).

Shamsutdinov, S.: The selection of cosmonauts, *Journal of the British Interplanetary Society*, **50**, 311–316 (1997).

Thiele, G., Schaarschmidt, H., Damann, V, Weerts, G., and Danesy, F.: *Report on the ESA Astronaut Selection*, ESA (2010).

Websites

Canadian Space Agency: *www.asc-csa.gc.ca*
Japan Aerospace Exploration Agency: *www.jaxa.jp/index_e.html*
Mission Control Center, Moscow: *www.mcc.rsa.ru*
NASA, The International Space Station Live: *http://spacestationlive.jsc.nasa.gov*
Yuri Gagarin Training Center: *www.gctu.ru*

8

A new image of Earth: The flight of European astronaut Gerhard Thiele on STS-99 in 2000

Gerhard Thiele

ABSTRACT

Dr. Gerhard Thiele participated as a mission specialist on the Shuttle Radar Topography Mission (SRTM) on board STS-99, February 11–22, 2000. The mission goal was the three-dimensional digital mapping of the Earth's surface. This chapter contains Dr. Thiele's own account of his space mission.

8.1 GENERAL REMARKS

The opportunity to work as an astronaut is a unique situation in one's life. Human spaceflight is still one of the noblest of endeavors, a challenge right at the edge of the seemingly impossible and yet a reality for more than 50 years. We have learned a lot since the first days when daring men and women flew into space in order to prove that a human being could actually survive in that hostile environment. Today we fly into space not to survive, but to live and work. The modern space traveler has little in common with the pioneering men and women of the early days, but is still associated with the spirit of those early attempts to venture into the unknown.

In this chapter, I do not seek to make an extended and careful review of the human path into space; Chapter 2 provides a review of the various phases in the history of human spaceflight. Rather, I will try to provide an insider's view of what it takes and what it means for an individual to travel into space. As I had the privilege to fly on the now retired U.S. Space Shuttle, much of my account is specific to this transportation system and the procedures employed by NASA, but I will try to highlight those issues that are still important more than a decade after my flight and are likely to remain so into the future.

Where technical details are necessary they will be explained at a level sufficient to enable the reader to follow my story without being overwhelmed with technicalities. Unless otherwise stated the technical references were my own notes, which I collected during my training as an astronaut, as well as training material provided for both the Space Shuttle and the payload, the so-called flight data file (FDF) of mission-specific information, and last but not least the Shuttle Crew Operations Manual (SCOM), for many the astronauts' bible. It should be noted that the documentation for my flight was out of date by the time the Shuttle was retired as a result of ongoing enhancements to the vehicles during the final decade of operations. Nevertheless, in giving a personal account it seemed best to use the information that applied at the time I was active.

The chapter is organized as follows. At the start I will briefly introduce the Shuttle Radar Topography Mission (SRTM) in terms of its scientific assignment. The main part of the text will explain how these objectives were translated into operational tasks and roles, and how the mission was flown. Because the astronauts play a very visible role in the mission, I shall introduce the crew first. Then I will describe my own perspective as mission specialist 1 (MS1), including the training period and specific preparations for the flight. The account of the mission follows the sequence of flight phases, beginning with launch, then on-orbit operations, and finally landing. Some scientific results will be presented to highlight the outcome of the mission. To wrap up, I provide a personal opinion of what was learned from the Shuttle during this marvelous machine's 30 years of operations.

8.2 THE MISSION GOAL AND THE CREW

The task of the SRTM was to obtain the data to create a digital three-dimensional map of most of the Earth's land mass at an unprecedented level of detail for such an extensive survey. The areal resolution of the final map was to be 1 second of arc in both latitude and longitude, corresponding to a linear resolution of roughly 30 m, and the vertical accuracy was to be about 16 m. These data now serve scientific disciplines like geology, hydrology, ecology, and even archeology. Beyond the scientific interest, topographic data and maps are an essential part of daily life. Any kind of infrastructural development relies on the information contained in maps. Telecommunications is one case of commercial exploitation, because wireless service providers need topographic data to identify the most appropriate locations for their antennas.

Let us turn now to the crew of the SRTM mission. Often people think that the key personnel in spaceflight are the astronauts. This certainly appears to be so at first sight. When the astronauts in their launch and entry suits board a Shuttle on launch day the international press is there to report the event, especially if the crew includes a non-U.S. citizen. In my case their reports appeared on the evening news on television in my home country and in newspapers the next day. But it is often overlooked that behind this very visible part of a space mission is a large team, some members of which may have spent years preparing for that flight. In

Fig. 8.1. Crew portrait STS-99 (source: NASA).

the case of SRTM some 300 scientists, engineers, and technicians were involved in the preparation of the mission over a 3-year period. In addition, a small team continued to work for another two and a half years to analyze the data that were obtained. Hence the astronauts are only a small part of a large team. They are the ones in the spotlight, similar to the tip of an iceberg that can be seen rising above sea level only because there is a large mass underneath the surface supporting it. In this sense, the crew picture should be viewed as a symbol for the entire SRTM team. Fig. 8.1 is our official crew picture. In the back, starting at the left are Janice Voss (JV),[1] mission specialist 3 (MS3) and payload commander (P/L CDR); Kevin R. Kregel, mission commander (CDR); Dominic L. Gorie, pilot (PLT); and Janet L. Kavandi, mission specialist 2 (MS2). The front row shows on the left Mamoru Mohri, mission specialist 4 (MS4) representing Japan's National Space Development Agency (NASDA); and myself, flying as MS1 and representing ESA.[2]

All crew members except myself already had spaceflight experience. Kevin flew his fourth space mission, and SRTM was his second as commander. JV was

[1] JV passed away on February 6, 2012, during the production of this chapter.

[2] If I refer in the text to a specific role rather than the astronaut as an individual I will use the abbreviations as given here in parentheses (e.g., MS2 instead of Janet).

payload commander of SRTM and with this her fifth flight she was the most experienced crew member. Dom and Janet had previously flown together on STS-91, the final Shuttle flight to dock with the Russian space station Mir. Mamoru flew in 1992 on STS-47 as payload specialist for NASDA.[3]

8.3 MISSION ASSIGNMENT AND PREPARATION

The reader should bear in mind that the descriptions of the tasks and responsibilities of the individual crew members given below are related to the specifics of the SRTM flight. Some of the descriptions may have evolved during the 30 or so missions that flew afterwards owing to technical enhancements ot the Space Shuttle, changes in procedures, or different mission profiles.

For a typical Shuttle mission the crew was assigned about a year ahead of the scheduled launch date. Assignment to a mission has a significant effect on the daily routine of an astronaut. When not on a crew, an astronaut's training is restricted to maintaining a certain qualification level from which one can become fully proficient in a short time. Most of the time an astronaut supports spaceflight activities in one or several of the many technical and operational areas. For example, before my flight I worked in the Computer Support Branch on issues regarding the laptop computers used on a Shuttle flight.[4] After the end of SRTM, I became a CAPCOM (capsule communicator, a term introduced during the Mercury program) and as such supported other flights from the Mission Control Center (MCC) in Houston. These are just two of many examples of tasks performed by astronauts on the ground. They are organized in several branches, so are called "branch jobs", and they fall away as mission training and preparation take precedence.

8.4 TRAINING

The details of the training program depended on the role of each astronaut during the mission. While some of the training was generic, due to the role as for example CDR or MS2, a significant portion of the training depended on the specific technical or scientific tasks of the mission. During the first meeting of our crew the commander defined the responsibilities of the crew members beyond the generic ones of their assignments. This enabled each person to focus on distinct areas.

[3] I vividly remember Mamoru's first spaceflight on STS-47 as I was starting my training in 1992 at NASA's Johnson Space Center in Houston, Texas, as a backup payload specialist for the German D2 mission.

[4] These laptop computers perform secondary functions (e.g., an application called "WORLDMAP" allows the display of the Orbiter's current position over the Earth's surface). These laptops must not be confused with the General Purpose Computers (GPCs) mentioned in Section 8.4 on training.

Interestingly, sessions where the entire crew trained together occurred significantly less than one might imagine. For instance, ascent and reentry training was mostly performed by the crew seated on the flight deck: CDR, PLT, MS1, and MS2. On the other hand, the four mission specialists were predominantly responsible for the scientific aspects of the mission, and so Kevin and Dom did not participate in payload-related training to the same level of detail as the mission specialists.

My assignment to SRTM as MS1 resulted in three groups of responsibilities: (1) the generic tasks of MS1 on a Shuttle flight; (2) the role of shift-lead during payload operations; and (3) the tasks of a contingency EVA crew member.

Let us focus first on the generic tasks as these pertained to the role of MS1 at the time of SRTM. During launch and reentry MS1 was seated on the flight deck together with CDR, PLT, and MS2.[5] Consequently, a sound understanding of all the flight deck activities was essential, especially for the dynamic phases of launch and reentry. This required a profound knowledge of the relevant Shuttle systems, and especially those that you might be required to operate. For MS1 this direct interaction was limited to the communication system (COMM).

In order to be ready for the ascent and reentry training with Kevin, Dom, and Janet, I took refresher lessons and even some of the so-called QUAL[6] classes for the major systems that would be especially important for these phases of the flight. This meant reviewing the data-processing system (DPS; in layman's terms the Orbiter's computer system), the environmental control and life support system (ECLSS), the main propulsion system (MPS), the orbital maneuvering system (OMS), the reaction control system (RCS), the auxiliary power unit and hydraulics system (APU/HYD), and the electrical power system (EPS). These refresher classes were mostly in the first two or three weeks of the training, and were a very good preparation for the joint training with the other crew members on the flight deck.

The second large block of training was payload related. Because the radar system operated quasi-continuously during the mission, the crew worked in two shifts named red and blue. While the commander carried the overall responsibility for the safe and successful execution of the mission, responsibility for payload operations was given to a specific member of the crew on each shift. As P/L CDR, Janice bore the overall responsibility for the payload. She served as shift-lead for the blue shift, working with Mamoru and Dom. I was shift-lead for the red shift, working with Janet and Kevin. Being shift-lead required a knowledge of payload activities well beyond the realm of nominal operations. Consequently a thorough familiarization with the scientific goals and the scientific radar equipment provided the mandatory background for the required understanding of the operational activities during the SRTM mission.

[5] During the landing of SRTM Mamoru was seated on the flight deck. Accordingly, he trained as MS1 during reentry simulations.

[6] QUAL stands for qualification. At the end of a given basic training phase, the trainee demonstrates during a QUAL that he has gained the required knowledge and skills. QUAL lessons are required for most of the various Shuttle systems.

Finally, Janet and I were the crew members assigned to EVA duties. No EVA was scheduled for the SRTM mission because the achievement of the specific mission objectives did not require it, but on any mission it might become necessary to go outside to deal with a faulty piece of equipment either to restore it to operation or to ensure the safe return of the Orbiter. One example of a contingency EVA would be to close and/or to manually latch the payload bay doors at the end of the mission. Hence two members of each crew were trained for an unscheduled EVA.

These three training responsibilities resulted in MS1 having the highest training load of the SRTM crew. Occasionally, surprising training plans were devised in order to fulfill the many obligations. As an EVA crew member I was obliged to do physical training, which was basically weight lifting for two hours twice per week. To squeeze this into the schedule, once the first sport class of the week appeared from 5 to 7 PM one afternoon and the second the next morning at 7 AM. Of course, having two of these training sessions so close together would undermine the objective. This just reflects how dense a training schedule could be at times in preparation for a Shuttle flight.

8.4.1 Training as MS1

The Space Shuttle was probably the most complex flying machine that was ever built. While this complexity offered options in the event that a certain system failed, it also imposed a high demand on the crew to maintain situational awareness and master the potential failure scenarios. Ascent and reentry therefore played a dominant role in the overall training schedule because in these dynamic phases of the flight decisions had to be taken very quickly. A system failure on orbit would not usually result in a critical situation for the Orbiter or the crew. Once the vehicle was circling the Earth at almost 8 km/s, it would continue to do so even if one of the currently operating computer systems were to fail or if a short were to cripple an electrical bus. There would be time to reconfigure the affected system and put the Orbiter back into a safe and even operational mode. This "fail operational/fail safe" design principle means that after a first system failure the crew would have been able to recover and carry out normal operations with reduced redundancy; and after a second failure the vehicle could have been configured to return safely home.

During ascent to orbit or during descent to Earth the situation was very dynamic and therefore downright different. There was hardly time to evaluate a failure, and the right decision had to be taken very quickly, sometimes within a few seconds. The quick and correct recognition of a failure was therefore an essential prerequisite to the execution of the necessary corrective actions. However, it was not sufficient for one crew member to recognize the situation and take action. CDR, PLT, and MS2 had to rapidly agree on the nature of the situation and the corrective action. In essence MS2 served as a flight engineer and sat behind and between the commander and pilot. My role as MS1 was somewhat limited here, since the seat was directly behind the pilot and provided only a restricted view of certain instruments. The task of MS1 was therefore initially that of an observer who monitored the execution of the malfunction procedures. It required a high level of situational awareness to keep track of which systems were still operating and what kind of additional failure could have potentially severe consequences.

The complexity of the task of the crew on the flight deck becomes apparent when one studies some of the systems listed above. For the ascent and reentry phases of the flight all five general purpose computers were operating (GPC1 to GPC5). Four of them were in charge of executing the necessary commands and running the so-called primary aviation systems software (PASS). The fifth computer provided redundancy in case a generic failure in the PASS required switching to its redundant backup flight software (BFS). The four GPCs commanded, amongst other equipment, three main engines, three electrical main buses (plus a number of secondary buses) powered independently by three fuel cells, three auxiliary power units and their related equipment, and a pair of cooling loops in the ECLSS. The interdependencies rapidly become very complex. A simple example illustrates this. On STS-93 Columbia delivered the Chandra X-Ray Observatory to orbit. An electrical bus suffered a short several seconds after lift-off. This electrical malfunction caused the shutdown of one of a pair of redundant main engine controllers on two of the three main engines of the Orbiter. As a result, the center and right engines were running on one controller. Fortunately, both redundant controllers on those two engines functioned normally, the engines continued to run, and the ascent to orbit was not jeopardized. However, another failure, not necessarily in the electrical system, could have affected one of the remaining controllers on those engines and caused it to shut down. The crew needed to be prepared in order to know what to do in that eventuality. In this specific case, the anticipated actions would change with time, because an engine failure early in a flight had a different consequence than one later in the ascent, or near MECO[7] when the engines had accomplished most of their work and would soon be shut down anyway.

For SRTM we simulated well over a hundred ascents, quickly advancing from rather simple failures to complex scenarios. We might have a cooling loop failure which allowed the temperature of the electronics in the Orbiter to rise and put equipment at risk. An engine might shut down early in the ascent, making it impossible to reach orbit and requiring us to select one of the available abort modes. An electrical failure might disable the engine controllers, as on STS-93. And last, but not least, the GPCs might cease to agree on the proper commands; if they voted down one of their number then it was a simple situation, but if they were to split with one pair voting against the other pair, who was to say which pair were right? Things could become very busy very quickly on the flight deck. However, hectic action is of little help in such a situation. The key was a sound understanding of the systems and their various interrelationships, and a clear, concise, and strict communications scheme. It was interesting and personally rewarding to

[7] MECO is main engine cutoff. It was a signal by the computer system that commanded the three main engines to shut down, normally sent after 8 minutes and 36 seconds.

watch the crew evolve into a well-functioning unit capable of dealing with the challenges set by the training team.

I will conclude this section with an important observation. The flight control team in Houston was indispensable for the execution of the mission. I had the privilege to work in the flight control team for several Shuttle missions during the assembly phase of the ISS and thus gained a view of a mission from the perspective of ground control. Behind every successful human space mission stands the flight control team of highly dedicated and very knowledgeable experts. Without their support, the often spectacular outcomes would not be possible.

8.4.2 Payload training

Payload-related training activities were naturally the most time consuming after those for the Orbiter flight deck systems. This began with an introduction into the scientific field of remote sensing in general and the goals of the SRTM mission in particular. A familiarization of the radar system introduced the technical details of the equipment, focusing on its functionality and the principles of operation. It made measurements at two wavelengths in the C-band and the X-band corresponding to roughly 6 cm and 3 cm. The C-band radar was developed by NASA's Jet Propulsion Laboratory (JPL) in Pasadena, California, and the X-band radar (X-SAR) by the German Aerospace Center in Oberpfaffenhofen.

The training in Pasadena resulted in a large amount of travel and time away from Houston, especially early in the training, but this was time well spent because it enabled us to become familiar not only with the scientific background and technical details of the mission but also with the people who "made it work", the scientists and engineers for whom the mission was being flown. Fortunately, many of the X-SAR scientists and engineers spent several months collocated with their JPL counterparts, making good contact with this group much easier.

From a payload point of view, the critical flight phases included the extension and retraction of the mast; the adjustment of the radar interferometer, which in essence meant the high-precision orientation between the primary antenna in the payload bay and the secondary antenna at the far end of the mast; and the calibration of the radar system. These activities took up a large part of the training. During this time, I learned a lot about novel concepts in remote sensing and radar technology. Of course, such training does not make one an expert capable of designing a new instrument, but the insights gained in this interesting field were for me a special reward.

8.4.3 EVA training

For many astronauts, performing an EVA protected solely by a space suit, which is like a personal spaceship giving an unfettered view of the Earth and space, is the crowning experience of their career. A number of scenarios could have required an unscheduled EVA during SRTM. One case would have been to extend the mast for the secondary radar manually if the drive motors failed. Likewise the X-SAR antenna in the payload bay could have been moved manually from its stowage configuration to the operating position. The EVA training for purely contingency reasons typically consisted of four training runs in the huge water tank of the Neutral Buoyancy Laboratory (NBL), but SRTM required roughly twice as much EVA training. Due to the complexity of the radar system in Endeavour's cargo bay, several scenarios for certain payload failures were identified and training devised to repair or secure the apparatus. Therefore Janet and I were trained for an additional five runs in the NBL to rehearse these tasks (Fig. 8.2). Although nine training days under water may not appear to the reader like a very heavy load, the overhead of familiarization with the EVA equipment and especially the suit made it quite pronounced.

8.4.4 Interjection

It is interesting that although a significant amount of training time was devoted to the dynamic flight phases of launch and reentry the ascent lasted only about 30 minutes from launch until the OMS-2 burn that achieved a circular orbit and the descent lasted about one hour from the deorbit burn until landing. This amounted to well under 1% of the 11-day flight time. The overall training time nevertheless exceeded some 30% for these activities. While it is not justified to compare the risks or potential dangers of launch or reentry with the payload activities, it is still legitimate to ask whether it is necessary to devote that much training time to such brief mission phases.

8.5 OTHER FLIGHT PREPARATIONS

Besides training, the preparations for a Shuttle mission involved a multitude of other activities, of which I will describe just a few.

The design of the mission patch was one of the first tasks for a crew. Since I was the only rookie on the flight I was responsible for the patch, a tradition that dates back to the Gemini program. The STS-99 patch (Fig. 8.3) depicts the main characteristics of the flight. The structure in the cargo bay resembles the main antennas of the radar. A long mast is extended 60 m out of the bay with the secondary antenna at its end. Radar beams are illustrated by the golden beams. A certain artistic license was taken, as in reality no active radar signal was sent by the secondary antenna; its role was to receive the signal from the primary antenna after it had been scattered by the Earth. Even if the Earth is covered by clouds the surface becomes visible because radar beams can penetrate the cloud layer, as illustrated in the patch. Since ours was a mission to map Earth, the patch includes latitude and longitude circles. But the most important feature is the stars. We astronauts believe that spaceflight is an essential step on our path to the future. It may not be the most important one, but it must not be neglected! Therefore we wanted to represent the future in our patch and the best symbol of the future that

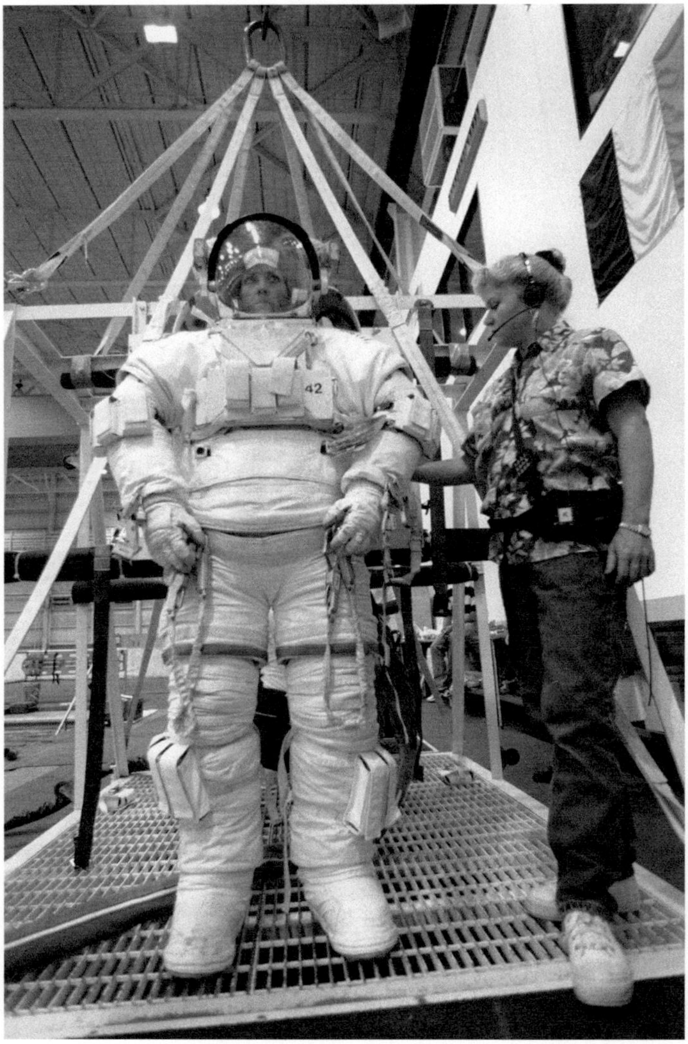

Fig. 8.2. Astronauts Janet L. Kavandi and Gerhard Thiele are about to be lowered into a deep pool at the Johnson Space Center for an underwater training session. Note that Thiele cannot be seen in this picture because he is positioned behind Kavandi. Kavandi has weights on the training version of her space suit (EMU) which help to provide neutral buoyancy in the pool (source: NASA).

we could think of was our children. Each star represents a child of a crew member. JV did not have a child, but she could not be omitted from this symbolism and so was given the rising star (with the thought that she might have a child at some time in the future). An interesting item is the placement of the names of the crew members, which is a remnant of an attempt to represent the border of the patch as a Möbius

Fig. 8.3. Mission patch STS-99 (source: NASA).

strip, a symbol for the infinite yet closed universe; eventually this feature was discarded from the patch.

The selection of clothing, food, hygiene, and personal items may not appear to be so important during flight preparation, but if one considers that the crew is required to undertake a challenging task on orbit it is apparent that they must not only be well trained but also enjoy an acceptable level of well-being. Here food and hygiene play a role that must not be underestimated.

Food selection was one of the most entertaining parts of flight preparation. Twice we were invited to the "Shuttle kitchen", a facility where Shuttle food was prepared. Each crew member tasted a small sample of the many food items available, and noted down on a list those items they especially liked or, in some cases, disliked. Then each astronaut submitted a proposal for the individual menus for the 11 days on orbit. A nutritionist assessed the menus for all the required vitamins and minerals and made some minor adjustments where necessary. My favorite food was shrimp in a hot horseradish sauce. I requested it for every other day, but it was actually included only three times.

My personal hygiene kit was a toothbrush and toothpaste, a no-rinse shampoo (as there is no shower on board), a comb, a razor and shaving cream, a deodorant and a hand cream. The kit led to an interesting discovery, which I will explain in the section on living in space!

The selection of clothing was pretty straightforward since the proper amount for each astronaut was determined by strict regulations. For each day a polo shirt could be selected, most of them with short sleeves but some with long sleeves. Likewise, short or long pants could be selected. Interestingly, astronauts seem to prefer warmer cabin temperatures at least during the first couple of days in space. As a result, and because the Orbiter would be "cold-soaked" on the day prior to reentry, making it quite chilly inside, a sweater was recommended. Shoes are of no use on orbit but socks are worn for protection and comfort. The astronauts were allowed to keep the shirts and sweater after the flight, which made great giveaways for friends and family.

8.6 LAUNCH PREPARATIONS

Each Shuttle crew training for a spaceflight was assigned a priority which essentially reflected its position in the launch schedule, progressively increasing until it became "number one", meaning that it would be the next mission to be launched. During this time small changes occurred. In particular, training intensity decreased and training in the simulator was assigned at more convenient times rather than starting very early or very late in the day. The most notable change was the period of quarantine that began seven days prior to launch. The crew said goodbye to their families and moved into crew quarters at the Johnson Space Center in Houston, which provided amenities similar in standard to a small mid-class hotel. Quarantine includes a strict health stabilization plan designed to minimize the risk of a crew member becoming ineligible to fly for medical reasons. Contacts with third parties are allowed only if those people have passed a medical examination to verify they are not carrying contagious diseases. These restrictions apply to family members as well, and children under the age of 16 are no longer allowed to come into physical contact with their parents in quarantine.

One milestone that has to be achieved prior to launching some missions is the so-called sleep-shifting. The day–night cycle for a crew is shifted such that launch and the first and mandatory flight activities occur in the equivalent late morning hours and the remainder of the first workday. SRTM was a two-shift flight and the red shift was lucky because no sleep-shift was necessary for Kevin, Janet, and me at all. The projected launch time was 12:30 PM, allowing us to wake up normally. The contrary was true for Dom, JV, and Mamoru of the blue shift. They had to be sleep-shifted by 12 hours over a period of only a few days and for them the launch occurred late in the day. To support this sleep-shifting, strong lights in the common room of the crew quarters at the Cape gave the impression of a bright day even in the middle of the night. Those already asleep in their rooms were obviously not affected by this.

On the third day before launch the crew flew in NASA's T-38 jets to the Kennedy Space Center, often referred to as the Cape. Some of the engineers and technicians who had spent months preparing our Orbiter for flight had gathered on the landing strip to welcome us and our CDR thanked them for their efforts. Normally not many people were able to attend this ceremony, since the majority were either still working on the vehicle or had already been reassigned to one of the other upcoming missions. This reflects how large the team was that worked on the Shuttle program.

The last days in the crew quarters were fairly quiet. We again reviewed checklists and procedures, the CDR and PLT participated in meetings in some cases relating to the latest work on the Orbiter (there were always last minute issues) and others about weather forecasting which most of the time put an interesting flavor on the attempted Shuttle launch. The Cape is located at a latitude of 28°N, and so is subtropical where weather can change quickly, especially in the summer. Therefore the weather plays an important role in determining what is called the launch probability. The first launch attempt for SRTM was scheduled for January 31, 2000. However, a warm front was expected to pass through the area. The predictions for meeting the weather minimums for a Shuttle launch were not at all favorable for that morning. Heavy rain, low clouds, and gusting winds signaled from the very start that we would be very lucky to leave Earth that morning. Nevertheless, we hoped the front would feather out and at the scheduled time we would be between two strips of rainy weather with blue sky above us, but it was not so. In any case the launch had to be scrubbed due to a problem with an electronics box called the MECA (Master Events Controller Assembly) that governed events in the solid rocket motors.

As the Shuttle launch managers decided to exchange the defective MECA in the solid rocket booster (SRB) rather than relying on the backup alone our launch was delayed for 11 days. The crew returned to Houston, then reentered quarantine seven days prior to the new launch date. We had no specific tasks to carry out during this additional time; indeed there ought not to have been any because we were ready to launch the first time around! The challenge was to stay focused on the mission and remain vigilant. In the case of SRTM this was not as difficult as it was, for example, for the crew of STS-51 in 1993, where the launch was delayed three times before finally lifting off and they went into quarantine on four occasions in a period of two months. On February 7 we returned to the Cape for our second attempt at launch.

"Night-viewing" is a fairly moving experience that takes place the evening prior to the launch attempt. Each member of our crew was able to name 22 people to be taken by bus at about 10 PM to a location some 300 m from the Shuttle. The buses came and the people spilled out, took photos, and milled around. What they did not know was that we were also there. Since we were in quarantine we could not greet them. It is strange to see vague figures of your family and friends, and attempt to make out the silhouette of your wife or children against the backdrop of the brilliantly illuminated Shuttle.

8.7 LAUNCH

To witness the launch of a Shuttle was a remarkable experience for any spectator. One not only saw it go, one *felt* it. Fig. 8.4 shows our launch photographed from a safe distance of about 8 km. Shortly after the picture was taken, the vehicle rolled around its main axis onto the desired azimuth then progressively tilted over and flew away from the observer. With the engine exhaust directed towards the viewing site, people would have felt the heat on their faces. Depending on the weather, and particularly in the summer when high humidity efficiently carries low acoustic frequencies, one could also feel one's stomach vibrating. This is an indication of the power released at launch. The white clouds seen around the launch pad are clouds of water vapor, which originated from a basin of water right beneath the launch pad. The water was used to damp acoustic shock waves that would otherwise have been reflected from the ground and could have destroyed the Orbiter.

Fig. 8.4. Launch of Space Shuttle Endeavour STS-99 from Launch Pad 39A, photographed from the roof of the Vehicle Assembly Building at the Kennedy Space Center, February 11, 2000 (source: NASA).

On launch day we followed the same protocols as for the first attempt less than two weeks earlier. We were very well prepared and remained ahead of the timeline, a very comforting feeling. The weather was perfect, there were no technical issues, and it was clear to us that we would lift off. A bus picked us up at the crew quarters. The head of the Flight Crew Operations Directorate (FCOD) accompanied us as far as the Launch Control Center (LCC) where he disembarked. For the final three miles to the launch pad the crew were alone.

At the launch site the astronauts were met by the Closeout Crew. They supervised the boarding of the Shuttle and ensured that we were properly strapped into our seats. The hatch of the Orbiter was closed and leak checks were carried out to verify that the cabin was airtight. We monitored the communication loops to keep track of events. The CDR and PLT performed some standard final reconfigurations to the vehicle's systems. For the mission specialists, the check-out of the various communication loops was one of the main activities.

In the final hours of the countdown the Shuttle entered two preplanned "holds", one at T-20 minutes for a duration of 20 minutes and another at T-9 minutes for 9 minutes.[8] "T" refers to the computer time, which is the sequential time used by the computer to command countdown activities. The actual time to launch was obviously longer by the durations of the holds. The reason for preplanned holds was to provide off-line time to undertake certain mechanical

[8] The holds always occur at the same time in the countdown. The length of each hold is flight specific (e.g., flights to the space station had normally a 40-minute hold at T-20).

activities. The most prominent example was the loading of liquid hydrogen and oxygen propellants into the vehicle's external tank, which began at T-3 hours, some 10 hours prior to the scheduled time of launch. The T-20 minutes and T-9 minutes holds were to provide time to overcome problems such as might occur during leak checks after closing the cabin hatch, or if the Closeout Crew needed extra time to leave the pad after completing their activities.

For SRTM, the T-9 minutes hold was extended by some 10 minutes as a result of three different issues, all minor. It was clear, however, that this would not jeopardize the launch. With two and a half minutes to go, we closed the visors of our helmets and activated the oxygen supply. CDR thanked the ground team for wishing us a safe and successful flight, and then he spoke to us on the private intercom loop. Finally, we fell silent.

At T-7 seconds the three main engines ignited in sequence within less than half a second. As they developed full thrust, a gimbal check verified the full range of motion because only if the thrust could be applied in a chosen direction could the Shuttle be steered on the desired trajectory. Inside, we could both feel and hear how the vehicle was coming alive. The engines at the tail of the Orbiter were not located directly on the centerline of the entire stack. This caused the Orbiter to "lean forward" slightly, a motion that could be clearly felt inside. But because the vehicle was firmly held to the ground by four huge bolts on the skirt of each of the SRBs, the system acted like a spring. After the maximum deflection the vehicle swung gently back, and as it passed through the upright position at $T = 0$ the SRBs ignited. The solid rocket motors were very powerful and when they were fired up there was no doubt we were definitely starting our journey! Many astronauts have attempted to describe the moment of SRB ignition, but words simply do not do it justice. Indeed, the SRBs provided in excess of 80% of the lift-off thrust. Each SRB weighed about 590 tons, 500 tons of which were the propellants consisting of ammonium perchlorate as oxidizer and aluminum as fuel. A polymer acted as glue to bind all the elements together and served as a secondary fuel. Once ignited, these motors could not be shut down like an engine running on liquid fuel, they burned until exhausted. The two SRBs were required to deliver exactly the same amount of thrust within very tight limits, so they were made as a pair using a single production run of propellant. As the thrust of the SRBs could not be regulated in flight, their thrust had to be ensured through the proper layering of the fuel. Without equal thrust on the two sides of the external tank, the stack would have become non-steerable. The nature of a solid rocket, and the unavoidable slight thrust difference between the two SRBs, meant we had a fairly rocky ride during the first two minutes. Sometimes this phase was so rough that crews found it difficult to read their cockpit displays.

Returning to activities in the cockpit, the first call almost right at lift-off was "102-102" by the PLT to indicate that both computer systems, the four GPCs, and the BFS, had properly switched from the pre-launch phase operating mode of 101 to the first flight phase. Soon after lift-off the Shuttle rolled on its back and flew away from the launch site. The CDR reported completion of this maneuver with "Roll program, Houston", and was acknowledged by MCC "Roger roll,

Endeavour". These calls served to verify that communications had been switched from LCC at the Cape to MCC in Houston.

After two minutes the SRBs burned out. They lost pressure very quickly which meant that the thrust they were providing reduced significantly in a short time. To us, the removal of this acceleration felt as if someone had slammed on the brakes. It was a sensation that I hadn't experienced in the simulator. The solid rocket motors separated, parachuted into the Atlantic, were recovered, and refurbished for reuse.[9] Once we had released the SRBs, the remaining time on the main engines of the Orbiter was very smooth and the cabin became remarkably quiet.

In addition to the various instruments showing the status of the systems, the crew observed key milestones like the so-called "no comm boundaries" that indicated abort options in case the ascent could not be completed as planned. "No comm" meant the crew must execute the relevant abort even if they were out of communication with the ground.

The first abort mode is called RTLS, an abbreviation for return to launch site. If a problem occurred in the early part of the flight and RTLS was required, the Shuttle would continue to fly downrange until SRB separation. Then a pitch-down maneuver would flip the stack over so that the engines of the Orbiter would first cancel out the downrange motion and then propel the vehicle back to the Cape. During this phase the desired attitude of the Shuttle is a careful trade-off to build up the speed to return to the Cape without falling back into the atmosphere. On finally jettisoning the external tank, the Orbiter would glide to a normal landing at the Cape about 25 minutes after having been launched.

Three minutes and forty seconds after lift-off the call "negative return" from MCC indicated that the Shuttle was too far away from the launch site and moving too fast to return. The preferred abort option now was TAL, meaning a transatlantic landing. Various landing sites in Europe and Africa were identified for different inclinations of the Shuttle orbit relative to the equator. For SRTM one of the TAL abort sites was as high in latitude as Shannon in Ireland. A TAL abort would lead to a landing about 40 minutes after lift-off. An engine failure during a relatively short period in the later part of the ascent would have resulted in the AOA option, meaning abort once around. This would be invoked if the Shuttle could not reach a stable orbit but had sufficient energy to be able to go around the Earth once and land in California about 90 minutes later. The call "press to ATO", meaning abort to orbit, could occur roughly 3 minutes prior to main engine cutoff, and would indicate that although the desired orbit could not be achieved a stable orbit at a lower altitude was possible. No mission was ever called upon to attempt RTLS, TAL, or AOA, but on STS-51F Challenger had to

[9] The overall time to recover the burned-out boosters, return them to the Cape, and analyze them to confirm that their burn had been nominal, was 11 days. This was therefore the minimum time allowed between two consecutive Shuttle launches, as it was necessary to ensure that the previous launch went as expected before committing to the launch of the next Shuttle.

perform an ATO after one of the three main engines failed 5 minutes and 45 seconds into the ascent, several minutes short of the planned main engine cutoff, MECO.

Our launch was perfect, and after 8 minutes and 36 seconds the engines were shut down at the desired moment. The external tank was jettisoned, Endeavour flipped over to place the tank in clear view of the overhead windows so that I could carry out my first task in space, which was to take photographs of the tank to verify that it had not been damaged during the launch. When taking off my helmet to get a better view with the camera, I tried to hang it on a hook at Dom's seat as I had practiced in the simulator. Janet smiled at the rookie and said "just let it go", and when I did so the weightless helmet floated in place. Photographing the tank against the backdrop of Earth was spectacular, but the time available was short because we were approaching the terminator which separates day from night.

8.8 THE FIRST HOURS IN ORBIT AND POST INSERTION

Strictly speaking, Endeavour was not in an orbit after MECO. The ascent achieved an elliptical orbit whose perigee was on the Earth's surface. A burn by the OMS was needed 30 to 40 minutes after lift-off to circularize the orbit. The necessity for such a burn can be understood by considering the fate of the external tank. At MECO the tank was traveling in the same orbit as the Shuttle. However, before completing half a revolution around the Earth the tank had tumbled into the Indian Ocean (it was the only part of the Shuttle that was not reusable). Without the OMS burn to raise its perigee above the atmosphere, the Orbiter would also have quickly reentered the atmosphere. The burn was a significant milestone in the mission because it achieved a sustainable orbit. In the case of SRTM this orbit was unusually low at 230 km to satisfy the payload requirements. Now the actual mission could begin.

The post-insertion phase began with stowing our space suits and essentially turning our rocket into a flying laboratory. Endeavour's systems had to be configured for on-orbit operations. One important task was to open the payload bay doors, whose inner surfaces carried radiators to shed to space the excess heat generated by all the on-board equipment. The Ku-band antenna was also deployed in order to establish a broadband link to MCC via geostationary relay satellites.

For the blue shift of Dom, JV, and Mamoru, the launch had occurred late in their day, so their first working day in space came to an end four hours into the mission at MET 0/04:00 (MET means "mission-elapsed time" and is given in the format d/hh:mm starting at lift-off). The red shift of Kevin, Janet, and I were responsible for payload operations during the first hours after the launch.

8.9 PAYLOAD OPERATIONS

Already during post insertion, which was successfully completed two and a half hours after launch, we began the preliminary work for activating the payload. The most vital task was to deploy the 60 m long mast carrying the secondary antenna of the radar interferometer. The mast consisted of 86 bays that were folded into a length about one twentieth of its operating length. The canister that housed the mast during launch and landing was just 4 m long and contained in addition a spool-nut assembly to erect the individual bays during the deployment procedure. The scientific objectives of the mission required keeping the relative positions of the main antenna in the bay and the antenna on the mast within very strict limits, which imposed exceptional requirements on the mast in terms of stiffness and stability. Given the large temperature differences between the day and night phases of the orbit, thermal stability was achieved by using the proper ratio of stainless steel joints and titanium cables (with a positive thermal expansion coefficient) and graphite/epoxy longerons (with a negative coefficient). A further development of this mast serves today as the truss of the ISS to which the pressurized modules, solar panels, and radiators are attached.

The procedure to deploy the mast started 5 hours into the flight, and took about 1 hour. First, the brackets that held the lid closely tied to the canister were removed, so that the lid was free to move. Would the mast move at all? Would it move at the desired speed, smoothly and continuously? Never before had such a long rigid lattice structure been deployed in space. When Janet activated the final switch, to our relief the mast deployment started just as in a training session. At a speed of about 5 cm/s bay after bay of the mast was erected and eased out of the canister by the deployment mechanism. We monitored all the time to identify any irregular or otherwise evident deviations in the movement of the mast, but everything worked as planned. Twenty minutes after the deployment started, the baseplate of the mast locked into place to signify that the mast had been fully deployed and rigidly anchored. Mario Runco, who was CAPCOM in MCC, acknowledged our report of the successful deployment with the words "It's a long train out there!" The sight of this structure that extended out from the portside of the spacecraft was indeed impressive. It was to remain in place for the next 10 days, at which time the task would be to retract it back into its canister.

As noted above, the deployment of the mast started some 5 hours into the flight. This is noteworthy in comparison with the early missions into space because 5 hours after lifting off John Glenn, the first American to orbit the Earth, was already back on its surface! During SRTM at this time we initiated one of the most critical prerequisites for a successful mission. This illustrates the advances made in the decades since the first daring men and women flew into space.

After deploying the mast, several tasks had to be accomplished before it would be possible to calibrate the radar interferometer. These included flipping the outboard antenna so that its surface would point at the ground and unstowing and adjusting the "milkstool", a device at the end of the mast used to adjust the antenna in pitch and yaw. Within 2 hours of the mast being deployed, all

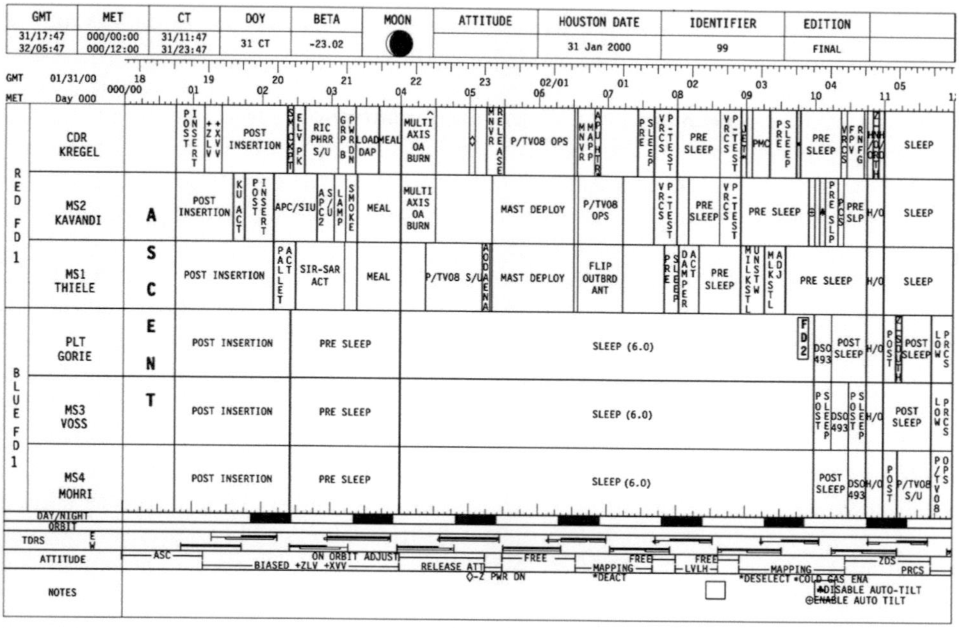

Fig. 8.5. Summary timeline for the first day of the STS-99 mission (source: NASA).

calibration activities were completed and the outboard antenna was properly aligned with the main antenna in the cargo bay. At MET 0/11:47 we began to gather the first scientific data of the SRTM mission, scanning the island of Diego Garcia in the central Indian Ocean.

Although astronauts are trained extensively in the preparation for a mission, one of the greatest concerns is the apprehension of making an eminently avoidable mistake. In addition to double-checking because four eyes see more than two eyes, the flight plan and various procedures guide the astronaut at work. Fig. 8.5 shows a page from the flight plan, also known for obvious reasons as a timeline, in this case for the first 12 hours of our mission. The main portion shows the activities for each member of the crew, with the red shift (CDR, MS1, MS2) in the upper half and the blue shift (PLT, MS3, MS4) in the bottom part. At the top of these sections two timescales, one in GMT (Greenwich Mean Time) and the second in MET, indicate the time when an activity was to be performed. For the first part, ascent and post insertion, the flight plan lacks detail because other documents govern these phases. As mentioned earlier, the blue shift retired after the post-insertion phase was completed at MET 0/04:00 and had a shortened sleep period of 6 hours. From the activities at MET 0/04:00 one can see that CDR and MS2 worked together to perform a multi-axis orbit adjust burn, the details of which were in a different book of procedures. At MET 05:20 MS1 and MS2 jointly deployed the mast. Highly critical activities were always performed by pairs of astronauts, but we attempted to be vigilant at all times and to monitor one another's activities. For example, although the flip of the outboard antenna and (a little later) the unstowing and adjustment of the milkstool were listed in the flight plan as being performed by MS1 alone, I had someone verifying the critical steps in the procedure. The main body of the flight plan also shows that the one and a half hour pre-sleep period was interrupted for all three crew members of the red shift several times. While this was not desirable in terms of resting, it reflected the fact that certain activities in the early part of the mission were very time critical and gave rise to a very dense schedule.

The fourth and fifth rows from the bottom of the main section of the flight plan show the day–night cycle and the current orbit. An orbit lasted about 90 minutes. Note that the night cycle was slightly shorter than the length of the day. The row named TDRS shows which of the Tracking and Data Relay Satellites, East or West, provide a communications link to MCC. When transferring from TDRS East to TDRS West, a small gap indicates the so-called Zone of Exclusion (ZOE) located roughly at the geographic latitude of India and the Indian Ocean. This blackout in communications could last anything up to 10 minutes, as shown by MET 02:15 to 02:25 and MET 10:05 to 10:15. The ATTITUDE row specifies the desired attitude of the Orbiter. Notice that during the mast deployment the Orbiter was in free drift, meaning that the reaction control system was completely inhibited to prevent it firing the primary thrusters or the verniers and subjecting the mast to any additional stress. The row for notes provided the opportunity to write in additional information where necessary.

SRTM was the last in a series of three Shuttle missions to use radar to study the Earth. The previous two missions known as Space Radar Laboratory 1 (SRL-1) and Space Radar Laboratory 2 (SRL-2) were flown within 6 months of one another in the spring and fall of 1994 in order to compare seasonal variations in their respective observations. Incidentally, all three radar missions were flown by Endeavour. Common to all these missions was an unusually low apogee of about 230 km. Normal Shuttle flights were made in the range 300 to 350 km. That said, the flights to maintain the Hubble Space Telescope (HST) were in excess of 550 km. The reason for the rather low orbit on the radar flights was a trade-off between getting a good signal-to-noise ratio (SNR) for a high-quality radar signal, on the one hand, and surface coverage and orbit stability, on the other hand. The lower the altitude the stronger the signals scattered back by the ground, giving a good SNR. But at a higher altitude the radar beam covered a larger area of the surface and it would be possible to cover the entire globe in fewer orbits. Another disadvantage of flying at low altitude was that the greater drag caused the orbit to decay more rapidly. Our mission required five correction burns to maintain the desired altitude.

The mission used C-band and X-band interferometric synthetic aperture radars to acquire topographic data of Earth's land surface. Fig. 8.6 depicts the operation of the radar system. The scattered signal was recorded in both the primary and the secondary antennas. By combining pairs of images using a technique called radar interferometry it was possible to measure the elevation of the terrain below. Fig 8.7 shows an image created by the SRTM system.

Fig. 8.6. With its C-band and X-band outboard antennae at work, the Earth-orbiting Space Shuttle Endeavour maps the topography of the Earth (source: NASA).

Our mission was a resounding success. The land mass in the latitude range 60°N to 56°S was surveyed at least once, in most cases twice, in some cases more often. This data set covered more than 119 million km^2 or roughly 80% of the land surface and 99.96% of the targeted land mass (Fig. 8.8). This success reflects the truly superb preparation and the excellent support by scientists, engineers, and technicians during the flight. But we also needed luck. We encountered a failure in

Fig. 8.7. Image of Ireland produced by shading and color-coding topographic height data from SRTM. The shade image was derived by computing the topographic slope in the north-west–southeast direction, so that northwest slopes appear bright and southeast slopes appear dark. Color-coding is directly related to topographic height, with green at the lower elevations, rising through yellow and tan, to white at the highest elevations (source: NASA/JPL/NGA).

the RCS that could have impaired the science. This system was responsible for maneuvering in orbit and to some extent for attitude control and hence was essential for maintaining the proper mapping attitude. Fortunately, the failure occurred just after we had passed over Japan for a long pass over the Pacific. Since we did not record the sea surface, no data were lost and we had sufficient time to reconfigure the RCS before we made landfall again.

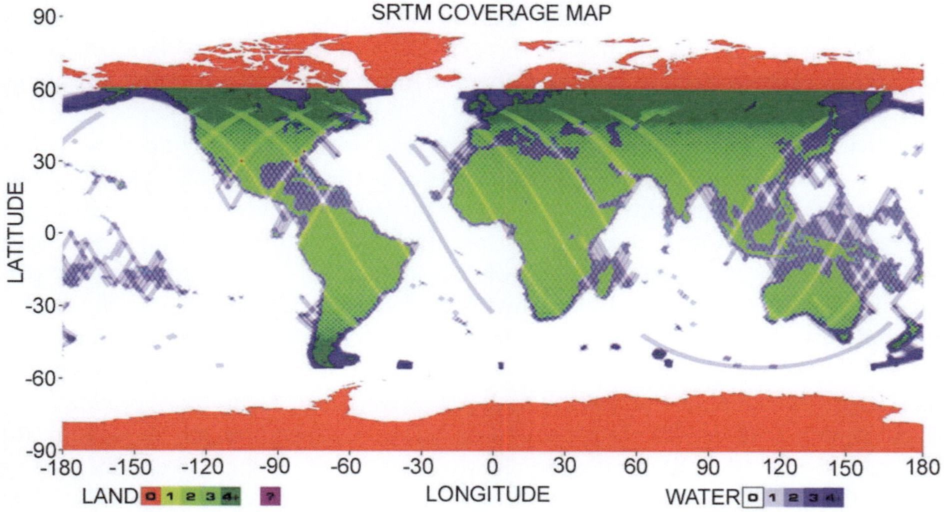

Fig. 8.8. Diagram to show the data coverage by SRTM. Green indicates the area was covered once, yellow-green twice, and so on, as indicated at bottom left. Data were mostly gathered over land (source: NASA).

8.10 OBSERVATIONS ABOUT LIVING IN SPACE

Living in space is a truly remarkable experience. What astonishes me to this day is that the human body apparently adapts to the condition of weightlessness in a fairly short period. When I flew into space I was 46 years of age, hence my body had known nothing but gravity for all that time. How long would it take to adjust to living in weightlessness? One of the biggest concerns at the very beginning was of course what is politely described as the space adaptation syndrome, which means in plain terms nothing less than space sickness. This could have created a potentially difficult situation considering the working intensity required to configure the radar equipment immediately after launch. To avoid the symptoms of space sickness we were advised to limit our head movements during the first days in space. Instead of turning the head left or right one would turn the entire body and keep the head aligned with the body axis. So for the first few days I was moving a little awkwardly whenever I wanted to turn to talk to someone. Once I made the mistake of tipping my head back in order to look up, and everything seemed to start spinning and there was nothing I could do to overcome the sensation. I was forced to take out my sick bag. After using it I felt quite alright again. The sensation of sickness can appear rapidly and disappear equally rapidly. If the best measure for well-being is a healthy appetite, then I must have felt splendid right from the beginning!

However, the first night in space was very difficult because I could hardly sleep. No matter what I did to find a suitable sleeping position, severe back pain meant rest eluded me. When I got up next morning, Janet noticed right away that

I had a problem. She massaged my back for about 10 minutes and the pain vanished. When we checked my height, the expansion of my vertebrae had caused me to grow by 3 cm overnight! I was rather fortunate, because for some colleagues back pain poses a long-term problem.

Obviously one needs to adjust when moving around in weightlessness. All it takes to get from one position to another is a gentle push in the right direction. There is no friction to slow a movement. In weightlessness one can experience Newton's laws of motion directly. For me, the more surprising adjustments were to do with orientation. After getting up one morning I was looking for my personal hygiene kit, which I had fastened with Velcro to a dedicated spot just like everyone else. However, the kit had disappeared. After I had searched the entire middeck of the Orbiter to no avail, Kevin suggested that I turn "upside down". In space there is no up or down, though we maintained a "normal orientation", which was the same orientation as the lockers and equipment stowed in the middeck of the Shuttle. Following Kevin's advice, I flipped over, the ceiling became the new floor and vice versa. And I immediately saw the hygiene kit. This may have been a lucky coincidence. Still, I used this orientation flip in other circumstances and usually with success.

Another memorable part of a Shuttle flight is of course the view. For me the most intriguing sight was the glittering blackness of space (as German astronaut Reinhard Furrer called it) but many astronauts cite the view of our home planet. The problem is that words just cannot do justice to this impressive sight. Since SRTM was a mapping mission which had to maintain a given attitude throughout, the Earth always appeared in the upper left of the window. In the image shown in Fig 8.9 one can see how the atmosphere appears as a thin blue line, roughly 10 km thick, separating the surface of the planet from space. A distance of 10 km seems quite far to us living on the surface but compared with the radius of the globe it is merely a fraction of a percent. All life that we know of lives in this thin blue layer and in the oceans beneath it. "One sees well only with the heart. The essential is invisible to the eyes."[10] By seeing the atmosphere from space you get indeed a new image of our Earth!

8.11 RETURNING TO EARTH

Coming home is for many reasons a very special moment in spaceflight. The mission is definitely over when the Shuttle touches down on the runway. Nothing more can be accomplished. In the case of SRTM, the amount of data gathered was near perfection in terms of areal coverage. Evaluations performed by the teams on the ground during the first days of the mission indicated the quality of the data was at least sufficient to achieve the scientific objectives.

From a personal perspective one of the questions one asks is: Will this be my last flight or will I be able to see this view again? I vividly remember that as the

[10] From *Le Petit Prince* by Antoine de Saint-Exupéry.

Fig. 8.9. Image of part of the aft cargo bay of the Space Shuttle Endeavour with the Earth in the background (source: NASA).

mission wound down I was not ready to return home. But the mapping was finished, the mast was stowed, and the energy resources of the Shuttle were almost depleted. We had no choice. Our flight predated the loss of Columbia and therefore the risks of reentering the atmosphere were probably not perceived as strongly as after 2003. The mood was good, and we were looking forward to landing and returning to our families.

In order to deorbit, the Shuttle had to decelerate. This burn was performed over Australia about one hour prior to landing at the Kennedy Space Center. The Orbiter adopted an attitude with the payload bay facing the Earth and the engines pointing in the direction of flight. The OMS engines were fired for about 4 minutes to lower the speed by 400 ft/s. The deceleration was so slight that it went unnoticed by the crew in terms of the rate at which the Earth passed by the windows. But because this burn lowered the perigee to *minus* 11 km, there could be no doubt that we were about to reenter the atmosphere.

During reentry the crew had to monitor two key parameters in order to land on the runway: the flight path, which is the line from the current position towards the runway, and total energy, which is a measure of the speed and height of the vehicle during reentry.

Let us examine the flight path first. After the deorbit burn the Shuttle had a slightly higher energy than necessary to reach the landing site when flying in a straight line (a great circle). We dissipated this excess energy by flying a trajectory that resembled a stretched "S". This deviation from a straight line was achieved by adopting the proper bank angle (i.e., the angle through which the Orbiter is rolled about the direction of the velocity vector). This introduced a small deviation from flying straight ahead, and the angle of deviation increased with time. The flight path was monitored via this delta azimuth angle, or ΔAz. It measured the deviation of the flight path from the direct line to the landing site, for which ΔAz is obviously zero. When ΔAz reached 17°, the Orbiter initiated a so-called S-turn by banking the other way. A second S-turn was performed once ΔAz again reached 17°, this time on the other side of the direct line to the runway. For SRTM three such S-turns were performed with the final ΔAz being 10°.

The key to maintaining the proper energy was the correct attitude angle, known as the angle of attack. It was the angle between the velocity vector and the x-axis of the vehicle's body; in essence the direction out of the nose of the Orbiter. Initially it was held at 41° with the very tight accuracy requirement of one degree. During the first half hour after the deorbit burn, whilst still traveling in space, the Shuttle barely slowed down. In our case we reached the coast of North America at the U.S.–Canadian border still going at a speed of Mach 25. However, our altitude was only half of what it had been at the time of the deorbit burn. The main deceleration occurred when the vehicle flew into the ever thickening atmosphere. As the Shuttle slowed down, the bank angle was continuously reduced. Maintaining the proper relationship between attitude and energy was of the greatest importance. To reenter at a shallower angle would build up heat and exceed thermal limits. A steeper angle would result in g-forces sufficient to violate the structural integrity of the Orbiter. During most of the reentry the attitude of the vehicle was maintained using the RCS jets because the aerosurfaces, aileron, and rudder could not exert the necessary steering forces until the altitude was down to 80,000 feet, at which time the vehicle would be some 60 nautical miles from the runway.

Inside the Shuttle the crew was subjected to steadily increasing forces due to the deceleration. The maximum loading was about 1.6–1.8 g, but after spending so long in weightlessness it felt like much more. Also, the temperature in the cabin was building up. Fortunately, the launch and entry suit had a cooling system that could be adjusted to maintain a comfortable body temperature.

Once the Shuttle was below Mach 2.5 the flight was controlled aerodynamically by the elevons and the rudder rather than the RCS jets. Around Mach 1 the vehicle was shaking quite significantly. A loud double "boom-boom" could be heard by the people at the landing strip as we broke the sound barrier. These two sounds were generated independently within a fraction of a second by the nose and the rudder of the vehicle. In approaching the runway the Orbiter descended at an angle of 19°, six times steeper than a commercial airliner. Dom

Fig. 8.10. Members of the STS-99 crew pose with NASA Administrator Dan Goldin underneath Space Shuttle Endeavour at the Shuttle Landing Facility after landing. From left are Commander Kevin Kregel, Mission Specialist Janet Kavandi, Pilot Dominic Gorie, Goldin, and Mission Specialists Gerhard Thiele and Mamoru Mohri. Mission Specialist Janice Voss is not in the photo (source: NASA).

constantly recited altitude, speed, and course to verify that we were on track. Kevin and Dom brought the mission to a successful end with a touchdown that was hardly noticeable.

After exiting from Endeavour we performed the final walk around the vehicle in a ritual that demonstrates the deep bond between our crew and the machine that kept us alive for 11 days in space (Fig. 8.10).

As at the start of the mission, upon returning to Earth the body needs to once again adjust to a change in environment. This time, however, the process goes in the reverse direction and it must get used to gravity again. This adjustment process is much faster because the body "knows" how to live in gravity. Nevertheless, some effects can be noticed for several days. As everything in space can be executed with much less force than on Earth, one must adjust the muscle force. When trying to throw a tennis ball to my partner I gave it an extra hard push but the ball didn't even make half the distance between us! Likewise reaction time appears to be slower when returning from a space mission. Maintaining posture can be challenging, especially with closed eyes or in the dark. It was difficult for the calf muscles to find the proper balance. After 2 days, the only symptom of having been in space was the muscle pain in the back of my legs due to the exertion required to steady myself in a gravity field.

8.12 REFLECTIONS

At 7 AM on Sunday, April 12, 1981, Columbia soared into the blue sky over Florida exactly 20 years after the first human spaceflight by Yuri Gagarin. The maiden flight of the new U.S. National Space Transportation System ended an almost 6-year period in which NASA's astronauts were grounded due to the lack of a spacecraft. This new Space Shuttle, as it was called, provided the backbone of the U.S. human spaceflight program for three decades. After 135 missions the program ended with the landing of Atlantis on July 21, 2011, and the "most amazing and complicated flying machine ever built"[11] became history. Its final flight provoked many comments, ranging from glorification to the verdict of being an outright failure. It may be too early to tell how history will judge the vehicle and its accomplishments, but some prominent aspects of the program may guide us in our attempt to classify the meaning of the Shuttle as a space vehicle.

There are numerous ways to categorize the various Space Shuttle missions. They were mostly civilian, but in the early years some were for the Department of Defense. In general there were three categories. One category served predominantly scientific goals. These include the Spacelab missions, maintaining the Hubble Space Telescope, and the SRTM even though the latter two were fairly technical in the execution and facilitated science rather than performed science. A second category was essentially technology demonstrations. Amongst other goals these included the deployment and/or retrieval of satellites. For example, six TDRS satellites were successfully deployed by Shuttles between 1983 and 1995. (Another one was aboard Challenger when that vehicle was lost in January 1986.) The third category of Shuttle missions are those to assemble and operate the ISS. A total of 37 missions were sent to the station between December 1998 and the end of the Shuttle program.

There can be no question that many extraordinary missions were accomplished by the Shuttle. In particular, several science-oriented missions yielded and still provide to this day extraordinary results that have changed the way we view Earth (SRTM) and the universe (Hubble and Chandra), with probably the most prominent ones being those to launch, repair, and service the Hubble Space Telescope. On the other hand, several critical aspects must be considered. The reusable space transportation system was expected to be capable of launching on a weekly basis, and provide an almost on-demand launch service at a much lower cost than previously. We now know that these goals were unreal-

[11] 12.12.2011: *www.nasa.gov/mission_pages/Shuttle/Shuttlemissions/sts135/launch/sts-135_mission-overview.html*

istic. On average, fewer than five flights per year were achieved; a factor of 10 less than was hoped. The Shuttle never provided regular, reliable, and affordable access to space, and despite being declared "operational" after four missions it remained a test vehicle throughout its lifetime.

It is a widely held view that the Shuttle was the most complex vehicle ever built. From a crew perspective the complexity is best seen when looking at the demands for training. A typical scenario for the majority of ascent or landing sessions featured a multitude of failures that were carefully constructed to allow the crew "a way out", meaning that the crew could save the Orbiter and themselves. However, none of these complex scenarios occurred in reality simply because they were highly unlikely. Why then did we lose two vehicles? Neither Challenger, which was lost during ascent, nor Columbia, which was lost during reentry owing to an incident at launch, could have been saved by the crew because neither was caused by a major failure of the Orbiter systems. The events that led to the loss of these Orbiters and the deaths of their crews were scenarios we never trained for and probably could not have trained for.

While the Shuttle's role in building the ISS is evident from the sheer number of flights to the station, one must not overlook a major drawback of this approach. The Orbiter and its payload weighed a total of about 120 tons on the pad. Of the mass that reached orbit, only 20 tons was left there. The 100 ton Orbiter returned to Earth. Yet if one lifts a huge mass into orbit for the purpose of constructing an infrastructure like the ISS, then one should leave the majority of it there! The use of the Shuttle to construct the ISS was inefficient to say the least. At one time Mike Griffin, the former NASA Administrator, pointed out that we should never again build an infrastructure like the ISS in 20-ton chunks.

Was the Shuttle, as many claim, ultimately a failure? I would not go that far. While it is true that the original goals and hopes could not be accomplished, we learned a lot during three decades of operation. If we apply the lessons learned to the design of the next space transportation system then we might one day be able to embark on space missions that today are impossible. No one now is concerned whether Columbus used the most suitable ships available in the 15th century when he set out to find a shortcut to India. It matters only that he set sail. When seen from this point of view, the Shuttle was probably a detour, perhaps an avoidable detour, and possibly a necessary one. However, when it contributed to the future of humans in space it was ultimately a success.

8.13 REFERENCES/RECOMMENDED LITERATURE

Gehman, H., Turcotte, S., Barry, J., Hess, K.W., Hallock, J.N., Wallace, S.B., Deal, D., Hubbard, S., Tetrault, R.E., Widnall, S. *et al.*: *Columbia Accident Investigation Board Report*, NASA (2003).

Hale, W., Lulla, K., and Chapline., G.: *Wings in Orbit: Scientific and Engineering Legacies of the Space Shuttle 1971–2010*, NASA (2011).

Logsdon, J.M.: The Space Shuttle program: A policy failure? *Science*, **232** 1099–1105 (1986).

Mullane, M.: *Riding Rockets: The Outrageous Tales of a Space Shuttle Astronaut*, Scribner (2006).
de Saint-Exupéry, A.: *Le Petit Prince*, Reynal & Hitchcock (1943).
Rogers, W.P., Armstrong, N.A., Acheson, D.C., Covert, E.E., Feynman, R.P., Hotz, R.B., Kutyna, D.J., Ride, S.K., Rummel, R.W., Sutter, J.F. *et al.*: *Report of the Presidential Commission on the Space Shuttle Challenger Accident*, U.S. Government Printing Office (1986).

Index

MIX
Papier aus verantwortungsvollen Quellen
Paper from responsible sources
FSC® C105338

If you have any concerns about our products,
you can contact us on
ProductSafety@springernature.com

In case Publisher is established outside the EU,
the EU authorized representative is:
Springer Nature Customer Service Center GmbH
Europaplatz 3, 69115 Heidelberg, Germany

Printed by Libri Plureos GmbH
in Hamburg, Germany